原版影印说明

1.《聚合物百科词典》(5册)是 Springer Reference *Encyclopedic Dictionary of Polymers*(2nd Edition)的影印版。为使用方便，由原版2卷改为5册：

第1册 收录 A–C 开头的词组；

第2册 收录 D–I 开头的词组；

第3册 收录 J–Q 开头的词组；

第4册 收录 R–Z 开头的词组；

第5册 为原书的附录部分及参考文献。

2. 缩写及符号、数学符号、字母对照表、元素符号等查阅说明各册均完整给出。

由 Jan W. Gooch 主编的《聚合物百科词典》是关于高分子科学与工程领域的参考书，2007 年出版第一版，2011 年再版。本书收录了 7 500 多个高分子材料方面的术语，涉及高分子材料的各个方面，如粘合剂、涂料、油墨、弹性体、塑料、纤维等，还包括生物化学和微生物学方面的术语，以及与新材料、新工艺相关的术语；并且不仅包括其物理、电子和磁学性能方面的术语，还增加了数据处理的统计和数值分析以及实验设计方面的术语。每个词条方便查找，并给出了简洁的定义，以及相互参照的相关术语。为了说明得更清晰，全书给出 1 160 个图、73 个表。有的词条还给出方程式、化学结构等。

材料科学与工程图书工作室

联系电话 0451–86412421

0451–86414559

邮　　箱 yh_bj@aliyun.com

xuyaying81823@gmail.com

zhxh6414559@aliyun.com

Springer 词典精选原版系列

聚合物百科词典

Jan W. Gooch

Encyclopedic Dictionary of Polymers

2nd Edition

VOLUME 5

（附录）

黑版贸审字08-2014-010号

Reprint from English language edition:
Encyclopedic Dictionary of Polymers
by Jan W.Gooch
Copyright © 2011 Springer New York
Springer New York is a part of Springer Science+Business Media
All Rights Reserved

This reprint has been authorized by Springer Science & Business Media for distribution in China Mainland only and not for export therefrom.

图书在版编目（CIP）数据

聚合物百科词典. 5, 附录 :英文/（美）古驰（Gooch, J. W.）主编.—哈尔滨:哈尔滨工业大学出版社，2014.3

（Springer词典精选原版系列）

ISBN 978-7-5603-4446-1

Ⅰ.①聚… Ⅱ.①古… Ⅲ.①聚合物-词典-英文 Ⅳ.①O63-61

中国版本图书馆CIP数据核字（2013）第292194号

责任编辑 许雅莹 张秀华 杨 桦
出版发行 哈尔滨工业大学出版社
社 址 哈尔滨市南岗区复华四道街10号 邮编 150006
传 真 0451-86414749
网 址 http://hitpress.hit.edu.cn
印 刷 哈尔滨市石桥印务有限公司
开 本 787mm × 1092mm 1/16 印张 13.75
版 次 2014 年 3 月第 1 版 2014 年 3 月第 1 次印刷
书 号 ISBN 978-7-5603-4446-1
定 价 108.00元

（如因印刷质量问题影响阅读，我社负责调换）

Acknowledgements

The editor wishes to express his gratitude to all individuals who made available their time and resources for the preparation of this book: James W. Larsen (Georgia Institute of Technology), for his innovations, scientific knowledge and computer programming expertise that were invaluable for the preparation of the Interactive Polymer Technology Programs that accompany this book; Judith Wiesman (graphics artist), for the many graphical presentations that assist the reader for interpreting the many complex entries in this publication; Kenneth Howell (Springer, New York), for his continued support for polymer science and engineering publications; and Daniel Quinones and Lydia Mueller (Springer, Heidelberg) for supporting the printed book and making available the electronic version and accompanying electronic interactive programs that are important to the scientific and engineering readers.

Preface

The second edition of Encyclopedic Dictionary of Polymers provides 40% more entries and information for the reader. A Polymers Properties section has been added to provide quick reference for thermal properties, crystallinity, density, solubility parameters, infrared and nuclear magnetic spectra. Interactive Polymer Technology is available in the electronic version, and provides templates for the user to insert values and instantly calculate unknowns for equations and hundreds of other polymer science and engineering relationships. The editor offers scientists, engineers, academia and others interested in adhesives, coatings, elastomers, inks, plastics and textiles a valuable communication tool within this book. In addition, the more recent innovations and biocompatible polymers and adhesives products have necessitated inclusion into any lexicon that addresses polymeric materials. Communication among scientific and engineering personnel has always been of critical importance, and as in any technical field, the terms and descriptions of materials and processes lag the availability of a manual or handbook that would benefit individuals working and studying in scientific and engineering disciplines. There is often a challenge when conveying an idea from one individual to another due to its complexity, and sometimes even the pronunciation of a word is different not only in different countries, but in industries. Colloquialisms and trivial terms that find their way into technical language for materials and products tend to create a communications fog, thus unacceptable in today's global markets and technical communities.

The editor wishes to make a distinction between this book and traditional dictionaries, which provide a word and definition. The present book provides for each term a complete expression, chemical structures and mathematic expression where applicable, phonetic pronunciation, etymology, translations into German, French and Spanish, and related figures if appropriate. This is a complete book of terminology never before attempted or published.

The information for each chemical entry is given as it is relevant to polymeric materials. Individual chemical species (e.g., ethanol) were taken from he *CRC Handbook of Chemistry and Physics*, 2004 Version, the Merck Index and other reference materials. The reader may refer to these references for additional physical properties and written chemical formulae. Extensive use was made of ChemDraw®, CambridgeSoft Corporation, for naming and drawing chemical structures (conversion of structure to name and vice versa) which are included with each chemical entry where possible. Special attention was given to the IUPAC name that is often given with the common name for the convenience of the reader.

The editor assembled notes over a combined career in the chemical industries and academic institutions regarding technical communication among numerous colleagues and helpful acquaintances concerning expressions and associated anomalies. Presently, multiple methods of nomenclature are employed to describe identical chemical compounds by common and IUPAC names (eg. acetone and 2-propanone) because the old systems (19th century European and trivial) methods of nomenclature exists with the modern International Union of Pure and Applied Chemistry, and the conflicts between them are not likely to relent in the near future including the weights and measures systems because some nations are reluctant to convert from English to metric and, and more recently, the International Systems of Units (SI). Conversion tables for converting other systems to the SI units are included in this book for this purpose. In addition, there are always the differences in verbal pronunciation, but the reasons not acceptable to prevent cogent communication between people sharing common interests.

In consideration of the many challenges confronting the reader who must economize time investment, the structure of this book is optimized with regard the convenience of the reader as follows:

- Comprehensive table of contents
- Abbreviations and symbols
- Mathematics signs
- English, Greek, Latin and Russian alphabets
- Pronunciation/phonetic symbols
- Main body of terms with entry term in English, French German and Italian
- Conversion factors

- Microbiology nomenclature and terminology
- References

The editor acknowledges the utilization of many international sources of information including journals, books, dictionaries, communications, and conversations with people experienced in materials, polymer science and engineering. A comprehensive reference section contains all of the sources of information used in this publication. Pronunciation, etymological, cross-reference and related information is presented in the style of the 11th Edition of the Meriam-Webster Dictionary, where known, for each term. The spelling for each term is presented in German, French, and Spanish where translation is possible. Each term in this book includes the following useful information:

- Spelling (in **bold** face) of each term and alternative spellings where more than one derivation is commonly used
- Phonetic spelling \-\ using internationally published phonetic symbols, and this is the first book that includes phonetic pronunciation information missing in technical dictionaries that allows the reader to pronounce the term
- Parts of speech in English following each phonetic spelling, eg. *n.*, *adj.*
- Cross-references in CAPITALS letters
- Also called *example* in italics
- Etymological information [-] for old and new terms that provides the reader the national origins of terms including root words, prefixes and suffixes; historical information is critical to the appreciation of a term and its true meaning
- French, German, Italian and Spanish spellings of the term { - }
- A comprehensive explanation of the term
- Mathematical expressions where applicable
- Figures and tables where applicable
- A comprehensive reference section is included for further research

References are included for individual entries where a publication(s) is directly attributable to a definition or description. Not all of the references listed in the Reference section are directly attributable to entries, but they were reviewed for information and listed for the reader's information. Published dictionaries and glossaries of materials were very helpful for collecting information in the many diverse and smaller technologies of the huge field of polymers. The editor is grateful that so much work has been done by other people interested in polymers.

The editor has attempted to utilize all relevant methods to convey the meaning of terms to the reader, because a term often requires more information than a standard entry in a textbook dictionary, so this book is dedicated to a complete expression. Terminology and correct pronunciation of technical terms is continuously evolving in scientific and industrial fields and too often undocumented or published, and therefore, not shared with others sometimes leading to misunderstandings. Engineering and scientific terms describe a material, procedure, test, theory or process, and communication between technical people must involve similar jargon or much will be lost in the translation as often has been the editor's experience. The editor has made an attempt to provide the reader who has an interested in the industries that have evolved from adhesives, coatings, inks, elastomers, plastics and textiles with the proper terminology to communicate with other parties whether or not directly involved in the industries. This publication is a single volume in the form of a desk-handbook that is hoped will be an invaluable tool for communicating in the spoken and written media.

Physics, electronic and magnetic terms because they are related to materials and processes (e.g., *ampere*).

Biomolecular materials and processes have in the recent decade overlapped with polymer science and engineering. Advancements in polymeric materials research for biomolecular and medical applications are rapidly becoming commercialized, examples include biocompatible adhesives for sutureless tissue bonding, liquid dressings for wounds and many other materials used for *in vitro* and *in vivo* medical applications. To keep pace with these advancements, the editor has included useful terms in the main body that are commonly used in the material sciences for these new industries.

A microbiology section has been included to assist the reader in becoming familiar with the proper nomenclature of bacteria, fungi, mildew, and yeasts – organisms that affect materials and processes because they are ubiquitous in our environment. Corrosion of materials by microorganisms is commonplace, and identification of a specific organism is critical to prevent its occurrence. Engineers and materials scientists will appreciate the extensive sections on different types of microorganisms together with a section dedicated to microbiology terminology that is useful for communicating in the jargon of biologists instead of referring to all organisms as "bugs."

New materials and processes, and therefore new terms, are constantly evolving with research, development and global commercialization. The editor will periodically update this publication for the convenience of the reader.

Statistics, numerical analysis other data processing and experimental design terms are addressed as individual terms and as a separate section in the appendix, but only as probability and statistics relate to polymer technology and not the broad field of this mathematical science. The interactive equations are listed in the Statistics section of the Interactive Polymer Technology program.

Interactive Polymer Technology Programs

Along with this book we are happy to provide a collection of unique and useful tools and interactive programs along with this Springer Reference. You will find short descriptions of the different functions below. Please download the software at the following website: http://extras.springer.com/2011/978-1-4419-6247-8

Please note that the file is more than 200 MB. Download the ZIP file and unzip it. It is strongly recommended to read the **ReadMe.txt** before installing. The software is started by opening the file InPolyTech.pdf and following the instructions. Detailed instructions can be found under 'Help Instructions'.

The software consists of 15 programs and tools that are briefly described in the appendix.

Abbreviations and Symbols

Abbreviations	Symbols
An	absorption (formerly extinction) (= log t_i^{-1})
A	Area
A	surface
A	Helmholtz energy ($A = U - TS$)
A	preexponential constant [in $k = A \exp(-E^{\ddagger}/RT)$]
A_2	second virial coefficient
a	exponent in the property/ molecular weight relationship ($E^{\ddagger} = KM^{a}$); always with an index, e.g., a_{η}, a_s, etc.
a	linear absorption coefficient, $a = l^{-1}$
absolute	abs
acre	spell out
acre-foot	acre-ft
air horsepower	air hp
alternating-current (as adjective)	a-c
A^m	molar Helmholtz energy
American Society for Testing and Materials	ASTM
amount of a substance (mole)	n
ampere	A or amp
ampere-hour	amp-hr
amplitude, an elliptic function	am.
angle	β
angle, especially angle of rotation in optical activity	∝
Angstrom unit	Å
antilogarithm	antilog
a_o	constant in the Moffit–Yang equation
Area	A
Atactic	at
atomic weight	at. wt
Association	Assn.
atmosphere	atm
average	avg
Avogadro number	N_L
avoirdupois	avdp
azimuth	az or α
barometer	bar.
barrel	bbl
Baumé	Bé
b_o	constant in the Mofit–Yang equation
board fee (feet board measure)	fbm
boiler pressure	spell out
boiling point	bp
Boltzmann constant	k
brake horsepower	bhp
brake horsepower-hour	bhp-hr
Brinell hardness number	Bhn
British Standards Institute	BSI
British thermal unit[1]	Btu or B
bushel	bu
C	heat capacity
c	specific heat capacity (formerly; specific heat); c_p = specific isobaric heat capacity, c_v = specific isochore heat capacity
c	"weight" concentration (= weight of solute divided by volume of solvent); IUPAC suggests the symbol ρ for this quantity, which could lead to confusion with the same IUPAC symbol for density
c	speed of light in a vacuum
c	speed of sound
calorie	cal
candle	c
candle-hour	c-hr
candlepower	cp
ceiling temperature of polymerization, °C	T_c

Abbreviations	Symbols
cent	c or ¢
center to center	c to c
centigram	cg
centiliter	cl
centimeter or centimeter	cm
centimeter-gram-second (system)	cgs
centipoise	cP
centistokes	cSt
characteristic temperature	Θ
chemical	chem.
chemical potential	μ
chemical shift	δ
chemically pure	cp
circa, about, approximate	ca.
circular	cir
circular mils	cir mils
cis-tactic	ct
C^m	molar heat capacity
coefficient	coef
cologarithm	colog
compare	cf.
concentrate	conc
conductivity	cond, λ
constant	const
continental housepower	cont hp
cord	cd
cosecant	csc
cosine	cos
cosine of the amplitude, an elliptic function	cn
cost, insurance, and freight	cif
cotangent	cot
coulomb	spell out
counter electromotive force	cemf
C_{tr}	transfer constant ($C_{tr} = k_{tr}/k_p$)
cubic	cu
cubic centimeter (liquid, meaning milliliter. ml)	cu, cm, cm^3
cubic centimeter	cm^3cubic expansion coefficient ∝
cubic foot	cu ft
cubic feet per minute	cfm
cubic feet per second	cfs

Abbreviations	Symbols
cubic inch	cu in.
cubic meter	cu m or m^3
cubic micron	cu μ or cu mu or μ^3
cubic millimeter	cu mm or mm^3
cubic yard	cu yd
current density	spell out
cycles per second	spell out or c
cylinder	cyl
D	diffusion coefficient
D_{rot}	rotational diffusion coefficient
day	spell out
decibel	db
decigram	d.g.
decomposition, °C	T_{dc}
degree	deg or °
degree Celsius	°C
degree centigrade	C
degree Fahrenheit	F or °
degree Kelvin	K or none
degree of crystallinity	∝
degree of polymerization	X
degree Réaumur	R
delta amplitude, an elliptic function	dn
depolymerization temperature	T_{dp}
density	ρ
diameter	diam
Dictionary of Architecture and Construction	DAC
diffusion coefficient	D
dipole moment	p
direct-current (as abjective)	d-c
dollar	$
dozen	doz
dram	dr
dynamic viscosity	η
E	energy (E_k = kinetic energy, E_p = potential energy, $E^{\ddagger}$ = energy of activation)
E	electronegativity
E	modulus of elasticity, Young's modulus ($E = \sigma_{ij}/\varepsilon_{ij}$)
E	general property

Abbreviations	Symbols
E	electrical field strength
e	elementary charge
e	parameter in the *Q-e* copolymerize-tion theory
e	cohesive energy density (always with an index)
edition	Ed.
Editor, edited	ed.
efficiency	eff
electric	elec
electric polarizability of a molecule	∝
electrical current strength	I
electrical potential	V
electrical resistance	R or X
electromotive force	emf
electronegativity	E
elevation	el
energy	E
enthalpy	H
entropy	S
equation	eq
equivalent weight	equiv wt
et alii (and others)	et al.
et cetera	etc.
excluded volume	u
excluded volume cluster integral	β
exempli gratia (for example)	e.g.
expansion coefficient	∝
external	ext
F	force
f	fraction (exclusing molar fraction, mass fraction, volume fraction)
f	molecular coefficient of friction (e.g., f_s, f_D, f_{rot})
f	functionality
farad	spell out or f
Federal	Fed.
feet board measure (board feet)	fbm
feet per minute	fpm
feet per second	fps
flash point	flp

Abbreviations	Symbols
fluid	fl
foot	ft
foot-candle	ft-c
foot-Lambert	ft-L
foot-pound	ft-lb
foot-pound-second (system)	fps
foot-second (see cubic feet per second)	
fraction	∫
franc	fr
free aboard ship	spell out
free alongside ship	spell out
free on board	fob
freezing point	fp
frequency	spell out
fusion point	fnp
G	Gibbs energy (formerly free energy or free enthalpy) ($G = H - TS$)
G	shear modulus ($G = \sigma_{ij}$/angle of shear)
G	statistical weight fraction ($G_i = g_i/\Sigma_i\, g_i$)
g	gravitational acceleration
g	statistical weight
g	*gauche* conformation
g	parameter for the dimensions of branched macromolecules
G^m	molar Gibbs energy
gallon	gal
gallons per minute	gpm
gallons per second	gps
gauche conformation	g
Gibbs energy	G
grain	spell out
gram	g
gram-calorie	g-cal
greatest common divisor	gcd
H	enthalpy
H^m	molar enthalpy
h	height
h	Plank constant
haversine	hav

Abbreviations	Symbols
heat	Q
heat capacity	C
hectare	ha
henry	H
high pressure (adjective)	h-p
hogshead	hhd
horsepower	hp
horsepower-hour	hp-hr
hour	h or hr
hundred	C
hundredweight (112 lb)	cwt
hydrogen ion concentration, negative logarithm of	pH
hyperbolic cosine	cosh
hyperbolic sine	sinh
hyperbolic tangent	tanh
I	electrical current strength
I	radiation intensity of a system
i	radiation intensity of a molecule
ibidem (in the same place)	ibid.
id est (that is)	i.e.
inch	in.
inch-pound	in-lb
inches per second	ips
indicated horsepower	ihp
indicated horsepower-hour	ihp-hr
infrared	IR
inside diameter	ID
intermediate-pressure (adjective)	i-p
internal	int
International Union of Pure and Applied Chemistry	IUPAC
isotactic	it
J	flow (of mass, volume, energy, etc.), always with a corresponding index
joule	J
K	general constant
K	equilibrium constant
K	compression modulus ($p = -K\Delta V/V_o$)
k	Boltzmann constant

Abbreviations	Symbols
k	rate constant for chemical reactions (always with an index)
Kelvin	K (Not °K)
kilocalorie	kcal
kilocycles per second	kc
kilogram	kg
kilogram-calorie	kg-al
kilogram-meter	kg-m
kilograms per cubic meter	kg per cu m or kg/m^3
kilograms per second	kgps
kiloliter	Kl
kilometer or kilometer	km
kilometers per second	kmps
kilovolt	kv
kilovolt-ampere	kva
kilowatt	kw
kilowatthour	kwhr
Knoop hardeness number	KHN
L	chain end-to-end distance
L	phenomenological coefficient
l	length
lambert	L
latitude	lat or ϕ
least common multiple	lcm
length	l
linear expansion coefficient	Y
linear foot	lin ft
liquid	liq
lira	spell out
liter	l
logarithm (common)	log
logarithm (natural)	log. or ln
kibgutyde	kibg. or λ
loss angle	δ
low-pressure (as adjuective)	l-p
lumen	1*
lumen-hour	1-hr*
luments per watt	lpw
M	"molecular weight" (IUPAC molar mass)
m	mass
mass	spell out or m
mass fraction	w

Abbreviations	Symbols
mathematics (ical)	math
maximum	max
mean effective pressure	mep
mean horizontal candlepower	mhcp
meacycle	mHz
megohm	MΩ
melting point, -temperature	mp, T_m
meter	m
meter-kilogram	m-kg
metre	m
mho	spell out
microsmpere	μa or mu a
microfarad	μf
microinch	μin.
micrometer (formerly micron)	μm
micromicrofarad	μμf
micromicron	μμ
micron	μ
microvolt	μv
microwatt	μw or mu w
mile	spell out
miles per hour	mph
miles per hour per second	mphps
milli	m
milliampere	ma
milliequivalent	meq
milligram	mg
millihenry	mh
millilambert	mL
milliliter or milliliter	ml
millimeter	mm
millimeter or mercury (pressure)	mm Hg
millimicron	mμ or m mu
million	spell out
million gallons per day	mgd
millivolt	mv
minimum	min
minute	min
minute (angular measure)	′

Abbreviations	Symbols
minute (time) (in astronomical tables)	m
mile	spell out
modal	m
modulus of elasticity	E
molar	M
molar enthalpy	H_m
molar Gibbs Energy	G_m
molar heat capacity	C_m
mole	mol
mole fraction	x
molecular weight	mol wt or M
month	spell out
N	number of elementary particles (e.g., molecules, groups, atoms, electrons)
N_L	Avogadro number (Loschmidt's number)
n	amount of a substance (mole)
n	refractive index
nanometer (formerly millimicron)	nm
National Association of Corrosion Engineers	NACE
National Electrical Code	NEC
newton	N
normal	N
number of elementary particles	N
Occupational Safety and Health Administration	OSHA
ohm	Ω
ohm-centimeter	ohm-cm
oil absorption	O.A.
ounce	oz
once-foot	oz-ft
ounce-inch	oz-in.
outside diameter	OD
osomotic pressure	∏
P	permeability of membranes
p	probability
p	dipole moment
$\mathbf{p}_i$	induced dipolar moment
p	pressure

Abbreviations	Symbols
p	extent of reaction
Paint Testing Manual	PTM
parameter	Q
partition function (system)	Q
parts per billion	ppb
parts per million	ppm
pascal	Pa
peck	pk
penny (pency – new British)	p.
pennyweight	dwt
per	diagonal line in expressions with unit symbols or (see Fundamental Rules)
percent	%
permeability of membranes	P
peso	spell out
pint	pt.
Planck's constant (in E = hv) (6.62517 +/– 0.00023 x 10^{-27} erg sec	h
polymolecularity index	Q
potential	spell out
potential difference	spell out
pound	lb
pound-foot	lb-ft
pound-inch	lb-in.
pound sterling	£
pounds-force per square inch	psi
pounds per brake horsepower-hour	lb per bhp-hr
pounds per cubi foot	lb per cut ft
pounds per square foot	psf
pounds per square inch	psi
pounds per square inch absolute	psia
power factor	spell out or pf
pressure	p
probability	p
Q	quantity of electricity, charge
Q	heat
Q	partition function (system)
Q	parameter in the Q–e copolymerize-tion equation

Abbreviations	Symbols
Q, Q	polydispersity, polymolecularity in-dex ($Q = \overline{M_w}/\overline{M_n}$)
q	partition function (particles)
quantity of electricity, charge	Q
quart	qt
quod vide (which see)	q.v.
R	molar gas constant
R	electrical resistance
R_G	radius of gyration
R_n	run number
R_ϑ	Rayleigh ratio
r	radius
r_o	initial molar ratio of reactive groups in polycondensations
radian	spell out
radius	r
radius of gyration	R_G
rate constant	k
Rayleigh ratio	R_9
reactive kilovolt-ampere	kvar
reactive volt-ampere	var
reference(s)	ref
refractive index	n
relaxation time	τ
resistivity	ρ
revolutions per minute	rpm
revolutions per second	rps
rod	spell out
root mean square	rms
S	entropy
S^m	molar entropy
S	solubility coefficient
s	sedimentation coefficient
s	selectivity coefficient in osmotic measurements)
Saybolt Universal seconds	SUS
secant	sec
second	s or sec
second (angular measure)	″
second-foot (see cubic feet per second)	

Abbreviations	Symbols
second (time) (in astronomical tables)	s
Second virial coefficient	A_2
shaft horsepower	shp
shilling	s
sine	sin
sine of the amplitude, an elliptic function	sn
society	Soc.
Soluble	sol
solubility coefficient	S
solubility parameter	δ
solution	soln
specific gravity	sp gr
specific heat	sp ht
specific heat capacity (formerly: specific heat)	c
specific optical rotation	$[\alpha]$
specific volume	sp vol
spherical candle power	scp
square	sq
square centimeter	sq cm or cm^2
square foot	sq ft
square inch	sq in.
square kilometer	sq km or km^2
square meter	sq m or m^2
square micron	sq μ or μ^2
square root of mean square	rms
standard	std
Standard	Stnd.
Standard deviation	σ
Staudinger index	$[\eta]$
stere	s
syndiotactic	st
T	temperature
t	time
t	*trans* conformation
tangent	tan
temperature	T or temp
tensile strength	ts
threodiisotactic	tit
thousand	M
thousand foot-pounds	kip-ft
thousand pound	kip

Abbreviations	Symbols
ton	spell out
ton-mile	spell out
trans conformation	t
trans-tactic	tt
U	voltage
U	internal energy
U^m	molar internal energy
u	excluded volume
ultraviolet	UV
United States	U.S.
V	volume
V	electrical potential
v	rate, rate of reaction
v	specific volume always with an in-dex
vapor pressure	vp
versed sine	vers
versus	vs
volt	v or V
volt-ampere	va
volt-coulomb	spell out
voltage	U
volume	V or vol.
Volume (of a publication)	Vol
W	weight
W	work
w	mass function
watt	w or W
watthour	whr
watts per candle	wpc
week	spell out
weight	W or w
weight concentration*	c
work	*y* yield
X	degree of polymerization
X	electrical resistance
x	mole fractio *y* yield
yard	yd
year	yr
Young's	E
Z	collision number
Z	*z* fraction
z	ionic charge

Abbreviations	Symbols
z	coordination number
z	dissymmetry (light scattering)
z	parameter in excluded volume theory
α	angle, especially angle of rotation in optical activity
α	cubic expandion coefficient [$\alpha = V^{-1} (\partial V/\partial T)_p$]
α	expansion coefficient (as reduced length, e.g., α_L in the chain end-to-end distance or α_R for the radius of gyration)
α	degree of crystallinity (always with an index)
α	electric polarizability of a molecule
$[\alpha]$	"specific" optical rotation
β	angle
β	coefficient of pressure
β	exclused volume cluster integral
Γ	preferential solvation
γ	angle
γ	surface tension
γ	linear expansion coefficient
δ	loss angle
δ	solubility parameter
δ	chemical shift
ε	linear expansion ($\varepsilon = \Delta l/l_o$)
ε	expectation
ε_r	relative permittivity (dielectric number)
η	dynamic viscosity
$[\eta]$	Staudinger index (called J_o in DIN 1342)
Θ	characteristic temperature, especial-ly theta temperature
θ	angle, especially angle of rotation
ϑ	angle, especially valence angle
κ	isothermal compressibility [$\kappa = V^{-1} (\partial V/\partial p)_T$]
κ	enthalpic interaction parameter in solution theory

Abbreviations	Symbols
λ	wavelength
λ	heat conductivity
λ	degree of coupling
μ	chemical potential
μ	moment
μ	permanent dipole moment
ν	mement, with respect to a reference value
ν	frequency
ν	kinetic chain length
ξ	shielding ratio in the theory of random coils
Ξ	partition function
Π	osmotic pressure
ρ	density
σ	mechanical stress (σ_{ii} = normal stress, σ_{ij} = shear stress)
σ	standard deviation
σ	hindrance parameter
τ	relaxation time
τ_i	internal transmittance (transmission factor) (represents the ratio of transmitted to absorbed light)
ϕ	volume fraction
$\varphi(r)$	potential between two segments separated by a distance r
Φ	constant in the viscosity-molecular-weight relationship
$[\Phi]$	"molar" optical rotation
χ	interaction parameter in solution theory
ψ	entropic interaction parameter in solution theory
ω	angular frequency, angular velocity
Ω	angle
Ω	probability
Ω	skewness of a distribution

*(= weight of solute divided by volume of solvent); IUPAC suggests the symbol ρ for this quantity, which could lead to confusion with the same IUPAC symbol for density.

Notations

The abbreviations for chemicals and polymer were taken from the "Manual of Symbols and Terminology for Physicochemical Quantities and Units," *Pure and Applied Chemistry* **21***1) (1970), but some were added because of generally accepted use.

The ISO (International Standardization Organization) has suggested that all extensive quantities should be described by capital letters and all intensive quantities by lower-case letters. IUPAC doe not follow this recommendation, however, but uses lower-case letters for specific quantities.

The following symbols are used above or after a letter.

Symbols Above Letters

— signifies an average, e.g., $\overline{M}$is the average molecular weight; more complicated averages are often indicated by ⟨⟩, e.g., $\langle R_G^2 \rangle$ is another way of writing $\overline{(R_G^2)}_z$

— stands for a partial quantity, e.g., $\tilde{v}_A$ is the partial specific volume of the compound A; V_A is the volume of A, wherea $\tilde{V}_A^m$xxx is the partial molar volume of A.

Superscripts

°	pure substance or standard state
∞	infinite dilution or infinitely high molecular weight
m	molar quantity (in cases where subscript letters are impractical)
(*q*)	the *q* order of a moment (always in parentheses)
‡	activated complex

Subscripts

Initial	State
1	solvent
2	solute
3	additional components (e.g., precipitant, salt, etc.)
am	amorphous
B	brittleness
bd	bond
cr	crystalline
crit	critical
cryst	crystallization
e	equilibrium
E	end group
G	glassy state
i	run number
i	initiation
i	isotactic diads
ii	isotactic triads
Is	heterotactic triads
j	run number
k	run number
m	molar
M	melting process
mon	monomer
n	number average
p	polymerization, especially propagation
pol	polymer
r	general for average
s	syndiotactic diads
ss	syndiotactic triads
st	start reaction
t	termination
tr	transfer
u	monomeric unit
w	weight average
z	*z* average
Prefixes	
at	atactic
ct	*cis*-tactic
eit	erythrodiisotactic
it	isotactic
st	syndiotactic
tit	threodiisotactic
tt	*trans*-tactic

Square brackets around a letter signify molar concentrations. (IUPAC prescribes the symbol *c* for molar councentrations, but to date this has consistently been used for the mass/volume unit.)

Angles are always given by °.

Apart from some exceptions, the meter is not used as a unit of length; the units cm and mm derived from it are used. Use of the meter in macromolecular science leads to very impractical units.

Mathematical Signs

Sign	Definition
Operations	
$+$	Addition
$-$	Subtraction
$\times$	Multiplication
$\cdot$	Multiplication
$\div$	Division
$/$	Division
$\circ$	Composition
$\cup$	Union
$\cap$	Intersection
$\pm$	Plus or minus
$\mp$	Minus or plus
Convolution	
$\oplus$	Direct sum, variation
$\ominus$	Various
$\otimes$	Various
$\odot$	Various
$:$	Ratio
$\amalg$	Amalgamation
Relations	
$=$	Equal to
$\neq$	Not equal to
$\approx$	Nearly equal to
$\cong$	Equals approximately, isomorphic
$<$	Less than
$<<$	Much less than
$>$	Greater than
$>>$	Much greater than
$\leq$	Less than or equal to
$\leq$	Les than or equal to
$\leqq$	Less than or equal to
$\geq$	Greater than or equal to
$\geq$	Grean than or equalt o
$\geqq$	Greater than or equal to
$\equiv$	Equivalent to, congruent to
$\not\equiv$	Not equivalent to, not congruent to
$\mid$	Divides, divisible by
$\sim$	Similar to, asymptotically equal to
$:=$	Assignment

Sign	Definition
$\in$	A member of
$\subset$	Subset of
$\subseteq$	Subset of or equal to
$\supset$	Superset of
$\supseteq$	Superset of or equal to
$\propto$	Varies as, proportional to
$\doteq$	Approaches a limit, definition
$\rightarrow$	Tends to, maps to
$\leftarrow$	Maps from
$\mapsto$	Maps to
$\hookrightarrow$ or $\hookleftarrow$	Maps into
$\Box$	d'Alembertian operator
Σ	Summation
Π	Product
$\int$	Integral
$\oint$	Contour integral
Logic	
$\wedge$	And, conjunction
$\vee$	Or, distunction
$\neg$	Negation
$\Rightarrow$	Implies
$\rightarrow$	Implies
$\Leftrightarrow$	If and only if
$\leftrightarrow$	If and only if
$\exists$	Existential quantifier
$\forall$	Universal quantifier
$\in$	A member o
$\notin$	Not a member of
$\vdash$	Assertion
$\therefore$	Hence, therefore
$\because$	Because
Radial units	
$'$	Minute
$''$	Second
$^\circ$	Degree
Constants	
π	pi ($\approx$3.14159265)
e	Base of natural logarithms ($\approx$2.71828183)

Sign	Definition
Geometry	
$\perp$	Perpendicular
$\parallel$	Parallel
$\nparallel$	Not parallel
$\angle$	Angle
$\sphericalangle$	Spherical angle
⩣	Equal angles
Miscellaneous	
i	Square root of -1
$'$	Prime
$''$	Double prime
$'''$	Triple prime
$\sqrt{}$	Square root, radical
$\sqrt[3]{}$	Cube root
$\sqrt[n]{}$	nth root
!	Factorial
!!	Double factorial
Ø	Empty set, null set
∞	Infinity
∂	Partial differential
Δ	Delta
∇	Nabla, del
∇^2, Δ	Laplacian operator

English-Greek–Latin Numerical Prefixes

English	Greek	Latin
2	bis	di
3	tris	tri
4	tetrakis	tetra
5	pentakis	penta
6	hexakis	hexa
7	heptakis	hepta
8	octakis	octa
9	nonakis	nona
10	decakis	deca

Greek-Russian-English Alphabets

Greek letter		Greek name	English equivalent	Russian letter		English equivalent
Α	α	Alpha	(ä)	А	а	(ä)
Β	β	Beta	(b)	Б	б	(b)
				В	в	(v)
Γ	γ	Gamma	(g)	Г	г	(g)
Δ	δ	Delta	(d)	Д	д	(d)
Ε	ϵ	Epsilon	(e)	Е	е	(ye)
Ζ	ζ	Zeta	(z)	Ж	ж	(zh)
				З	з	(z)
Η	η	Eta	(ā)	И	и	(i, ē)
Θ	θ	Theta	(th)	Й	й	(ē)
Ι	ι	Iota	(ē)	К	к	(k)
				Л	л	(l)
Κ	k	Kappa	(k)	М	м	(m)
Λ	λ	Lambda	(l)	Н	н	(n)
				О	о	(ô, o)
Μ	μ	Mu	(m)	О	о	(ô, o)
				П	П	(p)
Ν	ν	Nu	(n)	Р	р	(r)
Ξ	ξ	Xi	(ks)	С	с	(s)
				Т	т	(t)
Ο	ο	Omicron	α	У	у	ōō
Π	π	Pi	(P)	Ф	ф	(f)
				Х	х	(kh)
Ρ	ρ	Rho	(r)	Х	х	(kh)
				Ц	ц	(t_s)
Σ	σ	Sigma	(s)	Ч	ч	(ch)
Τ	τ	Tau	(t)	Ш	ш	(sh)
ϒ	ν	Upsilon	(ü, ōō)	Щ	щ	(shch)
				Ъ	ъ	8
Φ	ø	Phi	(f)	Ы	ы	(ë)
Χ	χ	Chi	(H)	ь	ь	(ë)
Ψ	ψ	Psi	(ps)	Э	э	(e)
				Ю	ю	(ū)
Ω	ω	Omega	(ō)	Я	я	(yä)

English-Greek-Latin Numbers

English	Greek	Latin
1	mono	uni
2	bis	di
3	tris	tri
4	tetrakis	tetra
5	pentakis	penta
6	hexakis	hexa
7	heptakis	hepta
8	octakis	octa
9	nonakis	nona
10	decakis	deca

International Union of Pure and Applied Chemistry: Rules Concerning Numerical Terms Used in Organic Chemical Nomenclature (specifically as prefixes for hydrocarbons)

1	mono-or hen-	10 deca-	100 hecta-	1000 kilia-
2	di- or do-	20 icosa-	200 dicta-	2000 dilia-
3	tri-	30 triaconta-	300 tricta-	3000 trilia-
4	tetra-	40 tetraconta-	400 tetracta	4000 tetralia-
5	penta-	50 pentaconta-	500 pentactra	5000 pentalia-
6	hexa-	60 hexaconta-	600 hexacta	6000 hexalia-
7	hepta-	70 hepaconta-	700 heptacta-	7000 hepalia-
8	octa-	80 octaconta-	800 ocacta-	8000 ocatlia-
9	nona-	90 nonaconta-	900 nonactta-	9000 nonalia-

Source: IUPAC, Commission on Nomenclature of Organic Chemistry (N. Lorzac'h and published in *Pure and Appl. Chem* 58: 1693–1696 (1986))

Elemental Symbols and Atomic Weights

Source: International Union of Pure and Applied Chemistry (IUPAC) 2001Values from the 2001 table *Pure Appl. Chem.*, **75**, 1107–1122 (2003). The values of zinc, krypton, molybdenum and dysprosium have been modified. The *approved name* for element 110 is included, see *Pure Appl. Chem.*, **75**, 1613–1615 (2003). The *proposed name* for element 111 is also included.

A number in parentheses indicates the uncertainty in the last digit of the atomic weight.

List of Elements in Atomic Number Order

At No	Symbol	Name	Atomic Wt	Notes
1	H	Hydrogen	1.00794(7)	1, 2, 3
2	He	Helium	4.002602(2)	1, 2
3	Li	Lithium	[6.941(2)]	1, 2, 3, 4
4	Be	Beryllium	9.012182(3)	
5	B	Boron	10.811(7)	1, 2, 3
6	C	Carbon	12.0107(8)	1, 2
7	N	Nitrogen	14.0067(2)	1, 2
8	O	Oxygen	15.9994(3)	1, 2
9	F	Fluorine	18.9984032(5)	
10	Ne	Neon	20.1797(6)	1, 3
11	Na	Sodium	22.989770(2)	
12	Mg	Magnesium	24.3050(6)	
13	Al	Aluminium	26.981538(2)	
14	Si	Silicon	28.0855(3)	2
15	P	Phosphorus	30.973761(2)	
16	S	Sulfur	32.065(5)	1, 2
17	Cl	Chlorine	35.453(2)	3
18	Ar	Argon	39.948(1)	1, 2
19	K	Potassium	39.0983(1)	1
20	Ca	Calcium	40.078(4)	1
21	Sc	Scandium	44.955910(8)	
22	Ti	Titanium	47.867(1)	
23	V	Vanadium	50.9415(1)	
24	Cr	Chromium	51.9961(6)	
25	Mn	Manganese	54.938049(9)	
26	Fe	Iron	55.845(2)	
27	Co	Cobalt	58.933200(9)	
28	Ni	Nickel	58.6934(2)	
29	Cu	Copper	63.546(3)	2
30	Zn	Zinc	65.409(4)	
31	Ga	Gallium	69.723(1)	
32	Ge	Germanium	72.64(1)	
33	As	Arsenic	74.92160(2)	
34	Se	Selenium	78.96(3)	
35	Br	Bromine	79.904(1)	
36	Kr	Krypton	83.798(2)	1, 3
37	Rb	Rubidium	85.4678(3)	1
38	Sr	Strontium	87.62(1)	1, 2
39	Y	Yttrium	88.90585(2)	
40	Zr	Zirconium	91.224(2)	1
41	Nb	Niobium	92.90638(2)	
42	Mo	Molybdenum	95.94(2)	1
43	Tc	Technetium	[98]	5
44	Ru	Ruthenium	101.07(2)	1
45	Rh	Rhodium	102.90550(2)	
46	Pd	Palladium	106.42(1)	1
47	Ag	Silver	107.8682(2)	1
48	Cd	Cadmium	112.411(8)	1
49	In	Indium	114.818(3)	
50	Sn	Tin	118.710(7)	1
51	Sb	Antimony	121.760(1)	1
52	Te	Tellurium	127.60(3)	1
53	I	Iodine	126.90447(3)	
54	Xe	Xenon	131.293(6)	1, 3
55	Cs	Caesium	132.90545(2)	
56	Ba	Barium	137.327(7)	
57	La	Lanthanum	138.9055(2)	1
58	Ce	Cerium	140.116(1)	1
59	Pr	Praseodymium	140.90765(2)	
60	Nd	Neodymium	144.24(3)	1
61	Pm	Promethium	[145]	5
62	Sm	Samarium	150.36(3)	1
63	Eu	Europium	151.964(1)	1
64	Gd	Gadolinium	157.25(3)	1
65	Tb	Terbium	158.92534(2)	
66	Dy	Dysprosium	162.500(1)	1
67	Ho	Holmium	164.93032(2)	
68	Er	Erbium	167.259(3)	1

At No	Symbol	Name	Atomic Wt	Notes
69	Tm	Thulium	168.93421(2)	
70	Yb	Ytterbium	173.04(3)	1
71	Lu	Lutetium	174.967(1)	1
72	Hf	Hafnium	178.49(2)	
73	Ta	Tantalum	180.9479(1)	
74	W	Tungsten	183.84(1)	
75	Re	Rhenium	186.207(1)	
76	Os	Osmium	190.23(3)	1
77	Ir	Iridium	192.217(3)	
78	Pt	Platinum	195.078(2)	
79	Au	Gold	196.96655(2)	
80	Hg	Mercury	200.59(2)	
81	Tl	Thallium	204.3833(2)	
82	Pb	Lead	207.2(1)	1, 2
83	Bi	Bismuth	208.98038(2)	
84	Po	Polonium	[209]	5
85	At	Astatine	[210]	5
86	Rn	Radon	[222]	5
87	Fr	Francium	[223]	5
88	Ra	Radium	[226]	5
89	Ac	Actinium	[227]	5
90	Th	Thorium	232.0381(1)	1, 5
91	Pa	Protactinium	231.03588(2)	5
92	U	Uranium	238.02891(3)	1, 3, 5
93	Np	Neptunium	[237]	5
94	Pu	Plutonium	[244]	5
95	Am	Americium	[243]	5
96	Cm	Curium	[247]	5
97	Bk	Berkelium	[247]	5
98	Cf	Californium	[251]	5
99	Es	Einsteinium	[252]	5
100	Fm	Fermium	[257]	5
101	Md	Mendelevium	[258]	5
102	No	Nobelium	[259]	5
103	Lr	Lawrencium	[262]	5
104	Rf	Rutherfordium	[261]	5, 6
105	Db	Dubnium	[262]	5, 6
106	Sg	Seaborgium	[266]	5, 6
107	Bh	Bohrium	[264]	5, 6
108	Hs	Hassium	[277]	5, 6
109	Mt	Meitnerium	[268]	5, 6
110	Ds	Darmstadtium	[281]	5, 6
111	Rg	Roentgenium	[272]	5, 6
112	Uub	Ununbium	[285]	5, 6
114	Uuq	Ununquadium	[289]	5, 6
116	Uuh	Ununhexium		see Note above
118	Uuo	Ununoctium		see Note above

1. Geological specimens are known in which the element has an isotopic composition outside the limits for normal material. The difference between the atomic weight of the element in such specimens and that given in the Table may exceed the stated uncertainty.
2. Range in isotopic composition of normal terrestrial material prevents a more precise value being given; the tabulated value should be applicable to any normal material.
3. Modified isotopic compositions may be found in commercially available material because it has been subject to an undisclosed or inadvertant isotopic fractionation. Substantial deviations in atomic weight of the element from that given in the Table can occur.
4. Commercially available Li materials have atomic weights that range between 6.939 and 6.996; if a more accurate value is required, it must be determined for the specific material [range quoted for 1995 table 6.94 and 6.99].
5. Element has no stable nuclides. The value enclosed in brackets, e.g. [209], indicates the mass number of the longest-lived isotope of the element. However three such elements (Th, Pa, and U) do have a characteristic terrestrial isotopic composition, and for these an atomic weight is tabulated.
6. The names and symbols for elements 112-118 are under review. The temporary system recommended by J Chatt, *Pure Appl. Chem.*, **51**, 381–384 (1979) is used above. The names of elements 101-109 were agreed in 1997 (See *Pure Appl. Chem.*, 1997, **69**, 2471–2473) and for element 110 in 2003 (see *Pure Appl. Chem.*, 2003, **75**, 1613–1615). The proposed name for element 111 is also included.

List of Elements in Name Order

At No	Symbol	Name	Atomic Wt	Notes
89	Ac	Actinium	[227]	5
13	Al	Aluminium	26.981538(2)	
95	Am	Americium	[243]	5
51	Sb	Antimony	121.760(1)	1

At No	Symbol	Name	Atomic Wt	Notes
18	Ar	Argon	39.948(1)	1, 2
33	As	Arsenic	74.92160(2)	
85	At	Astatine	[210]	5
56	Ba	Barium	137.327(7)	
97	Bk	Berkelium	[247]	5
4	Be	Beryllium	9.012182(3)	
83	Bi	Bismuth	208.98038(2)	
107	Bh	Bohrium	[264]	5, 6
5	B	Boron	10.811(7)	1, 2, 3
35	Br	Bromine	79.904(1)	
48	Cd	Cadmium	112.411(8)	1
55	Cs	Caesium	132.90545(2)	
20	Ca	Calcium	40.078(4)	1
98	Cf	Californium	[251]	5
6	C	Carbon	12.0107(8)	1, 2
58	Ce	Cerium	140.116(1)	1
17	Cl	Chlorine	35.453(2)	3
24	Cr	Chromium	51.9961(6)	
27	Co	Cobalt	58.933200(9)	
29	Cu	Copper	63.546(3)	2
96	Cm	Curium	[247]	5
110	Ds	Darmstadtium	[281]	5, 6
105	Db	Dubnium	[262]	5, 6
66	Dy	Dysprosium	162.500(1)	1
99	Es	Einsteinium	[252]	5
68	Er	Erbium	167.259(3)	1
63	Eu	Europium	151.964(1)	1
100	Fm	Fermium	[257]	5
9	F	Fluorine	18.9984032(5)	
87	Fr	Francium	[223]	5
64	Gd	Gadolinium	157.25(3)	1
31	Ga	Gallium	69.723(1)	
32	Ge	Germanium	72.64(1)	
79	Au	Gold	196.96655(2)	
72	Hf	Hafnium	178.49(2)	
108	Hs	Hassium	[277]	5, 6
2	He	Helium	4.002602(2)	1, 2
67	Ho	Holmium	164.93032(2)	
1	H	Hydrogen	1.00794(7)	1, 2, 3
49	In	Indium	114.818(3)	
53	I	Iodine	126.90447(3)	
77	Ir	Iridium	192.217(3)	
26	Fe	Iron	55.845(2)	

At No	Symbol	Name	Atomic Wt	Notes
36	Kr	Krypton	83.798(2)	1, 3
57	La	Lanthanum	138.9055(2)	1
103	Lr	Lawrencium	[262]	5
82	Pb	Lead	207.2(1)	1, 2
3	Li	Lithium	[6.941(2)]	1, 2, 3, 4
71	Lu	Lutetium	174.967(1)	1
12	Mg	Magnesium	24.3050(6)	
25	Mn	Manganese	54.938049(9)	
109	Mt	Meitnerium	[268]	5, 6
101	Md	Mendelevium	[258]	5
80	Hg	Mercury	200.59(2)	
42	Mo	Molybdenum	95.94(2)	1
60	Nd	Neodymium	144.24(3)	1
10	Ne	Neon	20.1797(6)	1, 3
93	Np	Neptunium	[237]	5
28	Ni	Nickel	58.6934(2)	
41	Nb	Niobium	92.90638(2)	
7	N	Nitrogen	14.0067(2)	1, 2
102	No	Nobelium	[259]	5
76	Os	Osmium	190.23(3)	1
8	O	Oxygen	15.9994(3)	1, 2
46	Pd	Palladium	106.42(1)	1
15	P	Phosphorus	30.973761(2)	
78	Pt	Platinum	195.078(2)	
94	Pu	Plutonium	[244]	5
84	Po	Polonium	[209]	5
19	K	Potassium	39.0983(1)	1
59	Pr	Praseodymium	140.90765(2)	
61	Pm	Promethium	[145]	5
91	Pa	Protactinium	231.03588(2)	5
88	Ra	Radium	[226]	5
86	Rn	Radon	[222]	5
75	Re	Rhenium	186.207(1)	
45	Rh	Rhodium	102.90550(2)	
111	Rg	Roentgenium	[272]	5, 6
37	Rb	Rubidium	85.4678(3)	1
44	Ru	Ruthenium	101.07(2)	1
104	Rf	Rutherfordium	[261]	5, 6
62	Sm	Samarium	150.36(3)	1
21	Sc	Scandium	44.955910(8)	
106	Sg	Seaborgium	[266]	5, 6
34	Se	Selenium	78.96(3)	
14	Si	Silicon	28.0855(3)	2

At No	Symbol	Name	Atomic Wt	Notes
47	Ag	Silver	107.8682(2)	1
11	Na	Sodium	22.989770(2)	
38	Sr	Strontium	87.62(1)	1, 2
16	S	Sulfur	32.065(5)	1, 2
73	Ta	Tantalum	180.9479(1)	
43	Tc	Technetium	[98]	5
52	Te	Tellurium	127.60(3)	1
65	Tb	Terbium	158.92534(2)	
81	Tl	Thallium	204.3833(2)	
90	Th	Thorium	232.0381(1)	1, 5
69	Tm	Thulium	168.93421(2)	
50	Sn	Tin	118.710(7)	1
22	Ti	Titanium	47.867(1)	
74	W	Tungsten	183.84(1)	

At No	Symbol	Name	Atomic Wt	Notes
112	Uub	Ununbium	[285]	5, 6
116	Uuh	Ununhexium		see Note above
118	Uuo	Ununoctium		see Note above
114	Uuq	Ununquadium	[289]	5, 6
92	U	Uranium	238.02891(3)	1, 3, 5
23	V	Vanadium	50.9415(1)	
54	Xe	Xenon	131.293(6)	1, 3
70	Yb	Ytterbium	173.04(3)	1
39	Y	Yttrium	88.90585(2)	
30	Zn	Zinc	65.409(4)	
40	Zr	Zirconium	91.224(2)	1

Pronounciation Symbols and Abbreviations

ə	Banana, collide, abut
ˈə, ˌə	Humdrum, abut
^{ə}	Immediately preceding \l\, \n\, \m\, \ŋ\, as in battle, mitten, eaten, and sometimes open
	\ˈō-p^{ə}m\, lock **and** key \-^{ə} ŋ-\; immediately following \l\, \m\, \r\, as often in French table, prisme, titre
ər	further, merger, bird
ˈə-, ˈə-r	As in two different pronunciations of hurry \ˈ hər-ē, \ˈ hə-rē\
a	mat, map, mad, gag, snap, patch
ā	day, fade, date, aorta, drape, cape
ä	bother, cot, and, with most American speakers, father, cart
á	father as pronounced by speakers who do not rhyme it with *bother*; French patte
aú	now, loud, out
b	baby, rib
ch	chin, nature \ˈnā-chər\
d	did, adder
e	bet, bed, peck
ˈē, ˌē	beat, nosebleed, evenly, easy
ē	easy, mealy
f	fifty, cuff
g	go, big, gift
h	hat, ahead
hw	whale as pronounced by those who do not have the same pronunciation for both *whale* and *wail*
i	tip, banish, active
ī	site, side, buy, tripe
j	job, gem, edge, join, judge
k	kin, cook, ache
k̲	German ich, Buch; one pronunciation of loch
l	lily, pool
m	murmur, dim, nymph
n	no, own
^{n}	Indicates that a preceeding vowel or diphthong is pronounced with the nasal passages open, as in French *un bon vin blanc* \œ^n-bō^{n}va^n-blä^n\
ŋ	sing \ˈ siŋ \, singer \ˈ siŋ-ər\, finger \ˈfiŋ-gər\, ink \ˈiŋk\
ō	bone, know, beau

ó	saw, all, gnaw, caught
o̅o̅	fool
o͝o	took
œ	French coeuf, German Hölle
œ̲	French feu, German Höhle
ói	coin, destroy
p	pepper, lip
r	red, car, rarity
s	source, less
sh	as in shy, mission, machine, special (actually, this is a single sound, not two); with a hyphen between, two sounds as in *grasshopper* \ˈgras-ˌhä-pər\
t	tie, attack, late, later, latter
th	as in thin, ether (actually, this is a single sound, not two); with a hyphen between, two sounds as in knighthood \ˈnīt-ˌh----d\
t̲h̲	then, either, this (actually, this is a single sound, not two)
ü	rule, youth, union \ˈyün-yən\, few \ˈfyü\
ú	pull, wood, book, curable \ˈky ú r-ə-bəl\, fury \ˈfy----r-ē\
ue	German füllen, hübsch
u̲e̲	French rue, German fühlen
v	vivid, give
w	we, away
y	yard, young, cue \ˈkyü\, mute \ˈmyüt\, union \ˈyün-yən\
^{y}	indicates that during the articulation of the sound represented by the preceding character the front of the tongue has substantially the position it has for the articulation of the first sound of yard, as in French *digne* \dēn^y\
Z	zone, raise
zh	as in vision, azure \ˈa-zhər\ (actually this is a single sound, not two).
\	reversed virgule used in pairs to mark the beginning and end of a transcription: \ˈpen\
ˈ	mark preceding a syllable with primary (strongest) stress: \ˈpen-mən-ˌship\
ˌ	mark preceding a syllable with secondary (medium) stress: \ˈpen-mən-ˌship\
-	mark of syllable division

( )	indicate that what is symbolized between is present in some utterances but not in others: *factory* \ ˈfak-t(ə-)rē
÷	indicates that many regard as unacceptable the pronunciation variant immediately following: *cupola* \ ˈkyü-pə-lə, ÷- ˌlō\

Explanatory Notes and Abbreviations

(date)	date that word was first recorded as having been used
[. . .]	etomology and origin(s) of word
{. . .}	usage and/or languages, including French, German, Italian and Spanish
adj	adjective
adv	adverb
B.C.	before Christ
Brit.	Britain, British
C	centigrade, Celsius
c	century
E	English
Eng.	England
F	French, Fahrenheit
Fr.	France
fr.	from
G	German
Gr.	Germany
L	Latin
ME	middle English
n	noun
neut.	neuter
NL	new Latin
OE	old English
OL	old Latin
pl	plural
prp.	present participle
R	Russian
sing.	singular
S	Spanish
U.K.	United Kingdom
v	verb

Source: From *Merriam-Webster's Collegiate© Dictionary*, Eleventh Editioh, ©2004 by Merriam-Webster, incorporated, (www.Merriam-Webster.com). With permission.

Languages

French, German and Spanish translations are enclosed in {--} and preceded by F, G, I and S, respectively; and gender is designated by f-feminine, m-masculine, n-neuter. For example: **Polymer**--{F polymere m} represents the French translation "polymere" of the English word polymer and it is in the masculine case. These translations were obtain from multi-language dictionaries including: *A Glossary of Plastics Terminology in 5 Languages*, 5th Ed., Glenz, W., (ed) Hanser Gardner Publications, Inc., Cinicinnati, 2001. By permission).

Appendices

Appendix A

Conversion factors

To convert	Into	Multiply by
A		
atm	cm of mercury	76.0
atm	ft of water (at 4°C)	33.90
atm	in. of mercury (at 0°C)	29.92
atm	kg/cm^2	1.0333
atm	$lb/in.^2$	14.70
atm	$tons/ft^2$	1.058
B		
btu	ft-lb	778.3
btu	g-cal	252.0
btu	J	1,054.8
btu	k-cal	0.2520
btu	kW-h	2.928×10^{-4}
btu/h	g-cal/s	0.0700
btu/h	W	0.2931
C		
cm	ft	3.281×10^{-2}
cm	in.	0.3937
cm	miles	6.214×10^{-6}
cm	mm	10.0
cm/s	ft/min	1.1969
cm/s	ft/s	0.03281
cm/s	k/h	0.036
cm/s	m/min	0.6
cc	ft^3	3.531×10^{-5}
cc	$in.^3$	0.06102
cc	m^3	10^{-6}
cc	gal (U.S. liq.)	2.642×10^{-4}
cc	'	0.001
ft^3	$in.^3$	1,728.0
ft^3	$m.^3$	0.02832
ft^3	gal (U.S. liq.)	7.48052
ft^3	1	28.32
ft^3/min	cm^3/s	472.0
ft^3/min	gal/s	0.1247
ft^3/min	1/s	0.4720

To convert	Into	Multiply by
$in.^3$	cc	16.39
$in.^3$	ft^3	5.787×10^{-4}
ft^3/lb	cm^3/g	62.43
$in.^3/oz.$	cc/g	0.577
$in.^3$	gal	4.329×10^{-3}
$in.^3$	1	0.01639
m^3	ft^3	35.31
m^3	$in.^3$	61,023.0
m^3	gal (U.S. liq.)	264.2
D		
days	min	1,440.0
days	s	86,400.0
degrees (angle)	rad	0.01745
deg/s	rad/s	0.01745
dynes	G	1.020×10^3
dynes	J/cm	10^{-7}
dynes	lb	2.248×10^6
F		
ft	km	3.048×10^4
ft	M	0.3048
ft of water	atm	0.02950
ft of water	in. of mercury	0.8826
ft of water	kg/cm^2	0.03048
ft of water	$lb/in.^2$	0.4335
ft/min	cm/s	0.5080
ft/min	m/min	0.3048
ft/s	cm/s	30.48
ft/s	m/min	18.29
ft-lb	btu	1.286×10^{-3}
ft-lb	g-cal	0.3238
ft-lb	hp-h	5.050×10^{-7}
ft-lb	J	1.356
ft-lb	k-h	3.766×10^{-7}
ft-lb/min	btu/min	1.286×10^{-3}
ft-lb/min	hp	3.030×10^{-5}
ft-lb/min	kg-cal/min	3.24×10^{-4}
ft-lb/min	kW	2.260×10^{-5}

To convert	Into	Multiply by
G		
gal	cc	3,785.0
gal	ft^3	0.1337
gal	$in.^3$	231.0
gal (liq. Br. Impt)	gal (U.S. liq.)	1.20095
gal (U.S.)	gal (Imp.)	0.83267
gal of water	lb of water	8.3453
gal/min	ft^3/s	2.228×10^{-3}
gal/min	1/s	0.06308
gal/min	1/min	3.785
g	oz (avdp)	0.03527
g	lb	2.205×10^{-3}
g/cm	lb/in.	5.600×10^{-3}
g/cc	lb/ft^3	62.43
g/cc	lb/ft^3	0.03613
g/cc	$oz/in.^3$	0.5781
g/cc	$lb/in.^3$	0.03613
gr/cm^2	lb/ft^2	2.0481
H		
hp	btu/min	42.44
hp	ft-lb/s	550.0
hp	W	745.7
hp-h	btu	2,547
hp-h	kg-cal	641.1
hp-h	k-h	0.7457
I		
in.	cm	2.540
in. of mercury	atm	0.03342
in. of mercury	kg/cm^2	0.03453
in. of mercury	$lb/in.^2$	0.4912
in. of water (at 4°C)	atm	2.458×10^{-3}
in. of water (at 4°C)	in. of mercury	0.07355
in. of water (at 4°C)	kg/cm^2	2.540×10^{-3}
in. of water (at 4°C)	$lb/in.^2$	0.03613
J		
J	btu	9.480×10^{-4}
J/cm	G	1.020×10^{4}
J/cm	lb	22.48
K		
kg/cm^2	atm	0.9678
kg/cm^2	ft of water	32.81
kg/cm^2	in. of mercury	28.96
kg/cm^2	$lb/in.^2$	14.22
kg-cal	btu	3.968
kg-cal	J	4,186
kg-cal	kW-h	1.163×10^{-3}
kg-m	btu	9.294×10^{-3}
kg-m	k-h	2.723×10^{-6}
kW	btu/min	56.92
kW	hp	1.341
kW	kg-cal/min	14.34
kW-h	btu	3,413
kW-h	hp	1.341
kW-h	kg-cal	860.5
L		
L	$in.^3$	61.02
L	gal (U.S. liq.)	0.2642
1/min	ft^2/s	5.886×10^{-4}
1/min	gal/s	4.403×10^{-3}
M		
m/min	cm/s	1.667
m/min	miles/h	0.03728
m/s	ft/s	2.281
m/s	k/min	0.06
m/s	miles/h	2.237
m-kg	cm-dynes	9.807×10^{7}
m-kg	lb-ft	7.233
miles (statute)	ft	5,280
miles (statute)	K	1.609
miles (statute)	yards	1,760
miles/h	cm/s	44.70
miles/h	ft/min	88
miles/h	m/min	26.82
miles/min	cm/s	2,682
miles/min	natural knots/min	0.8684
mm	ft	3.281×10^{-3}
mm	in.	0.03937
mils	cm	2.540×10^{-3}
mils	in.	0.001
O		
oz	g	28.3495
oz	lb	0.0625
oz (fluid)	$in.^3$	1.805
oz (fluid)	L	0.02957
$oz/in.^2$	$lb/in.^2$	0.0625
$oz/in.^2$	g/cc	1.733

To convert	Into	Multiply by
P		
lb	dynes	44.4823×10^4
lb	g	453.59
lb of water	ft^3	0.01602
lb of water	$in.^3$	27.68
lb of water	gal	0.1198
lb of water/min	ft^3/s	2.670×10^{-4}
lt-ft	cm-dynes	1.356×10^7
lb-ft	m-kg	0.1383
lb/ft^3	g/cc	0.01602
lb/ft^3	kg/m^3	16.02
$lb/in.^3$	g/cc	27.68
lb/in.	g/cm	178.6
$lb/in.^2$	atm	0.06804
$lb/in.^2$	ft of water	2.307
$lb/in.^2$	in. of mercury	2.036
$kb/in.^2$	kg/m^2	703.1
$lb/in.^2$	kg/cm^2	0.07031
Q		
quarts (liq.)	cc	946.4
quarts (liq.)	$in.^3$	57.75
quarts (liq.)	L	0.9463
S		
cm^2	circular mils	1.973×10^5
cm^2	ft^2	1.076×10^{-3}
cm^2	$in.^2$	0.1550
ft^2	cm^2	929.0

To convert	Into	Multiply by
$in.^2$	ft^2	6.452
m^2	cm^2	10.76
m^2	$in.^2$	1,550
m^2	$yards^2$	1.196
mm^2	circular mils	1,973
mm^2	$in.^2$	1.550×10^{-3}
$yards^2$	cm^2	8,361
T		
tons (long)	lb	2,240
tons (metric)	k	1,000
tons (metric)	lb	2,205
tons (short)	kg	907.18
tons (short)	lb	2,000
tons (short)	tons (long)	0.89287
tons (short)	tons (metric)	0.9078
W		
W	btu/h	3,413
W	erg/s	107
W	ft-lb/min	44.27
W	hp	1.341×10^{-3}
W	kg-cal/min	0.01433
W-h	btu	3.413
W-h	g-cal	859.85
W-h	hp-h	1.341×10^{-3}
W-h	kg-m	367.2
Y		
yards	cm	91.4

Appendix B

International Standards Organization (ISO) units

The International System of units (SI) was adopted by the 11th General Conference on Weights and Measures (CGPM) in 1960. It is a coherent system of units built from seven SI base units, one for each of the seven dimensionally independent base quantities: they are the meter, kilogram, second, ampere, Kelvin, mole, and candela, for the dimensions length, mass, time, electric current, thermodynamic temperature, amount of substance, and luminous intensity, respectively. The definitions of the SI base units are given below. The SI derived units are expressed as products of powers of the base units, analogous to the corresponding relations between physical quantities but with numerical factors equal to unity.

In the International System there is only one SI unit for each physical quantity. This is either the appropriate SI base unit itself or the appropriate SI derived unit. However, any of the approved decimal prefixes, called SI prefixes, may be used to construct decimal multiples or submultiples of SI units.

It is recommended that only SI units be used in science and technology (with SI prefixes where appropriate). Where there are special reasons for making an exception to this rule, it is recommended always to define the units used in terms of SI units. This section was reprinted with the permission of IUPAXC.

Definitions of SI base units

Meter – The meter is the length of path traveled by light in vacuum during a time interval of 1/299,792,458 of a second (17th CGPM, 1983).

Kilogram – The kilogram is the unit of mass; it is equal to the mass of the international prototype of the kilogram (third CGPM, 1901).

Second – The second is the duration of 9,192,631,770 periods of the radiation corresponding to the transition between the two hyperfine levels of the ground state of the cesium-133 atom (13th CGPM, 1967).

Ampere – The ampere is that constant current which, if maintained in two straight parallel conductors of infinite length, of negligible circular cross-section, and placed 1 m apart in vacuum, would produce between these conductors a force equal to 2×10^{-7} newton per meter of length (ninth CGPM, 1948).

Kelvin – The Kelvin, unit of thermodynamic temperature, is the fraction 1/273.16 of the thermodynamic temperature of the triple point of water (13th CGPM, 1967).

Mole – The mole is the amount of substance of a system which contains as many elementary entities as there are atoms in 0.012 kg of cargon-12. When the mole is used, the elementary entities must be specific and may be atoms, molecules, ions, electrons, other particles, or specified groups of such particle (14th GCPM, 1971).

Examples of the use of the mole:

1 mol of H_2 contains about 6.022×10^{23} H^2 molecules, or 12.044×10^{23} H atoms
1 mol of HgCl has a mass of 236.04 g
1 mol of $Hg_{2_{2+}}cl_2$ has a mass of 472.08 g
1 mol of Hg_2 has a mass of 401.18 g and a charge of 192.97 kC
1 mol of $Fe_{0.91}S$ has a mass of 82.88 g
1 mol of e^- has a mass of 548.60 μg and a charge of −96.49 kC
1 mol of photons whose frequency is 10^{14} Hz has energy of about 39.90 kJ

Candela – The candela is the luminous intensity, in a given direction, of a source that emits monochromatic radiation of frequency 540×10^{12} Hz (0.556 μm = λ) and that has a radiant intensity in that direction of (1/683) watt per steradian (16th CGPM, 1979).

Names and symbols for the SI base units

Physical quantity	Name of SI unit	Symbol for SI unit
Length	Meter	m
Mass	Kilogram	kg
Time	Second	s
Electric current	Ampere	A
Thermodynamic temperature	Kelvin	K
Amount of substance	Mole	mol

Physical quantity	Name of SI unit	Symbol for SI unit	
Luminous intensity	Candela	cd	
Frequency[1]	Hertz	Hz	s^{-1}
Force	Newton	N	$m\ kg\ s^{-2}$
Pressure, stress	Pascal	Pa	$Nm^{-2} = m^{-1}\ kg\ s^{-2}$
Energy, work, heat	Joule	J	$Nm = m^2\ kg\ s^{-2}$
Power, radiant flux	Watt	W	$Js^{-1} = m^2\ kg\ s^{-3}$
Electric charge	Coulomb	C	A s
Electric potential, Electromotive force	Volt	V	$JC^{-1} = m^2\ kg\ s^{-3}\ A^{-1}$
Electric resistance	Ohm	Ω	$VA^{-1} = m^2\ kg\ s^{-3}\ A^{-2}$
Electric conductance	Siemens	S	$\Omega^{-1} = m^{-2}\ kg^{-1}\ s^3\ A^2$
Electric capacitance	Farad	F	$CV^{-1} = m^{-2}\ kg^{-1}\ s^4\ A^2$
Magnetic flux density	Tesla	T	$V\ s\ m^{-2} = kg\ s^{-2}\ A^{-1}$
Magnetic flux	Weber	Wb	$Vs = m^2\ kg\ s^{-2}\ A^{-1}$
Inductance	Henry	H	$VA^{-1}\ s = m^2\ kg\ s^{-2}\ A^{-2}$
Celsius temperature[2]	Degree celsius	°C	K
Luminous flux	Lumen	1m	cd sr
Illuminance	Lux	1x	$cd\ sr\ m^{-2}$
Activity[3] (radioactive)	Becquerel	Bq	s^{-1}
Absorbed dose[3] (of radiation)	Gray	Gy	$J\ kg^{-1} = m^2\ s^{-2}$
Dose equivalent[3] (dose equivalent index)	Sievert	Sv	$J\ kg^{-1} = m^2\ s^{-2}$
Plane angle[4]	Radian	rad	$1 = m\ m^{-1}$
Solid angle[4]	Stradian	sr	$1 = m^2\ m^{-2}$

[1]For radial (circular) frequency and for angular velocity the unit rad s^{-1}, or simply s^{-1}, should be used, and this may not be simplified to Hz. The unit Hz should be used only for frequency in the sense of cycles per second.

[2]The celsius temperature θ is defined by the equation: $\theta/°C = T/K - 273.15$

The SI unit of Celsius temperature interval is the degree celsius, °C, which is equal to the Kelvin, K. °C should be treated as a single symbol, with no space between the ° sign and the letter C. (The symbol °K, and the symbol °, should no longer be used.)

[3]The units gray and sievert are admitted for reasons of safeguarding human health.

[4]The units radian and steradian are described as "SI supplementary units." However, in chemistry, as well as in physics, they are usually treated as dimensionless derived units, and this was recognized by CIPM in 1980. Since they are then of dimension 1, this leaves open the possibility of including them or omitting them in expressions of SI derived units. In practice this means that rad and sr may be used when appropriate and may be omitted if clarity is not lost thereby.

SI prefixes

To signify decimal multiples and submultiples of SI units the following prefixes may be used.

Factor	Prefix	Symbol	Factor	Prefix	Symbol
10^{24}	yotta	Y	10^{-1}	deci	d
10^{21}	zetta	Z	10^{-2}	centi	c
10^{18}	exa	E	10^{-3}	milli	m
10^{15}	peta	P	10^{-6}	micro	u
10^{12}	tera	T	10^{-9}	nano	n
10^9	giga	G	10^{-12}	pico	p
10^6	meta	M	10^{-15}	femto	f
10^3	kilo	k	10^{-18}	atto	a
10^2	hecto	h	10^{-21}	zepto	z
10^1	deka	da	10^{-24}	yocto	y

Prefix symbols should be printed in roman (upright) type with no space between the prefix and the unit symbol.
Example: kilometer, km
When a prefix is used with a unit symbol, the combination is taken as a new symbol that can be raised to any power without the use of parentheses.
Examples: $1\ cm^3 = (0.01\ m)^3 = 10^{-6}\ m^3$
$1\ \mu s^{-1} = (10^{-6}\ s)^{-1} = 10^6\ s^{-1}$
1 V/cm = 100 V/m
$1\ mmol/dm^3 = mol\ m^{-3}$
A prefix should never be used on its own, and prefixes are not to be combined into compound prefixes.
Example: pm, not μμm
The names and symbols of decimal multiples and submultiples of the SI base unit of mass, the kg, which already contains a prefix, are constructed by adding the appropriate prefix to the word grarn and symbol g.
Examples: mg, not μkg; Mg, not kkg
The SI prefixes are not be used with °C.

Units in use together with the SI

These units are not part of the SI, but it is recognized that they will continue to be used in appropriate contexts. SI prefixes may be attached to some of these units, such as milliliter, ml; millibar, mbar; megaelectronvolt, McV; kilotonne, ktonne.

Physical Quantity	Name of unit	Symbol for unit	Value in SI units
Time	Minute	min	60 s
Time	Hour	h	3,600 s
Time	Day	d	86,400 s
Plane angle	Degree	o	$(\pi/180)$ rad
Plane angle	Minute	′	$(\pi/10{,}800)$ rad
Plane angle	Second	″	$(\pi/648{,}000)$ rad
Length	Ängström[1]	Å	10^{-10} m
Area	Barn	b	10^{-28} m^2
Volume	Liter	l, L	dm^3 = 10^{-3} m^3
Mass	Tonne	t	Mg = 10^3 kg
Pressure	Bar[1]	bar	10^5 Pa = 10^5 Nm^{-2}
Energy	Electrovolt[2]	eV (= e × V)	$(\approx 1.60218 \times 10^{-19})$
Mass	Unified atomic mass unit[2.3]	u(= m_2 (^{12}C)/12)	$\approx 1.66054 \times 10^{-27}$ kg

[1]The ängström and the bar are approved by CIPM for "temporary use with SI units," until CIPM makes a further recommendation. However, they should not be introduced where they are not used at present.
[2]The values of these units in terms of the corresponding SI units are not exact, since they depend on the values of the physical constants e (for the electrovolt) and N_A (for the unified atomic mass unit), which are determined by experiment.
[3]The unified atomic mass unit is also sometimes called the Dalton, with symbol Da, although the name and symbol have not been approved by CGPM.

Atomic units

For the purpose of quantum mechanical calculations of electronic wavefunctions, it is convenient to regard certain fundamental constants (and combinations of such constants) as though they were units. They are customarily called atomic units (abbreviated: au), and they may be regarded as forming a coherent system of units for the calculation of electronic properties in theoretical chemistry, although there is no authority from CGPM for treating them as units. The first five atomic units in the table below have special names and symbols. Only four of these are independent; all others may be derived by multiplication and division in the usual way, and the table includes a number of examples.

The relation of atomic units to the corresponding SI units involves the values of the fundamental physical constants, and is therefore not exact. The numerical values in the table are based on the 1986 CODATA values of the fundamental constants. The numerical results of calculations in theoretical chemistry are frequently quoted in atomic units, or as numerical values in the form (physical quantity)/(atomic unit), so that the reader may make the conversion using the current best estimates of the physical constants.

Physical Quantity	Name of unit	Symbol for unit	Definition and value of unit in SI
Mass	Electron rest mass	m_e	$m_e \approx 9.1095 \times 10^{-31}$ kg
Charge	Elementary charge	e	$e \approx 1.6022 \times 10^{-19}$ C
Action	Planck constant/2π	h	$h = h/2\pi \approx 1.0546 \times 10^{-34}$ Js
Length	Bohr	a_o	$4\pi \varepsilon_o h^2/m_e e^2 \approx 5.2918 \times 10^{-11}$ m
Energy	Hartree	E_h	$h^2/m_e a_o^2 \approx 4.3598 \times 10^{-18}$ J
Time	au of time	h/E_h	$\approx 2.44189 \times 10^{-17}$ s
Velocity[1]	au of velocity	a_oE_h/h	$\approx 2.1877 \times 10^6$ m s^{-1}
Force	au of force	E_h/a_o	$\approx 8.2389 \times 10^{-8}$ N
Momentum Linear	au of momentum	h/a_o	$\approx 1.9929 \times 10^{-24}$ N s
Electric current	au of current	eE_h/h	$\approx 6.6236 \times 10^{-3}$ A
Electric field	au of electric field	E_h/ea_o	$\approx 5.1422 \times 10^{11}$ V m^{-1}
Electric dipole moment	au of electric dipole	ea_o	$\approx 8.4784 \times 10^{-30}$ C m
Magnetic flux density	au of magnetic flux density	h/ea_o^2	$\approx 2.3505 \times 10^5$ T
Magnetic dipole moment[2]	au of magnetic dipole	mh/m_e	$= 2\mu_B \approx 1.8548 \times 10^{-23}$ JT^{-1}

[1]The numerical value of the speed of light, when expressed in atomic units, is equal to the reciprocal of the fine structure constant α; c/(au of velocity) = $ch/a_oE_h = \alpha^{-1}$ 137.04.
[2]The atomic unit of magnetic dipole moment is twice the Bohr magnetron, μB.

Conversion to SI units

The following tables gives conversion factors from various units of measure to SI units. It is reproduced from NIST Special Publication 811, Guide for the Use of the International System of Units (Superintendent of Documents, U.S. Government Printing Office, 1991), which in turn was derived from IEEE Std 268-1982, IEEE Standard Metric Practice (© 1982 by the Institute of Electrical and Electronics Engineers, Inc.).

The SI values are expressed in terms of the base, supplementary, and derived units of SI in order to provide a coherent presentation of the conversion factors and facilitate computations (see the table "International System of Units" in this section). Powers of ten can be avoided by using SI prefixes and shifting the decimal point if necessary. Conversion from a non-SI unit to a different non-SI unit may be carried out by using this table in two stages, e.g.,
1 cal (thermochemical) = 4.184 J
1 Btu (mean) = 1.05587 E + 03J
Thus, 1 Btu (mean)
= (1.05587 E + 03/4.184 cal (thermochemical)
= 252.359 cal (thermochemical).

Conversion factors are presented for ready adaptation to computer readout and electronic data transmission. The factors are written as a number equal to or greater than one and less than ten with six or less decimal places. This number is followed by the letter E (for exponent), plus or minus symbol, and two digits which indicate the power of 10 by which the number must be multiplied to obtain the correct value. For example

3.523907 E − 02 is 3.523907×10^{-2}

or

0.03523907

Similarly:

3.386389 E + 03 is 3.386389×10^{3}

or

3 386.389

An asterisk (*) after the sixth decimal place indicates that the conversion factor is exact and that all subsequent digits are zero. All other conversion factors have been rounded to the figures given in accordance with accepted practice. Where less than six decimal places are shown, more precision is not warranted.

To convert from	To	Multiply by
abampere	ampere (A)	1.000000 *E + 01
abcoulomb	coulomb (C)	1.000000 *E + 01

(Continued)

To convert from	To	Multiply by
abfaarad	farad (F)	1.000000 *E + 09
abhenry	henry (H)	1.000000 *E − 09
abmho	siemens (S)	1.000000 *E + 09
abohm	ohm (Ω)	1.000000 *E − 09
abvolt	volt (V)	1.000000 *E − 08
acre foot	meter^3 (m^3)	1.2335 E + 03
acre	meter^2 (m^2)	4.046873 E + 03
ampere hour	coulomb (C)	3.600000 *E + 03
angstrom	meter (m)	1.000000 *E − 10
are	meter^2 (m^2)	1.000000 *E + 02
astronomical unit	meter (m)	1.495979 E + 11
atmosphere (standard)	pascal (Pa)	1.013250 *E + 05
atmosphere (technical = 1 kgf/cm^2)	pascal (Pa)	9.806650 *E + 04
bar	pascal (Pa)	1.000000 *E + 05
barn	meter^2 (m^2)	1.000000 *E − 28
barrel (for petroleum, 42 gal)	meter^3 (m3)	
board foot	meter^3 (m^3)	2.359737 E − 03
British thermal unit (International Table)	joule (J)	1.055056 E + 03
British thermal unit (mean)	joule (J)	1.05587 E + 03
British thermal unit (thermochemical)	joule (J)	1.054350 E + 03
British thermal unit (39°F)	joule (J)	1.05967 E + 03
British thermal unit (59°)	joule (J)	1.05468 E + 03
British thermal unit (60°)	joule (J)	1.05468 E + 03
Btu (International Table)/ft/(h · ft^2 · °F) (thermal conductivity)	watt per meter kelvin [W/(m · k)]	1.730735 E + 00
Btu (thermochemical)/(ft/h · ft^2 · °F) (thermal conductivity)	watt per meter kelvin [W/(m · k)]	1.729577 E + 00
Btu (International Table)/in/(h · ft^2 · °F) (thermal conductivity)	watt per meter kelvin [W/(m · k)]	1.442279 E − 01
Btu (thermochemical)/in/(h · ft^2 · °F) (thermal conductivity)	watt per meter kelvin [W/(m · k)]	1.441314 E − 01
Btu (International Table)/in/(s · ft^2 · °F) (thermal conductivity)	watt per meter kelvin [W/(m · k)]	5.192204 E + 02

(Continued)

To convert from	To	Multiply by
Btu (thermal chemical)/ in/(s · ft^2 · °F) (thermal conductivity)	watt per meter kelvin [W/(m · k)]	5.188732 E + 02
Btu (International Table)/h	watt (W)	2.930711 E − 01
Btu (International Table)/s	watt (W)	1.055056 E + 03
Btu (thermochemical)/h	watt (W)	2.928751 E − 01
Btu (thermochemical)/min	watt (W)	1.757250 E + 01
Btu (thermochemical)/s	watt (W)	1.054350 E + 03
Btu (International Table)/ft^2	joule per $meter^2$ (j/m^2)	1.135653 E + 04
Btu (thermochemical)/ft^2	joule per $meter^2$ (J/m^2)	1.134893 E + 04
Btu (International Table)/(ft^2 · h)	watt per $meter^2$ (W/m^2)	3.154591 E + 00
Btu (International Table)/(ft^2 · s)	watt per $meter^2$ (W/m^2)	1.135653 E + 04
Btu (thermochemical)/ (ft^2 · h)	watt per $meter^2$ (W/m^2)	3.152481 E + 00
Btu (thermochemical)/ (ft^2 · min)	watt per $meter^2$ (W/m^2)	1.891489 E + 02
Btu (thermochemical)/ (ft^2 · s)	watt per $meter^2$ (W/m^2)	1.134893 E + 04
Btu (thermochemical)/ (in^2 · s)	watt per meter (W/m^2)	1.634246 E + 06
Btu (International Table)/(h · ft^2 · °F)	watt per $meter^2$ kelvin [W/(m^2 · K)]	5.678263 E + 00
Btu (thermalcohemical)/ (h · ft^2 · °F)	watt per $meter^2$ kelvin [W/m^2 · K)]	5.674466 E + 00
Btu (International Table)/(s · ft^2 · °F)	watt per $meter^2$ kelvin [W/(m^2 · K)]	2.044175 E + 04
Btu (thermochemical)/ (s · ft^2 · °F)	watt per $meter^2$ kelvin [W/(m^2 · K)]	2.042808 E + 04
Btu (International Table)/lb	joule per kilogram (J/kg)	2.326000 *E + 03
Btu (thermochemical)/ lb	joule per kilogram (J/kg)	2.324444 E + 03

(Continued)

To convert from	To	Multiply by
Btu (International Table)/(lb · °F) (specific heat capability)	joule per kilogram kelvin [J/(kg · K)]	4.186800 *E + 03
Btu (thermochemical)/ (lb · °F)/(specific heat capacity)	joule per kilogram kelvin [(J/(kg · K)]	4.184000 *E + 03
Btu (International Table)/ft^3	joule per $meter^3$ (J/m^3)	3.725895 E + 04
Btu (thermochemical)/ ft^3	joule per $meter^3$ (J/m^3)	3.723402 E + 04
bushel	$meter^3$ (m^3)	3.523907 E − 02
calorie (International Table)	joule (J)	4.186800 *E + 00
calorie (mean)	joule (J)	4.19002 E + 00
calorie (thermochemical)	joule (J)	4.184000 *E + 00
calorie (15°C)	joule (J)	4.18580 E + 00
calorie (20°C)	joule (J)	4.18190 E + 00
calorie (kilogram, International Table)	joule (J)	4.186800 *E + 03
calorie (kilogram, mean)	joule (J)	4.19002 E + 03
calorie (kilogram, thermochemical)	joule (J)	4.184000 *E + 03
cal (thermochemical)/cm^2	joule per $meter^2$ (J/m^2)	4.184000 *E + 04
cal (International Table)/g	joule per kilogram (J/kg)	4.186800 *E + 03
cal (thermochemical)/g	joule per kilogram (J/kg)	4.184000 *E + 03
cal (International Table)/(g · °C)	joule per kilogram kelvin [J/(kg · K)]	4.186800 *E + 03
cal (thermochemical)/ (g · °C)	joule per kilogram Kelvin [J/(kg · K)]	4.184000 *E + 03
cal (thermochemical)/ min	watt (W)	6.973333 E − 02
cal (thermochemical)/s	watt (W)	4.184000 *E + 00
cal (thermochemical)/ (cm^2 · min)	watt per $meter^2$ (W/m^2)	6.973333 E − 02
cal (thermochemical)/ (cm^2 · s)	watt per $meter^2$ (W/m^2)	4.184000 E + 04
cal (thermochemical)/ (cm · s · °C)	watt per meter Kelvin [W/(m · K)]	4.184000 E + 02

(Continued)

To convert from	To	Multiply by
cd/in^2	candela per meter^2 (cd/m^2)	1.550003 E + 03
carat (metric)	kilogram (kg)	2.000000 *E – 04
centimeter of mercury (0°C)	pascal (Pa)	1.33322 E + 03
centimeter of water (4°C)	pascal (Pa)	9.80638 E + 01
centipoises	pascal second (Pa.s)	1.000000 *E – 03
centistokes	meter^2 per second (m^2/s)	1.000000 *E – 06
chain	meter^2 (m^2)	5.067075 E + 01
circular mill	meter^2 (m^2)	5.067075 E – 10
clo	Kelvin meter^2 per watt (K – m^2/W)	2.003712 E – 01
cup	milliliter (mL)	2.366 E + 02
curie	becquerel (BQ)	3.700000 *E + 10
darcy^2	meter^2 (m^2)	9.869233 E – 13
day	second (s)	8.640000 *E + 04
day (sidereal)	second (s)	8.616409 E + 04
degree (angle)	radian (rad)	1.745329 E – 02
degree Celsius	Kelvin (K)	$T_K = t_C + 273.15$
degree centigrade	[see note below]	
degree Fahrenheit	degree Celsius (°C)	$t_C = (t_F - 32)/1.8$
degree Fahrenheit	Kelvin (K)	$T_k = t_F /1.8$
degree Rankine	Kelvin (K)	$T_K = T_R/1.8$
°F·h·ft^2/Btu (International table)	kelvin meter^2 per watt (K·m^2/W)	1.761102 E – 01
°F·h·ft^2/Btu (thermochemical)	kelvin meter^2 per watt (K·m^2/W)	1.762280 E – 01
°F·h·ft^2/[Btu (International Table)·in] (thermal resistivity)	kelvin meter^2 per watt (K·m^2/W)	6.933472 E + 00
°F·h·ft^2/[Btu (thermochemical) ·in] (thermal resistivity)	kelvin meter^2 per watt (K·m^2/W)	6.938112 E + 00
denier	kilogram per meter (kg/m)	1.111111 E – 07
dyne	newton (N)	1.000000* E – 05
dyne/cm	N/m	1.000000 E – 03
dyne/cm	m N/m	1.000000

(Continued)

To convert from	To	Multiply by
dyne – cm	newton meter (N – m)	1.000000 *E – 07
dyne/cm^2	pascal (Pa)	1.000000 *E – 01
electronvolt	joule (J)	1.60219 E – 19
EMU of capacitance	farad (F)	1.000000 *E + 09
EMU of current	ampere (A)	1.000000 *E – 01
EMU of electric potential	volt (V)	1.000000 *E – 08
EMU of inductance	henry (H)	1.000000 *E – 09
EMU of resistance	ohm (m)	1.000000 *E – 09
erg	joule (J)	1.000000 *E – 07
erg/cm^2 · s	watt per meter^2 (W/m^2)	1.000000 *E – 03
erg/s	watt (W)	1.000000 *E – 07
faraday (based on carbon-12)	coulomb (C)	9.64870 E + 04
faraday (chemical)	coulomb (C)	9.64957 E + 04
faraday (physical)	coulomb (C)	9.65219 E + 04
fathom	meter (m)	1.8288 E + 00
fermi (femtometer)	meter (m)	1.000000 *E – 15
fluid ounce (US)	meter^3 (m^3)	2.957353 E – 05
foot	meter (m)	3.048000 *E – 01
foot (US survey)	meter (m)	3.048006 E – 01
foot of water (39.2°F)	pascal (Pa)	2.98898 E + 03
ft^2	meter^2 (m^2)	9.290304 E – 02
ft^2/h (thermal diffusivity)	meter^2 per second (m^2/s)	2.580640 *E – 05
ft^2/s	meter^2 per second (m^2/s)	9.290340 E – 02
ft^3 (volume, section modulus)	meter^3 (m^3)	2.831685 E – 02
ft^3/min	meter^3 per second (m^3/s)	4.719474 E – 04
ft^3/s	meter^3 per second (m^3/s)	2.831685 E – 02
ft^4 (second moment of area)[b]	meter^4 (m^4)	8.630975 E – 03
ft/h	meter per second (m/s)	8.466667 E – 05
ft/min	meter per second (m/s)	5.080000 *E – 03

(Continued)

To convert from	To	Multiply by
ft/s	meter per second (m/s)	3.048000 *E − 01
ft/s^2	meter per $second^2$ (m/s^2)	3.048000 E − 01
footcandle	lux (lx)	1.076391 E + 01
footlambert	candela per $meter^2$ (cd/m^2)	3.426259 E + 00
ft · lbf	joule (J)	1.355818 E + 00
ft · lbf/h	watt (W)	3.766161 E − 04
ft · lbf/min	watt (W)	2.259697 E − 02
ft · lbf/s	watt (W)	1.355818 E + 00
ft·poundal	joule (J)	4.214011 E − 02
g, standard acceleration of free fall	meter per $second^2$ (m/s^2)	9.806650 *E + 00
gal	meter per $second^2$ (m/s^2)	1.000000 *E − 02
gallon (Canadian liquid)	$meter^3$ (m^3)	4.546090 *E − 03
gallon (UK liquid)	$meter^3$ (m^3)	4.546090 *E − 03
gallons (US liquid)	$meter^3$ (m^3)	3.785412 E − 03
gallon (US liquid) per day	$meter^3$ per second (m^3/s)	4.381264 E − 08
gallon (US liquid) per minute	$meter^3$ per second (m^3/s)	6.309020 E − 05
gallon (US liquid) per (hp · h) (SFC), (specific fuel consumption)	$meter^3$ per joule (m^3/J)	1.410089 E − 09
gamma	tesla (T)	1.000000 *E − 09
gauss	tesla (T)	1.000000 *E − 04
gilbert	ampere (A)	7.957747 E − 01
gill (UK)	$meter^3$ (m^3)	1.420654 E − 04
gill (US)	$meter^3$ (m^3)	1.182941 E − 04
grade	degree (angular)	9.000000 E − 01
grade	radian (rad)	1.570796 E − 02
grain	kilogram (kg)	6.497891 *E − 05
grain/gal (US liquid)	kilogram per $meter^3$ (kg/m^3)	1.711806 E − 02
gram	kilogram (kg)	1.000000 *E − 03

(Continued)

To convert from	To	Multiply by
g/cm^3	kilogram per $meter^3$ (kg/m^3)	1.000000 *E + 01
$gram/force/cm^2$	pascal (PA)	9.801650 *E + 03
hectare	$meter^2$ (m^2)	1.000000 *E + 04
horsepower (550 ft· lbf/s)	watt (W)	7.456999 E + 02
horsepower (boiler)	watt (W)	9.80950 E + 03
horsepower (electric)	watt (W)	7.460000 *E + 02
horsepower (metric)	watt (W)	7.35499 E + 02
horsepower (water)	watt (W)	7.46043 E + 02
horsepower (UK)	watt (W)	7.4570 E + 02
hour	second (s)	3.600000 *E + 03
hour (sidereal)	second (s)	3.590170 E + 03
hundredweight (long)	kilogram	5.080235 E + 01
hundredweight (short)	kilogram (kg)	4.535924 E + 01
inch	meter (m)	2.540000 *E − 02
inch of mercury (32°F)[C]	pascal (Pa)	3.38638 E + 03
inch of mercury (60°F)[C]	pascal (Pa)	3.37685 E + 03
inch of water (39.2°F)	pascal (Pa)	2.49082 E + 02
inch of water (60°F)	pascal (Pa)	2.4884 E + 02
in^2	$meter^2$ (m^2)	6.451600 *E − 04
in^2 (volume; section modulus)[d]	$meter^3$ (m^3)	1.638706 E − 05
in^3/min	$meter^3$ per second (m^3/s)	2.731177 E − 07
in^4 (second moment of area)[b]	$meter^4$ (m^4)	4.162314 E − 07
in/s	meter per second (m/s)	2.540000 *E − 02
in/s^2	meter per $second^2$ (m/s^2)	2.540000 *E − 02
kayser	1 per meter (1/m)	1.000000 *E + 02
kelvin	degree Celsius	$t_{\cdot C} = T_K - 273.15$
kilocalorie (International Table)	joule (J)	4.186800* E + 03
kilocalorie (mean)	joule (J)	4.19002 E + 03
kilocalorie (thermochemial)	joule (J)	4.184000 *E + 03
kilocalorie (thermochemical)/ min	joule (J)	6.973333 E + 01
kilocalorie (thermochemical)/s	watt (W)	4.184000 *E + 03
kilogram-force (kgf)	newton (N)	9.806650 *E + 00

(Continued)

To convert from	To	Multiply by
kgf · m	newton meter (N · m)	9.806650 *E + 00
kgf · s^2/m (mass)	kilogram (kg)	9.806650 *E + 00
kgf/cm^2	pascal (Pa)	9.806650 *E + 04
kgf/m^2	pascal (Pa)	9.806650 *E + 00
kgf/mm^2	pascal (Pa)	9.806650 *E + 06
Km/h	meter per second (m/s)	2.777778 E – 01
kilopond (1 kp = 1 kgf)	newton (N)	9.806650 *E + 00
kW · h	pascal (Pa)	6.894757 E + 06
kip (1000 lbf)	newton (N)	4.448222 E + 03
kip/in^2 (ksi)	pascal (Pa)	6.894757 E + 06
knot (international)	meter per second (m/s)	5.144444 E – 01
lambert	candela per $meter^2$ (cd/m^2)	1/π *E + 04
langley	joule per $meter^2$ (J/m^2)	4.184000 *E + 04
light year[e]	meter (m)	9.46073 E + 15
liter	$meter^3$ (m^3)	1.000000 *E – 03
lumen per ft^2	lumen per $meter^2$ (lm/m^2)	1.076391 E + 01
maxwell	weber (Wb)	1.000000 *E – 08
mho	siemens (S)	1.000000 *E + 00
microinch	meter (m)	2.540000 *E – 08
micron	meter (m)	1.000000 *E – 06
mil	meter (m)	2.540000 *E – 05
mile (international)	meter (m)	1.609344 *E + 03
mile (US statute)	meter (m)	1.6093 E + 03
mile (international nautical)	meter (m)	1.852000 *E + 03
m^2 (international)	$meter^2$ (m^2)	2.589988 E + 06
mi^2 (US statute)	$meter^2$ (m^2)	2.589998 E + 06
mi/h (international)	meter per second (m/s)	4.470400 *E – 01
mi/h (international)	kilometer per hour (km/h)	1.609344 *E + 00
mi/min (international)	meters per second (m/s)	2.682240 *E + 01
mi/s (international)	meter per second (m/s)	1.609344 *E + 03
millibar	pascal (Pa)	1.000000 *E + 02
millimeter of mercury (0°C)[C]	pascal (Pa)	1.333220 E + 02

(Continued)

To convert from	To	Multiply by
millimeter	meter (m)	1.000000 E – 03
minute (angle)	radian (rad)	2.908882 E – 04
minute	second (s)	6.000000 *E + 01
minute (sidereal)	second (s)	5.983617 E + 01
nanometer	meter (m)	1.000000 E – 09
oersted	ampere per meter (A/m)	7.957747 E + 01
ohm centimeter	ohm meter (Ω · m)	1.000000 *E – 02
ohm circular-mil per ft	ohm meter (Ω · m)	1.662426 E – 09
ounce (avoirdupois)	kilogram (kg)	2.834952 E – 02
ounce (troy or apothecary)	kilogram (kg)	3.110348 E – 02
ounce (UK fluid)	$meter^3$ (m^3)	2.841307 E – 05
ounce (US fluid)	$meter^3$ (m^3)	2.957353 E – 05
ounce-force	newton (N)	2.780139 E – 01
ozf · in	newton meter (N · m)	7.061552 E – 03
oz (avoirdupois)/gal (UK liquid)	kilogram per $meter^3$ (kg/m^3)	6.236023 E + 00
oz (avoirdupois)/gal (US liquid)	kilogram per $meter^3$ (kg/m^3)	7.489152 E – 00
oz (avoirdupois)/in	kilogram per $meter^3$	1.729994 E – 03
oz (avoirdupois)/ft^2	kilogram per $meter^2$ (kg/m^2)	3.051517 E – 01
oz (avoirdupois)/yd^2	kilogram per $meter^2$ (kg/m^2)	3.390575 E – 02
parsec	meter (m)	3.085678 E + 16
peck (US)	$meter^3$ (m^3)	8.809768 E – 03
pennyweight	kilogram (kg)	1.555174 E – 03
perm (0°C)	kilogram per pascal second $meter^2$ [(kg/Pa · s · m^2)]	5.72135 E – 11
perm (23°C)	kilogram per pascal second $meter^2$ [(kg/Pa · s · m^2)]	5.74525 E – 11
perm · in (0°C)	kilogram per pascal second meter [(kg/Pa · s · m)]	1.45322 E – 12

(Continued)

To convert from	To	Multiply by
perm · in (23°C)	kilogram per pascal second meter [(kg/Pa · s · m)]	1.45929 E − 12
phot	lumen per meter^2 (lm/m^2)	1.000000 *E + 04
pica (printer's)	meter (m)	4.217518 E − 03
pint (US dry)	meter^3 (m^3)	5.506105 E − 04
ping (US liquid)	meter^3 (m^3)	4.731765 E − 04
point (printer's)	meter (m)	3.514598 *E − 04
poise (absolute viscosity)	pascal second (Pa · s)	1.000000 *E − 01
pound (avoirdupois)[g]	kilogram (kg)	4.535924 E − 01
pound (troy or apothecary)	kilogram (kg)	3.732417 E − 01
lb/ft	kilogram per meter (kg/m)	1.488164 E + 00
lb · ft^2 (moment of inertia)	kilogram meter^2 (kg · m^2)	4.214011 E − 02
lb · ft^2 (moment of inertia)	kilogram meter^2 (kg · m^2)	2.926397 E − 04
lb/ft · h	pascal second (Pa · s)	4.133789 E − 04
lb/ft · s	pascal second (Pa · s)	1.488164 E − 00
lb/ft^2	kilogram per meter^2 (kg/m^2)	4.882428 E + 00
lb/ft^3	kilogram per meter^3 (kg/m^3)	1.601846 E + 01
lb/gal (UK liquid)	kilogram per meter^3 (kg/m^3)	9.977633 E + 01
lb/gal (US liquid)	kilogram per meter^3 (kg/m^3)	1.198264 E + 02
lb/h	kilogram per second (kg/s)	1.259979 E − 04
lb/hp · h (SFC, specific fuel consumption)	kilogram per joule (kg/J)	1.689659 E − 07
lb/in	kilogram per meter (kg.m)	1.785797 E − 01
lb/in^3	kilogram per meter^3 (kg/m^3)	2.767990 E + 04
lb/min	kilogram per second (kg/s)	7.559873 E − 03

(Continued)

To convert from	To	Multiply by
lb/s	kilogram per second (kg/s)	4.535924 E − 01
lb/yd^3	kilogram per meter^3 (kg/m^3)	5.932 E − 01
poundal	newton (N)	1.382550 E − 01
poundal/ft^2	pascal (Pa)	1.488164 E + 00
poundal · s/ft^2	pascal second (Pa · s)	1.488164 E + 00
pound-force (lbf)[h]	newton (N)	4.448222 E + 00
lbf · ft	Newton meter (N · m)	1.355818 E + 00
lbf · ft/in	Newton meter per meter (N · m/m)	5.337866 E + 01
lbf · in	newton meter (N · m)	1.129848 E − 01
lbf · in/in	newton meter per meter (N · m/m)	4.448222 E + 00
lbf · s/ft^2	pascal second (Pa · s)	4.788026 E + 01
lbf · s/in^2	pascal second (Pa · s)	6.894757 E + 03
lbf/ft	newton per meter (N/m)	1.459390 E + 01
lbf/ft^2	pascal (Pa)	4.788026 E + 01
lbf/in	newton per meter (N/m)	1.751268 E + 02
lbf/in^2 (psi)	pascal (Pa)	6.894757 E + 03
lbf/lb (thrust/ weight [mass] ratio)	newton per kilogram (N/kg)	9.806650 E + 00
quad	joule (J)	1.055 E + 18
quart (US dry)	meter^3 (m^3)	1.101221 E − 03
quart (US liquid)	meter^3 (m^3)	9.463529 E − 04
rad (absorbed dose)	gray (Gy)	1.000000 *E − 02
rem (dose equivalent)	sievert (Sv)	1.000000 *E − 02
rhe	1 per pascal second [1/Pa · s)]	1.000000 *E + 01
rod	meter (m)	5.029210 E + 00
roentgen	coulomb per kilogram (C/kg)	2.58 E − 04
second (angle)	radian (rad)	4.848137 E − 06
second (sidereal)	second (s)	9.972696 E − 01
shake	second (s)	1.000000 *E − 08
slug	kilogram (kg)	1.459390 E + 01

(Continued)

To convert from	To	Multiply by
slug/ft · s	pascal second (Pa · s)	4.788026 E + 01
slug/ft^3	kilogram per meter (kg/m^3)	5.153788 E + 02
statampere	ampere (A)	3.335641 E – 10
statcoulomb	coulomb (C)	3.335641 E – 10
stratfarad	farad (F)	1.112650 E – 12
stathenry	henry (H)	8.987552 E + 11
statmho	siemens (S)	1.112650 E – 12
statohm	ohm (Ω)	8.987552 E + 11
statvolt	volt (V)	2.997925 E + 02
stere	meter^3 (m^3)	1.000000 *E + 00
stilb	candela per meter^2 (cd/m^2)	1.000000 *E + 04
stokes (kinematic viscosity)	meter^2 per second (cd/m^2)	1.000000 *E – 04
tablespoon	milliliter (mL)	1.479 E + 01
teaspoon	meter^3 (m^3) milliliter (mL)	4.929 E + 01
tex	kilogram per meter (kg/m)	1.000000 *E – 06
therm (EEG)[i]	joule (J)	1.055060 *E + 08
therm (US)[i]	joule (J)	1.054804 *E + 08
ton (assay)	kilogram (kg)	2.916667 E – 02
ton (long, 2240 lb)	kilogram (kg)	1.016047 E + 03
ton (metric)	kilogram (kg)	1.000000 *E + 03
ton (explosive energy of one tone of TNT)	joule (J)	4.184 E + 09[j]
ton of refrigeration (12,000 Btu/h)	watt (W)	3.517 E + 03
ton (register)	meter^3 (m^3)	2.831685 E + 00
ton (short, 2,000 lb)	kilogram (kg)	9.071847 E + 02
ton (long)/yd^3	kilogram per meter^3 (kg/m^3)	1.328939 E + 03
ton (short)/yd^3	kilogram per meter^3 (kg/m^3)	1.186553 E + 03
ton (short)/h	kilogram per second (kg/s)	2.519958 E – 01
ton-force (2,000 lbf)	newton (N)	8.896443 E + 03
tonne	kilogram (kg)	1.000000 *E + 03
torr (mmHg, 0°C)[c]	pascal (Pa)	1.33322 E + 02
unit pole	weber (Wb)	1.256637 E – 07
W · h	joule (J)	3.600000 *E + 03

(Continued)

To convert from	To	Multiply by
W · s	joule (J)	1.000000 *E + 00
W/cm^2	watt per meter^2 (W/m^2)	1.000000 *E + 04
W/in^2	watt per meter^2 (W/m^2)	1.550003 *E + 03
yard	meter (m)	9.144000 *E – 01
yd^2	meter^2 (m^2)	8.361274 E – 01
yd^3	meter^3 (m^3)	7.645549 E – 01
yd^3/min	meter^3 per second m^3/s)	1.274258 E – 02
year (365 days)	second (s)	3.153600 *E + 07
year (sidereal)	second (s)	3.155815 E + 07
year (tropical)	second (s)	3.155693 E + 07

Note: the centigrade temperature scale is obsolete. The unit, degree centigrade, is only approximately equal to the degree Celsius.

[a]The darcy is a unit for measuring permeability of porous solids.

[b]This is sometimes called the moment of section or area moment of ineria of a plane section about a specified axis.

[c]Conversion factors for mercury manometer pressure units are calculated using the standard value for the acceleration of gravity and the density of mercury at the stated temperature. Higher levels of precision are not justified because the definitions of the units do not take into account the compressibility of mercury or the density value change caused by the revised practical temperature scale, ITS-90.

[d]The exact conversion factor is 1.638 706 4*E–05.

[e]This conversion factor is based on the astronomical unit of time of one day (86,400 s); an interval of 36,525 days is one Julian century. (See the Astronomical Almanac for the Year 1991, page K6, U.S. Government Printing Office, Washington, DC).

[f]In 1964 the General Conference on Weights and Measures reestablished the name liter as a special name for the cubic decimeter. Between 1901 and 1964, the liter was slightly larger (1.000028 dm^3); in the use of high-accuracy volume data of that time interval, this fact must be kept in mind.

[g]The exact conversion factor is 4.535 923 7*E – 01.

[h]The exact conversion factor is 4.448 221 615 260 5*E + 00.

[i]The therm (EEC) is legally defined in the Council Directive of 20 December 1979, Council of the European Communities. The therm (US) is legally defined in the Federal Register of July 27, 1968. Although the therm (EEC), which is based on the International Table Btu, is frequently used by engineers in the US, the therm (US) is the legal unit used by the US natural gas industry.

[j]Defined (not measured) value.

Note concerning the foot:

The U.S. Metric Law of 1866 gave the relationship, 1 m equals 39.37 in. Since 1893 the U.S. yard has been derived from the meter. In 1959 a refinement was made in the definition of the yard to bring the U.S. yard and the yard used in others countries into agreement. The U.S. yard was changed from 3,600/3,937 m to 0.9144 m exactly. The new length is shorter by exactly two parts in a million. At the same time it was decided that any data in feet derived from and published as a result of geodetic surveys within the US would remain with the old

standard (1 ft = 1,200/3,937 m) until further decision. This foot is named the US survey foot and has the following relationships:
1 rod (pole or perch) = 16½ feet
1 chain = 66 ft
1 mile (US statute) = 5,280 ft

Conversion of temperatures

From	To	
°Celsius	°Fahrenheit	$t_F = (t_c \times 1.8) + 32$
	Kelvin	$T_k = t_c + 273.15$
	°Rankin	$T_R = (t_c + 273.15) \times 1.8$
°Fahrenheit	°Celsius	$t_c = \frac{(t_F - 32)}{1.8}$
	Kelvin	$t_k = \frac{(t_F - 32)}{1.8} + 273.15$
	°Rankin	$T_R = t_F + 459.67$
Kelvin	°Celsius	$t_c = T_K - 273.15$
	°Rankin	$T_R = T_K \times 1.8$
°Rankin	°Fahrenheit	$t_F = T_R - 459.67$
	Kelvin	$t_k = \frac{t_R}{1.8}$

Designation of large numbers

	U.S.A	Other countries
10^6	Million	Million
10^9	Billion	Milliard
10^{12}	Trillion	Billion
10^{15}	Quadrillion	Billard
10^{18}	Quintillion	Trillion

Conversion factors for pressure units

Pa	kPa	MPa	bar	
Pa	1	0.001	0.000001	0.00001
kPa	1,000	1	0.001	0.01
MPa	10,00,000	1,000	1	10
bar	1,00,000	100	0.1	1
atmos	1,01,325	101.325	0.101325	1.01325
Torr	133.322	0.133322	0.000133322	0.00133322
µmHz	0.133322	0.000133322	1.33322×10^7	1.33322×10^{-6}
psi	6,894.757	6.894757	0.006894757	0.06894757

atmos	Torr	µmHz	psi	
Pa	9.8692×10^{-6}	0.0075006	7.5006	0.0001450377
kPa	0.0098692	7.5006	7,500.6	0.1450377
MPa	9.8692	7500.0	75,00,600	145.0377
bar	0.98692	750.06	7,50,060	14.50377
atmos	1	760	7,60,000	14.69594

atmos	Torr	µmHz	psi	
Torr	0.00131579	1	1000	0.01933672
µmHz	1.31579×10^{-6}	0.001	1	1.933672×10^{-5}
Psi	0.0680456	51.7151	51715.1	1

To convert a pressure value from a unit in the left-hand column to a new unit, multiply the value by the factor appearing in the column for the new unit. For example:

1 kPa = 9.8692 × 10–3 atmos
1 Torr = 1.33322 × 10–4 MPa

Notes: µmHz is often referred to as "micron"
Torr is essentially identical to mmHg
psi is pounds per square inch
Sub-units of the meter
1 meter = 1.0 E + 0 m
= 1.0 E – 1 dm
= 1.0 E – 2 cm
= 1.0 E – 3 mm
= 1.0 E – 6 µm
= 1.0 E – 9 nm
= 1.0 E – 10 Å

Commonly used units

Area
Density
Enegy
Flow rate
Force
Frequency
Heat
Heat capacity
Heat flux
Length
Mass

Perms, permenance, and permeability
Power
Pressure
Strength
Stress
Temperature
Thermal conductivity
Thermal diffusivity
Time
Torque
Velocity
Viscosity
Volume
Work

Appendix C: Polymer Properties

Thermal Transition of Polymers

Percent crystallinity, glass transition and melting temperatures of polymers

Repeating Unit, poly (repeating unit)	% Crystalline	T_g (°C)	T_m (°C)
Acenaphythylene		214	
Acetal	high	91–110	175–181
Acetaldehyde		−32	165
4-Acetoxystyrene		116	
Acrylamide		165	
Acrylic acid		105	
Acrylonitrile-butadiene-styrene	low	110	125
Acrylonitrile, syndiotactic		125	319
Allyl glycidyl ether		−78	
Benzyl acrylate		6	
Benzyle methacrylate		54	
Bisphenol A-alt-epichlorohydrin		100	
Bisphenol A terephthalate		205	
Bisphenol carbonate		174	
Bisphenol F carbonate		147	
Bisphenol Z carbonate		175	
4-Bromostyrene		118	
cis-Butadiene		102	1
trans-Butadiene		−58	148
1-Butene		−24	171
N-tert-Butylacrylamide		128	
Butyl acrylate		−54	
sec-Butyl acrylate		−26	
tert-Butyl acrylate		43–107	193
2-tert-Butylaminoethyl methacrylate		33	
Butyl glydicyl ether		−79	
Butyl methacrylate		20	
tert-Butyl methacrylate		118	
4-tert-Butylstyrene		127	
tert-Butyl vinyl ether		88	250
Butyl vinyl ether		−55	64
ε-Caprolactone		−60	

Percent crystallinity, glass transition and melting temperatures of polymers (Continued)

Repeating Unit, poly (repeating unit)	% Crystalline	T_g (°C)	T_m (°C)
Cellulose	high		decom-poses
Cellulose nitrate		53	
Cellulose tripropionate			
cis-Chlorobutadiene		−20	80
trans-Chlorobutadiene		−40	101
2-Chlorostyrene		119	
3-Chlorostyrene		90	
4-Chlorostyrene		110	
Chlorotrifluoroethylene		52	214
2-Cyanoethyl acrylate		4	
Cyclohexyl acrylate		19	
Cyclohexyl methacrylate		92	
Cyclohexyl vinyl ether		81	
2,6-Dichlorostyrene		167	
Diethylaminoethyl methacrylate		20	
N,N- Dimethylacrylamide		89	
Dimethylaminoethyl methacrylate		19	
2,6-Dimethyl-1,4-phenylene oxide		167	
Dimethylsiloxane		−127	−40
2,4-Dimethylstyrene		112	
2,5-Dimethylstyrene		143	
3,5-Dimethylstyrene		104	
Dodecyl acrylate		−3	
Dodecyl methacrylate		−65	
Dodecyl vinyl ether		−62	
Epibromohydrin		−14	
Ephichlorohydrin		−22	
1,2-Epoxybutane		−70	
1,2-Epoxydecane		−70	
1,2-Epoxyoctane		−67	
2-Ethoxyethyl acrylate		−50	
4-Ethoxystyrene		86	
Ethyl acrylate		−24	
Ethyl cellulose		43	
Ethylene, HDPE		−125	130

Percent crystallinity, glass transition and melting temperatures of polymers (Continued)

Repeating Unit, poly (repeating unit)	% Crystalline	T_g (°C)	T_m (°C)
Ethylene adipate		−46	54
Ethylene-trans-1,4-cyclohexyldicarboxylate		18	-
Ethylene isophthatate		51	
Ethylene malonate		−29	
Ethylene 2,6-napthalenedicarboxylate		113	
Ethylene oxide		−66	66
Ethylene terephthalate		72	265
Ethylene-vinyl acetate	high		65–110
2-Ethylhexyl acrylate		−50	
2-Ethylhexyl methacrylate		−10	
2-Ethylhexyl vinyl ether		−66	
Ethyl methacrylate		65	
Ethyl vinyl ether		−43	86
Fluorinated ethylene propylene	high		275
4-Fluorostyrene		95	
Formaldehyde		−82	181
Hexadecyl acrylate		35	
Hexadecyl methacrylate		15	
Hexyl acrylate		57	
Hexyl methacrylate		−5	
High-density polyethylene	95	−125	130–135
2-Hydropropyl methacrylate		76	
Hydroquinone-alt-epichlorohydrin		60	
2-Hydroxyethyl methacrylate		57	
Indene		85	
Isobornyl acrylate		94	
Isobornyl methacrylate		110	
Isobutyl acrylate		−24	
Isobutylene		−73	
Isobutyl methacrylate		53	
Isobutyl vinyl ether		−19	165
cis-Isoprene		−63	28
trans-Isoprene		−66	65
N-Isopropylacrylamide		85–130	
Isopropyl acrylate, isotactic		−11	162
Isopropyl methacrylate		81	
Low-density polyethylene	60	−25	109–125
Methacrylic acid		228	
Methacrylic anhydride		159	

Percent crystallinity, glass transition and melting temperatures of polymers (Continued)

Repeating Unit, poly (repeating unit)	% Crystalline	T_g (°C)	T_m (°C)
Methacrylonitrile		120	
2-Methoxyethyl acrylate		−50	
4-Methoxystyrene		113	
Methyl acrylate		10	
Methyl cellulose			
Methyl glycidyl ether		−62	
Methyl methacrylate, atactic		105, 120	
Methyl methacrylate, syndiotactic		115	200
4-Methylpentene		29	250
Methylphenlsiloxane		−86	
Methylstyrene		20	
3-Methylstyrene		97	
4-Methylstyrene		97	
Methyl vinyl ether		−31	144
Natural rubber	low		30
Nylon 4,6 (tetramethylene adipamide)		43	
Nylon 6		75	215–220
Nylon 6 (-caprolactam)		52	225
Nylon 6–6		57	250–260
Nylon 6,6 (hexamethylene adipamide)		50	265
Nylon 6,9 (hexamethylene azelamide)		58	
Nylon 6–10		50	215
Nylon 6,10 (hexamethylene sebacamide)		50	227
Nylon 6,12 (hexamethylene dodecanediamide)		46	
Nylon 11	high		185–195
Nylon 11 (ω-undecanamide)		42	189
Nylon 12 (ω-dodecanamide)		41	179
1-Octadecene		55	
Octadecyl methacrylate		−100	
1-Octene		−63	
Octyl methacrylate		−20	
Oxy-4,4′-biphenyleneoxy-1,4-phenylenesulfonyl-1,4-phenylene		230	290
Oxy-1,4-phenylenesesulfonyl-1,4-phenyleneoxy-1,4-phenyleneisopropylidene-1,4-phenylene		165	190

Percent crystallinity, glass transition and melting temperatures of polymers (Continued)

Repeating Unit, poly (repeating unit)	% Crystalline	T_g (°C)	T_m (°C)
Oxy-1,4-phenylenesulfonyl-1,4-phenylene ether		214	230
p-Phenylene isophthalamide		225	380
p-Phenylene terephthalamide		345	
Phenylene vinylene		80	380
Phenyl methacrylate		110	
Phenyl vinyl ketone		74	
Acrylonitrile	low	140	317
Butene 1		124–142	124–142
Butylene		126	126
Butylene terephthalate	high	40	232–267
Carbonate	low	145–150	215–230
Chlorotrifluorethylene	high	45	220
Ether ether ketone	high	143	334
Ether imide		217	
Ethylene terephthalate	high	70	265
Hexene 1			55
Methylbutene 1			300
Methylene	100		136
Methyl methacrylate	low	50	90–105
Oxymethylene	75–80	−85	175–180
Pentene 1			130
3-Phenylbutene 1			360
Phenylene oxide/ polystyrene	low	100–135	110–135
Phenylene sulphide	high	88–93	277–282
Propylene	65	−20	170
Styrene	low	>80	230
Sulphone	low	190	190
Tetrafluoroethylene	100	125	327
Vinyl chloride	5–15	80–85	75–105 (212)
Vinylidine chloride	High	−18	210
Vinylidine fluoride	high	−30 to −20	160–170
p-Xylene			>400
Potassium acrylate		194	
Propylene, atactic		−13	
Propylene, isotactic		−8	176
Propylene, syndiotactic		−8	
Propylene oxide		−75	66
Propyl vinyl ether		−49	76
1-Octene		−63	

Percent crystallinity, glass transition and melting temperatures of polymers (Continued)

Repeating Unit, poly (repeating unit)	% Crystalline	T_g (°C)	T_m (°C)
Octyl methacrylate		−20	
Oxy-4,4′-biphenyleneoxy-1,4-phenylenesulfony l-1,4-phenylene		230	290
Oxy-1,4-phenylenesesulfonyl-1,4-phenyleneoxy-1,4-phenyleneisopropylidene-1,4-phenylene		165	190
Oxy-1,4-phenylenesulfonyl-1,4-phenylene ether		214	230
p-Phenylene isophthalamide		225	380
p-Phenylene terephthalamide		345	
Phenylene vinylene		80	380
Phenyl methacrylate		110	
Phenyl vinyl ketone		74	
Acrylonitrile	low	140	317
Butene 1		124–142	124–142
Butylene		126	126
Butylene terephthalate	high	40	232–267
Carbonate	low	145–150	215–230
Chlorotrifluorethylene	high	45	220
Ether ether ketone	high	143	334
Ether imide		217	
Ethylene terephthalate	high	70	265
Hexene 1			55
Methylbutene 1			300
Methylene	100		136
Methyl methacrylate	low	50	90–105
Oxymethylene	75–80	−85	175–180
Pentene 1			130
3-Phenylbutene 1			360
Phenylene oxide/ polystyrene	low	100–135	110–135
Phenylene sulphide	high	88–93	277–282
Propylene	65	−20	170
Styrene	low	>80	230
Sulphone	low	190	190
Tetrafluoroethylene	100	125	327
Vinyl chloride	5–15	80–85	75–105 (212)
Vinylidine chloride	High	−18	210
Vinylidine fluoride	high	−30 to −20	160–170
p-Xylene			>400

Percent crystallinity, glass transition and melting temperatures of polymers (Continued)

Repeating Unit, poly (repeating unit)	% Crystalline	T_g (°C)	T_m (°C)
Potassium acrylate		194	
Propylene, atactic		−13	
Propylene, isotactic		−8	176
Propylene, syndiotactic		−8	
Propylene oxide		−75	66
Propyl vinyl ether		−49	76
Sodium acrylate		230	
Sodium methacrylate		310	
Styrene, atactic		100	
Styrene, isotactic		100	240
Styrene-acrylonitrile	low	100–120	120
Tetrabromobisphenol A carbonate		157	
Tetrafluoroethylene		117	327
Tetrahydrofuran		−84	
Tetramethylene adipate		−118	
Tetramethylene terephthalate		17	232
Thio-1,4-phenylene		97	285
2,2,2-Trifluoroethyl acrylate		−10	
Trimethylene oxide		−78	
Trimethylsilyl methacrylate		68	
2,4,6-Trimethylstyrene		162	
Vinyl acetal		355	82
Vinyl acetate		30	
Vinyl alcohol		85	220
Vinyl benzoate		71	
Vinyl 4-tert-butylbenzoate		101	-
Vinyl butyral		322	49
Vinyl carbazole		227	320
Vinyl chloride		81	227
Vinyl cyclohexanoate		76	
Vinylferrocene		189	
Vinyl fluoride		41	200
Vinyl formal		105	
Vinylidene chloride		−18	200
Vinylidene fluoride		−40	171
2-Vinyl naphthalene		151	
Vinyl pivalate		86	
Vinyl propionate		10	
2-Vinylpyridine		104	
4-Vinylpyridine		142	
1-Vinyl-2-pyrrolidone		54	
Vinyl trifluoroacetate		46	

Spectra

Infrared

Theory

Electromagnetic Radiation

Infrared waves like x-rays, light and radio waves are classified as electromagnetic radiation since they consist of both alternating electric and magnetic fields. Each of these types of radiation has a different amount of energy associated with it. The classification of radiation according to its energy gives rise to the electromagnetic spectrum. This is illustrated in the following Electromagnetic Spectrum.

The classical description of electromagnetic radiation is that it is continuous and has a sinusoidal wave motion. The wave motion of electromagnetic radiation can be described in terms of wavelength and frequency. The wavelength, λ, of radiation is the distance between two successive maxima or minima of the wave motion as illustrated in Sinusoidal Wave. The frequency, ν, is defined as the number of cycles which pass a given point per second. Wavelength and frequency are interrelated according to the following equation:

(Insert Electromagnetic Spectrum)

(Insert Sinusoidal Wave)

$$\lambda = \frac{c}{\nu} \tag{1}$$

where: λ = wavelength (cm)

c = velocity of light (3×10^{10} cm/sec)

ν = frequency (cycles/sec).

Frequency and energy are interrelated according to the following equation:

$$\mathrm{E} = \mathrm{h}\nu \tag{2}$$

where E = energy of a quantum in ergs.

H = Planck's constant (6.62×10^{-27} erg•sec)

This classical description of electromagnetic radiation satisfactorily explained the properties of visible radiation in the ultraviolet region. According to equations derived for the behavior of electromagnetic radiation based upon the concept of continuous wave motion, the energy emitted by a blackbody source at finite temperature increases without limit as the wavelength of the radiation approaches zero. This behavior is not experimentally observed. However, if one assumes, as did Planck (1901), that radiation is emitted by a source in discrete units called quanta or photons and not continuously, then the behavior of radiation in the ultraviolet region can be satisfactorily explained. This concept of

Decompositon rates of polymerization initiators

Initiator	Solvent	T(°C)	$k_d(s^{-1})$	10th Half-life °C (Solvent)
Tert-Amyl peroxybenzoate				99 (benzene)
4,4-Azobis(4-cyanovaleric acid)	Acetone Water	70	4.6×10^{-5}	69 (water)
	Water	69	1.9×10^{-5}	
	Water	80	9.0×10^{-5}	
1.1[a]-Azobis(cyclohexanecarbonitrile)	Toluene	80	6.5×10^{-6}	88 (toluene)
		95	5.4×10^{-5}	
		102	1.3×10^{-4}	
2,2[a]-Azobisisobutyronitrile (AIBN)	Benzene	50	2.2×10^{-6}	65 (toluene)
		70	3.2×10^{-5}	
		100	1.5×10^{-3}	
Benzoyl peroxide[b]	Benzene	60	2.0×10^{-6}	70 (benzene)
		78	2.3×10^{-5}	
		100	5.0×10^{-4}	
2,2-Bis(tert-butylperoxy)butane				100 (benzene)
1.1-Bis(tert-butylperoxy)cyclohexane	Benzene	93	1.9×10^{-5}	
2,5-Bis(tert-butylperoxy)-2,5-dimethylhexane	Benzene	115	1.1×10^{-5}	120 (benzene)
		145	4.7×10^{-4}	
2,5-Bis(tert-Butylperoxy)-2,5-dimethyl-3-hexyne				125 (benzene)
Bis(1-(tert-butylperoxy)-1-methylethyl)benzene				115 (benzene)
1,1-Bis(tert-butylperoxy)-3,3,5-trimethylcyclohexane				85 (dibutyl phthalate)
tert-Butyl hydroperoxide	Benzene	130	3×10^{-7}	170 (benzene)
		160	6.6×10^{-6}	
		170	2.0×10^{-5}	
		183	3.1×10^{-5}	
tert-Butyl peracetate	Benzene	85	1.2×10^{-6}	100 (benzene)
		100	1.5×10^{-5}	
		130	5.7×10^{-4}	
tert-Butyl peroxide	Benzene	80	7.8×10^{-8}	125 (benzene)
	Benzene	100	8.8×10^{-7}	
	Benzene	130	3.0×10^{-5}	
tert-Butyl peroxybenzoate	Benzene	100	1.1×10^{-5}	103 (benzene)
		130	3.5×10^{-4}	
tert-Butylperoxy isopropyl carbonate				98 (aliphatic hydrocarbons)
Cumene hydroperoxide	Benzene	115	4.0×10^{-7}	135 (toluene)
		145	6.6×10^{-6}	
Cyclohexanone peroxide	Benzene			90 (benzene)
Dicumyl peroxide	Benzene			115 (benzene)

[a]"Polymer Handbook", Eds. Brandrup, J; Immergut, E.H.; Grulke, E.A., 4th Edition, John Wiley, New York, 1999, 11/2–69

[b]Amines can significantly increase the decomposition rates of peroxides, e.g., addition of *N,N*-dimethyl aniline to benzoyl peroxide causes the latter to decompose rapidly at room temperature

γ-Rays	X-rays	Ultraviolet	Infrared	Microwave	Radio Television & Radar Waves
Nuclear Transitions	Inner Orbital Electron Transitions	Outer Orbital Electron Transitions	Molecular Vibrations	Molecular Rotations	

0.1A 0.5A 50A 4000A 700mμ 500μ 0.3m
400mμ 0.7μ

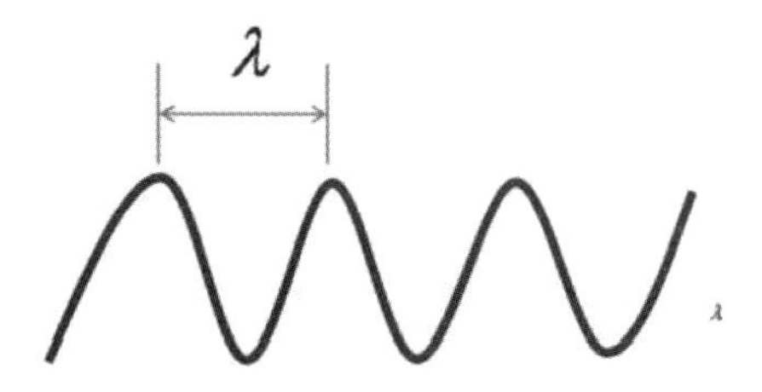

Sinusoidal wave and wavelength

discrete units for electromagnetic radiation is the basis of quantum theory which is necessary for a fundamental understanding of infrared spectroscopy.

Molecular Energy

When a molecule absorbs electromagnetic radiation, there is an increase in its thermal energy. According to quantum theory only a quantum of radiation of a specific energy can be absorbed by a molecule, thereby raising its energy from the ground state to an excited state. Absorption in Electromagnetic Energy shows this process schematically where E_1 and E_2 represent the energy of the molecule in the ground state and excited state, respectively, and E_p is the energy of the photon. It is important to note that the energy of transition, ΔE, is equal to the energy of the interacting photon, and thus, equal to hv. This can be stated mathematically according to equation (3).

$$\Delta E = E_2 - E_1 = E_p = hv \quad (3)$$

The total thermal energy, E_T, of a gas-phase molecule is comprised of electronic, vibrational, rotational, and translational energies:

$$E_T = E_e + E_v + E_r + E_t \quad (4)$$

where E_T = total thermal energy of a gas-phase molecule
E_e = electronic energy
E_v = vibrational energy
E_r = rotational energy
E_t = translational energy.

The energy involved in the transition from the electronic ground state to the first excited electronic state of a molecule usually requires radiation of greater energy than that available in the infrared region of the electromagnetic spectrum. Therefore, electronic energies and transitions usually do not have to be considered in discussing the theory of infrared spectroscopy.

Degrees of Freedom and Molecular Action

The remaining thermal energies of a molecule are due to vibrational, rotational and translational motions associated with the molecule. These molecular motions are commonly referred to as degrees of freedom. For every atom in a molecule there are three degrees of freedom corresponding to motions along the three mutually perpendicular x, y and z coordinates in space. For a non-linear molecule containing N atoms, there will be 3N degrees of freedom; three translational, three rotational, and the remainder, 3N-6, will be vibrational. These degrees of freedom are illustrated in the following figure, 3N-6 Degrees of Vibrational Freedom of a Molecule.

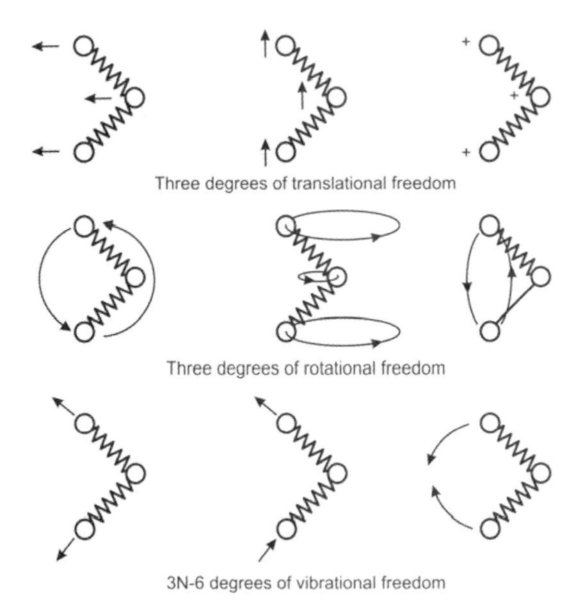

For a linear molecule there are 3N-5 vibrational degrees of freedom since a linear molecule has only two degrees of rotational freedom. No change occurs in a linear molecule because of rotation about its bond axis. Of these vibrational degrees of freedom, N-1 modes will be due to stretching vibrations and 2N-4 modes will be due to deformation. For a non-linear molecule, 2N-5 modes will be due to deformation.

Infrared radiation is of sufficient energy to cause transitions within the translational, rotational and vibrational energy levels of a molecule. Infrared spectroscopy is the study of transitions within these various energy levels due to the absorption of infrared radiation. Therefore, of primary concern are transitions between vibrational energy levels of a molecule and the various types of molecular vibrations.

For infrared absorption to occur, two major conditions must be fulfilled, and they are:

- The energy of the radiation must coincide with the energy difference between the excited and ground states of the molecule; radiant energy will then be absorbed by the molecule, increasing its natural vibration intensity.
- The vibration must entail a change in the electrical *dipole moment.* (The bond dipole moment uses the idea of electric dipole moment to measure the polarity of a chemical bond within a molecule. The bond dipole μ is given by: $\mu = \delta d$. The bond dipole is modeled as $+\delta$ — δ- with a distance d between the partial charges $+\delta$ and δ-. It is a vector, parallel to the bond axis, pointing from minus to plus, as is conventional for electric dipole moment vectors. (Some chemists draw the vector the other way around, pointing from plus to minus, but only in situations where the direction doesn't matter.)[1] This vector can be physically interpreted as the movement undergone by electrons when the two atoms are placed a distance d apart and allowed to interact, the electrons will move from their free state positions to be localised more around the more electronegative atom. The SI unit for electric dipole moment is the coulomb-meter, but that is much too large to be practical on the molecular scale. Bond dipole moments are commonly measured in debyes, represented by the symbol D, which is obtained by measuring the charge δ in units of 10^{-10} statcoulomb and measure the distance d in Angstroms. Note that 10^{-10} statcoulomb is 0.48 units of elementary charge. Another useful conversion factor is 1 C m = 2.9979×10^{29} D. Typical dipole moments for simple diatomic molecules are in the range of 0 to 11D. At one extreme, a symmetrical molecule such as chlorine, Cl_2, has zero dipole moment, while near the other extreme, gas phase potassium bromide, KBr, which is highly ionic, has a dipole moment of 10.5D. For a complete molecule the total molecular dipole moment may be approximated as the vector sum of individual bond dipole moments. Often bond dipoles are obtained by the reverse process: a known total dipole of a molecule can be decomposed into bond dipoles. The reason for doing this is the transfer of bond dipole moments to molecules that have the same bonds, but for which the total dipole moment is not yet known. The vector sum of the transferred bond dipoles gives an estimate for the total (unknown) dipole of the molecule. The bond dipole is two atoms in a bond, such that the electronegativity of one atom causes electrons to be drawn towards the other, in turn causing a partial negative charge. There is therefore a difference in polarity across the bond, which causes a dipole moment), a restriction which distinguishes infrared from Raman spectroscopy.

Fundamental Vibrations

Vibrational transitions or fundamental modes of vibration are classified as stretching modes and deformation modes. Stretching modes are described as changes in bond lengths and deformation modes as giving rise primarily to changes in bond angles as summarized below.

- Stretching: a change in the length of a bond, such as C-H or C-C.

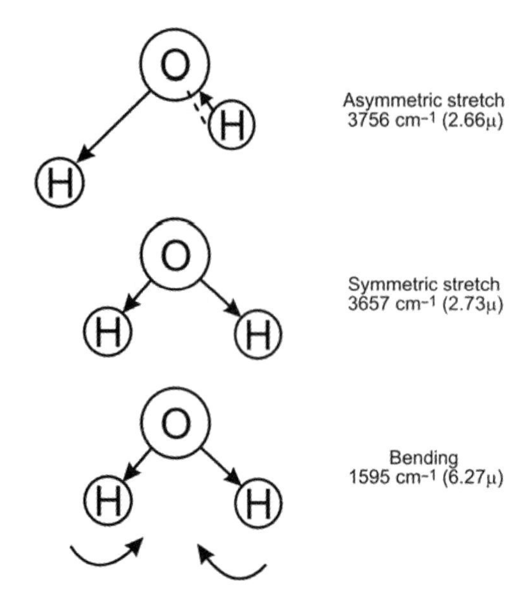

- Bending: a change in the angle between two bonds, such as the HCH angle in a methylene group.
- Rocking: a change in angle between a group of atoms, such as a methylene group and the rest of the molecule.
- Wagging: a change in angle between the plane of a group of atoms, such as a methylene group and a plane through the rest of the molecule.
- Twisting: a change in the angle between the planes of two groups of atoms, such as a change in the angle between the two methylene groups.
- Out-of-plane: Not present in ethene, but an example is in BF_3 when the boron atom moves in and out of the plane of the three fluorine atoms.

The fundamental modes of vibration of the methylene group are shown in the figure of Vibration of the Methylene Group. The approximate spectral positions where these vibrations will absorb infrared radiation are also given.

Infrared Spectral Regions

These same types of molecular motions occur even when other atoms or functional groups are involved in the vibrational transitions. However, depending on the types of atoms involved and their environments in the molecule, each transition will have a specific energy associated with it. Each of these vibrational modes will give rise to the absorption of infrared spectrum. This forms the basis for qualitative analysis and structural determinations by infrared spectroscopy. For organic molecules, hydrogen stretching vibrations (changes in bond length between hydrogen and any other atom) give rise to the absorption of infrared radiation in the 3,950–2,560 cm^{-1} region (2.53–3.91 μm). Asymmetrical stretching vibrations usually occur at higher frequencies than the corresponding symmetrical stretching vibrations. Multiple bond stretching of all types usually occurs between 2,260 and 1,150 cm^{-1} (4.42–8.70 μm). Deformation vibrations occur in the 1,670–600 cm^{-1} region (5.99–16.67 μ). Skeletal vibrations, motions of a major portion of a molecule, usually are found in the region below 1,250 cm^{-1} (8.00 μm). The specific wavelength where a functional group will absorb infrared radiation can also be calculated or predicted.

The water molecule can be used as a simple example for predicting the number and position of fundamental vibrations for a molecule. The water molecule with three atoms would be expected to have 3N or 9 degrees of freedom. Three of these degrees of freedom are translational, and since the water molecule is nonlinear, three are rotational. This leaves 3N-6 (or 3) degrees of vibrational freedom. Of these degrees of vibrational freedom, N-1 (or 2) are due to stretching vibrations, and 2N-5 (or 1) is due to a bending vibration. These vibrations are the asymmetrical stretching which occurs at 3,756 cm^{-1} (2.66 μm), the symmetrical stretching which occurs

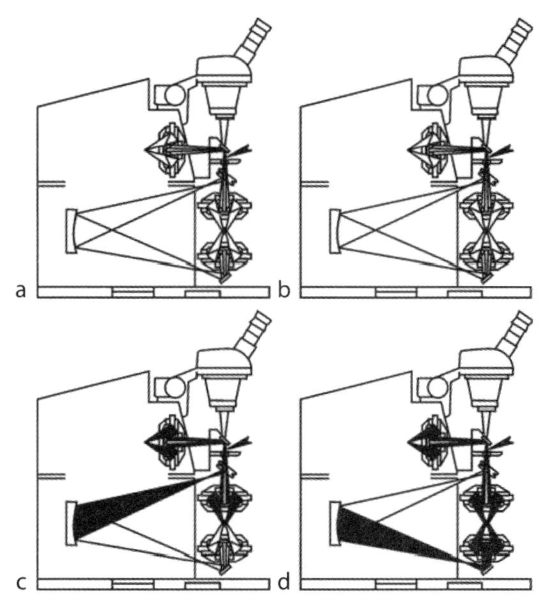

Perkin-Elmer FT-IR microscope. (a) Optical path-sample preparation, (b) Optical path-sample viewing, (c) Optical path-reflectance infrared, and (d) Optical path-transmittance infrared. (Arrowhead indicates sample position.)

at 3,657 cm^{-1} (2.73 μm) and the bending vibration which occurs at 1,595 cm^{-1} (6.27 μm) in the gas phase. These are illustrated in the figure of Fundamentals of the Water Molecule in the Gas Phase.

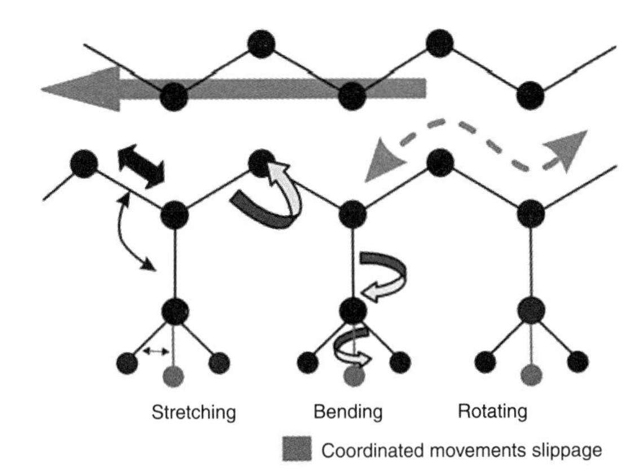

It is also possible to calculate the spectral position due to the stretching vibration of a two-atom pair using the following relationship:

$$v^*{}_{em}{}^{-1} = \frac{1}{2\pi c}\sqrt{\frac{k}{\mu}} \quad (5)$$

where $v^*{}_{em}{}^{-1}$ = spectral position in wavenumbers
k = molecular force constant in 5.0 × 10^5 dynes/cm (Willard et al. 1974)
μ = reduced mass of the atom pair in gram
c = speed of light in cm/sec.

The force constant is a measure of the bond energy of the two-atom bond and can be calculated empirically or obtained from the literature. Force constants usually have a magnitude on the order of 10^5 dynes/cm. The reduced mass of the atom pair is calculated from the expression:

$$\mu = \frac{m_1 \cdot m_2}{m_1 + m_2} \bullet \frac{1}{N} \quad (6)$$

where m_1 and m_2 = the atomic masses of the two atoms involved in the vibration

N = Avogadro's Number (6.023 × 10^{23} atoms/mole).

The reduced mass is always less than the mass of either of the atoms involved in the vibration. The carbonyl group can be used as an example of the calculation of the spectral position due to a stretching vibration. Where m_c and m_o are the atomic masses of carbon and oxygen, then the reduced mass of this atom pair (denoted in subscript by $_{c=o}$) is:

$$\mu_{c=o} = \frac{m_c \cdot m_2}{m_c + m_o} \bullet \frac{1}{N}$$

$$\mu_{c=o} = \frac{\frac{12.00\,\text{g atoms}}{\text{mole}} + \frac{15.99\,\text{g atoms}}{\text{mole}}}{\left(\frac{12.00\,\text{g atoms}}{\text{mole}} + \frac{15.99\,\text{g atoms}}{\text{mole}}\right)} \frac{1}{\left(\frac{6.023 \times 1023\,\text{atoms}}{\text{mole}}\right)}$$

$$\mu_{c=o} = \frac{191.9\,g^2\text{atoms}^2/\text{mole}^2}{27.99\,\text{g atoms/mole}} \frac{1}{6.023 \times 1023 \cdot \text{atoms/mole}}$$

$$\mu_{c=o} = 1.14 \times 10^{-23}\text{g}$$

The force constant for the carbonyl (e.g., C=O group as in acidic acid) stretching vibration is approximately 12 × 10^5 dynes/cm. From this value and the reduced mass for the carbonyl atom pair, the spectral position for this stretching vibration can be calculated using equation (5).

$$v^*{}_{c=o} = \frac{1}{2\pi c}\sqrt{\frac{k}{\mu}}$$

$$v^*{}_{c=o} = \frac{1}{2(3.14)(3 \times 10^{10}\,\text{cm/sec})}\sqrt{\frac{12 \times 10^5\,\text{dynes/cm}}{1.14 \times 10^{-23}\text{g}}}$$

$$v^*{}_{c=o} = 1722\,\text{cm}^{-1}$$

$$\lambda_{c=o} = \frac{10^4\mu/\text{cm}}{1722\,\text{cm}^{-1}} = 5.81\,\mu\text{m}$$

The actual position for the stretching vibration of most carbonyl groups is in the spectral region between 1,850 and 1,650 cm^{-1} (5.41 and 6.06 μm). Having obtained the spectral position of this vibration, it is possible to calculate the vibrational frequency from the expression:

$$v = cv^* \quad (7)$$

Where: v = the frequency of the vibration in cycles/sec.

Therefore, $v_{c=o}$ = 3 × 10^{10} cm/sec × 1,722 cm^{-1} = 5.2 × 10^{13} cps.

Using equation (2) it is possible to calculate the energy required for the transition between the ground state and the first excited state for the carbonyl stretching vibration.

E = hv

where E = transitional energy in ergs/atom

$E_{c=o}$ = (6.625 × 10^{-27} ergs•sec/atom) (5.2 × $10^{13}sec^{-1}$)

$E_{c=o}$ = 3.45 × 10^{-13} ergs/atom.

In terms of kilocalories per mole, this is equal to:

$E_{c=o}$ = (3.45 × 10^{-13} ergs/atom) (2.39 × 10^{-S} cal/erg)• (6.023 × 10^{23} atoms/mole)

$E_{c=o}$ = 4,960 cal/mole = 4.96K cal/mole.

This type of transition is depicted schematically in Potential Energy Diagram for Vibrational Transitions.

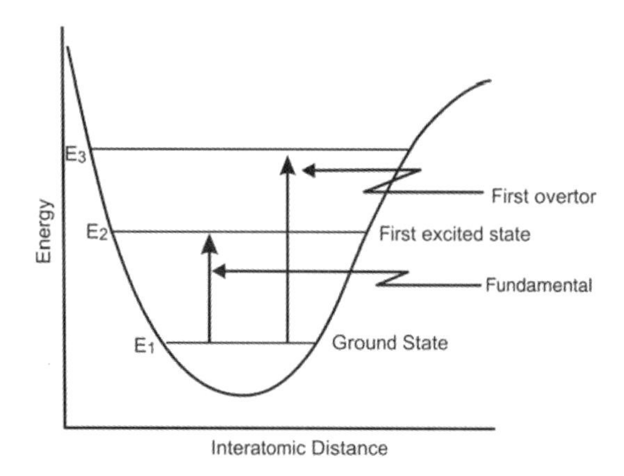

Additional Theoretical Considerations

As a heteronuclear molecule vibrates the interatomic distance between atoms changes as does its dipole moment. The vibrating molecule produces an alternating dipolar electric field changes periodically with time at a frequency equal to the vibrational frequency. It is this dipolar electric field which interacts with the electrical component of the electromagnetic radiation giving rise to the absorption of energy. This absorption of electromagnetic radiation is what is observed when recording an infrared spectrum.

If a diatomic molecule is homonuclear, there is no change in dipole moment with the vibration and no alternating dipolar electric field is produced. Thus, the molecule does not interact with or absorb electromagnetic radiation.

Whenever a change in dipole moment occurs, the fundamental vibration is said to be infrared active. When a vibrating molecule produces no change in its dipole moment because of symmetry, the fundamental vibration is said to be infrared inactive.

In molecules with a high degree of symmetry, many of the vibrations may occur between groups of atoms with the same reduced mass and same force constant. These vibrations will have the same frequency and will be superimposed on the spectrum. When this occurs, the vibrations are said to be degenerate. In symmetrical molecules several modes of vibration may occur at identical frequencies, such as the two identical frequencies such as the CH_2 vibrations in linear polyethylene. In cases like these, symmetry does not require the frequencies to coincide; however, due to chemical considerations the absorption bands overlap. Thus, a highly symmetrical molecule of many atoms, such as benzene, will give a simple infrared spectrum because many of its modes of vibration are degenerate. Also, the spectra of polymers can appear simpler than expected due to the accidental degeneracy of the chemically similar groups.

It is also possible to obtain a more complex infrared spectrum than that predicted from the 3N-6 equation. This will usually be due to overtones of a fundamental vibration and the combination of two or more fundamental vibrations. Overtone vibrations are due to transitions between the vibrational ground state and the second or their quantum energy states (See Potential Energy Diagram for Vibrational Transitions). The intensity of the first overtone of a fundamental is usually at least an order of magnitude lower than the fundamental vibration. The position of an overtone frequently occurs at slightly less than whole number multiples of very intense fundamental vibrations. If electromagnetic radiation of energy equal to the sum (or difference) of energies of two fundamental transitions is absorbed, an absorption band will appear in the spectrum at a frequency equal to this sum (or difference). The absorption band resulting from the sum of two or more fundamentals is called a combination band and that resulting from the difference of two fundamentals is known as a difference band.

Another phenomenon which adds to the complexity of an infrared spectrum is known as Fermi resonance. Fermi resonance is the result of an interaction of a fundamental vibration with an overtone or combination band which has nearly the same frequency. It is a consequence of quantum mechanical mixing. The phenomenon was explained by the Italian-American physicist Enrico Fermi. The result is the appearance of two bands equally displaced on both sides of the predicted interacting frequencies. The intensities of the two bands resulting from this interaction are not ordinarily the same. An example of Fermi resonance is the doublet centered at 1,765 cm^{-1} for benzoyl chloride.

It is possible to "predict" completely the positions and types of the fundamental vibrations for molecules. This derivation, however, involves the use of quantum mechanics and group theory and is beyond the scope of this book.

If vibrations within a molecule were independent of the rest of the molecule, then each mode of vibration would be limited to a definite frequency. As an example, all carbon-carbon stretching vibrations regardless of the molecule would have the same frequency. However, for the real molecules vibrational modes can occur over a range of frequencies. The frequency depends upon the mass of the adjacent atoms, resonance and inductive effects, and coupling of this vibration to other vibrations. Other factors such as physical state and matrix may also influence the frequency.

The effect of mass on the fundamental stretching vibration can be computed from equation above. Thus, if the force constants for two diatomic molecules are the same, the molecule having the heavier atoms will have its stretching vibrational mode at a lower frequency. However, the change in position of a vibrational mode when one of the atoms is changed cannot be totally attributed to mass effects. The change in position is partially due to the change in the electron distribution brought about by a substituted atom or group of atoms. As an example, these inductive and resonance effects account for most of the shift in the carbonyl stretching frequency as one changes the atom adjacent to the carbon atom of the carbonyl group. The carbonyl stretching frequency for an acid chloride is higher than that of an ester which is in turn higher than that of a ketone or amide. The high frequency associated with an acid chloride is due primarily to the inductive effects of the highly electronegative chlorine atom. The low frequency of the carbonyl for an amide is due predominantly to the resonance within the molecule. The position of the carbonyl stretching frequency for an ester depends upon both inductive and resonance

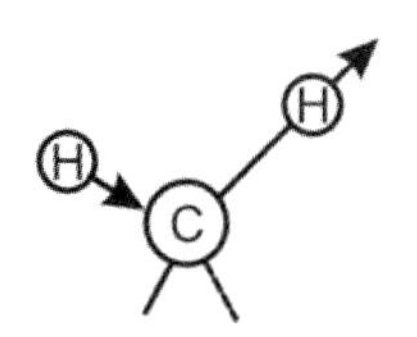

Asymmetric stretching
2926 cm-1 (3.42μ)

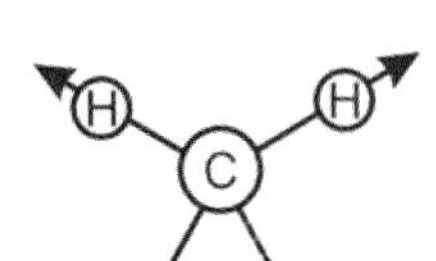

Symmetric stretching
2853 cm-1 (3.51μ)

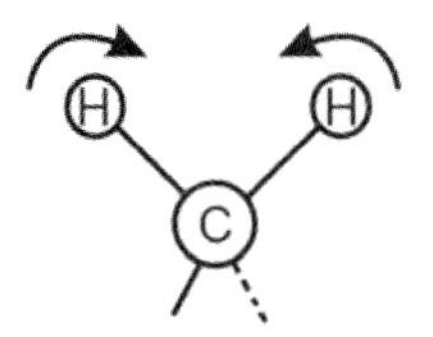

Scissoring deformation
1468 cm-1 (6.81μ)

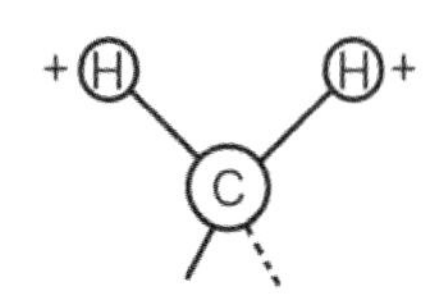

Wagging deformation
1305 cm-1 (7.55μ)

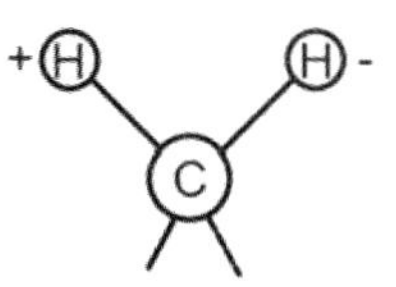

Twisting deformation
2405 cm-1 (7.66μ)

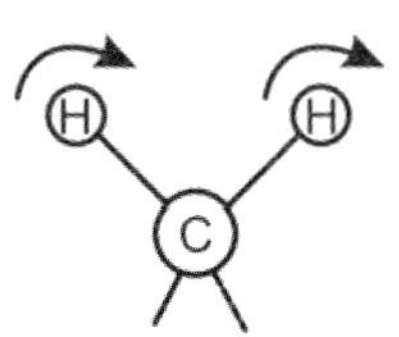

Rocking deformation
720 cm-1 (13.89μ)

effects. The inductive effect tends to raise the frequency of a vibrational mode while resonance tends to lower the frequency.

The interaction or coupling of vibrational modes can also give rise to a shift in the frequency for a given type of vibration. Thus, two carbon-carbon bonds having a carbon atom in common will give rise to two frequencies – one above and one below the frequency of a single carbon-carbon bond. The vibration at the higher frequency is due to the asymmetrical stretch while the vibration at the lower frequency is due to the symmetrical C–C–C stretch. Anhydrides are another class of compounds which exhibit coupling. The two carbonyl stretching frequencies are coupled giving rise to asymmetrical and symmetrical stretching frequencies. However, the coupling effect is not very strong since the frequencies are only separated by 70 cm^{-1}.

The intensities and shapes of vibrational bands are of considerable importance in spectra interpretation and infrared quantitative analysis. The intensity of an infrared band is directly proportional to the probability for the transition between the ground state and a vibrational excited state. This probability, in turn, depends primarily upon the square of the rate of change in the dipole moment for a particular vibration. An intense band indicates a large change in the dipole moment of the atoms involved during a vibration, or that many atoms or functional groups within the molecule have the same vibrational energy.

Band shapes may be sharp or broad depending upon the chemical environment that the molecule experiences. Intermolecular hydrogen bonding will increase the intensity and broaden the O-H stretching frequency in compounds such as alcohols and phenols, whereas the unassociated hydroxyl group has a sharp band of medium intensity. Band broadening may also occur when the spectrophotometer slit width is greater than the width of the absorption band being measured.

Bands whose shapes are not Gaussian often appear in the infrared spectrum of a material in a solid matrix. This effect is known as the "Christiansen effect" and is due to the particle size and refractive index differences between the material and matrix. The radiation loss by scattering is a function of this refractive index difference. At frequencies slightly higher than the absorption maximum of a band, the refractive index of the sample decreases rapidly and approaches the refractive index of the matrix. This results in less scattering and a rapid increase in transmission. At frequencies slightly lower than the absorption maximum, the refractive index of the sample is markedly different than that of the matrix and the transmission is much lower than would be caused by the absorption band alone. The resulting phenomena is represented in the following Christiansen Effect.

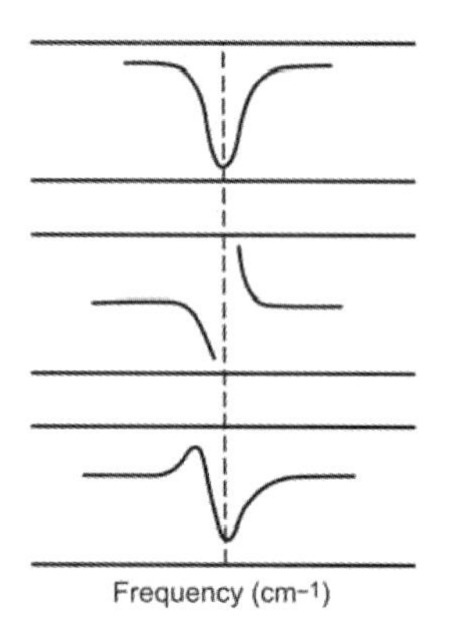

"True" band

Refractive index in region of absorption band (note discontinuity)

Resulting "distorted" absorption band

Characteristic infrared absorption frequencies[a]

Bond	Compound type	Frequency range, cm^{-1}
CH_2	Alkanes	1,450
CH_3	Alkanes	1,325–1,470
C–H	Alkanes	2,850–2,960
C–H	Alkenes	3,020–3,080 (*m*)
C–H	Alkynes	3,300
C=C	Alkenes	1,640–1,680(*v*)
C≡C	Alkynes	2,100–2,260(*v*)
$C\overset{\cdots}{=}$	Aromatic rings	1,500–1,600 (*v*)
C–O	Alcohol, ethers, carboxylic acids, esters	1,080–1,300
C=O	Aldehydes, ketones, carboxylic acids	1,690–1,760
O–H	Monomeric alcohols, phenols	3,610–3,640(*v*)
	Hydrogen-bonded alcohols, phenols	3,200–3,600 (*broad*)
	Carboxylic acids	2,500–3,000 (*broad*)
N–H	Amines	3,300–3,500 (*m*)
C–N	Amines	1,180–1,360
C≡N	Nitriles	2,210–2,260 (*v*)
$-NO_2$	Nitro compounds	1,515–1,560
		1,345–1,385

[a]All bands strong unless marked: *m*:moderate; *w*:weak; *v*:variable

References

Kazuo Nakamoto, Infrared *and Raman Spectra of Inorganic and Coordination Compounds: Theory and Applications in Inorganic Chemistry* (Volume A), John Wiley, 1997.

Planck, Max (1901), "Ueber das Gesetz der Energieverteilung im Normalspectrum", *Ann. Phys.* **309** (3): 553–63.

Willard, Merritt and Dean, *Instrumental Methods of Analysis*, 5th ed., D. Van Nostrand Company, New York, 1974.

Optical Properties

Transmission of Radiation

Refractive index of polymers

Abbr.	Polymer	Refractive Index
PHFPO	Poly(hexafluoropropylene oxide)	1.3010
	Water	1.33
	Alginic acid, sodium salt	1.3343
	Hydroxypropyl cellulose	1.3370
	Poly(tetrafluoroethylene-co-hexafluoropropylene)	1.3380
FEP	Fluorinated Ethylene Propylene	1.3380
	Poly(pentadecafluorooctyl acrylate)	1.3390
	Poly(tetrafluoro-3-(heptafluoropropoxy)propyl acrylate)	1.3460
	Poly(tetrafluoro-3-(pentafluoroethoxy)propyl acrylate)	1.3480
PTFE	Poly(tetrafluoroethylene)	1.3500
THV	Tetrafluoroethylene hexafluoropropylene vinylidene fluoride	1.3500
	Poly(undecafluorohexyl acrylate)	1.3560
PFA	Perfluoroalkoxy	1.3400
ETFE	Ethylene Tetrafluoroethylene	1.4000
	Poly(nonafluoropentyl acrylate)	1.3600
	Poly(tetrafluoro-3-(trifluoromethoxy)propyl acrylate)	1.3600
	Poly(pentafluorovinyl propionate)	1.3640
	Poly(heptafluorobutyl acrylate)	1.3670
	Poly(trifluorovinyl acetate)	1.3750
	Poly(octafluoropentyl acrylate)	1.3800

Refractive index of polymers (Continued)

Abbr.	Polymer	Refractive Index
	Poly(methyl 3,3,3-trifluoropropyl siloxane)	1.3830
	Poly(pentafluoropropyl acrylate)	1.3850
	Poly(2-heptafluorobutoxy)ethyl acrylate)	1.3900
PCTFE	Poly(chlorotrifluoroethylene)	1.3900
	Poly(2,2,3,4,4-hexafluorobutyl acrylate)	1.3920
	Poly(methyl hydro siloxane)	1.3970
	Poly(methacrylic acid), sodium salt	1.4010
	Poly(dimethyl siloxane)	1.4035
	Poly(trifluoroethyl acrylate)	1.4070
	Poly (2-(1,1,2,2-tetrafluoroethoxy)ethyl acrylate)	1.4120
	Poly(trifluoroisopropyl methacrylate)	1.4177
	Poly(2,2,2-trifluoro-1-methylethyl methacrylate)	1.4185
	Poly(2-trifluoroethoxyethyl acrylate)	1.4190
PVDF	Poly(vinylidene fluoride)	1.4200
ECTFE	Ethylene Chlorotrifluorotheylene	1.4470
	Poly(trifluoroethyl methacrylate)	1.4370
	Poly(methyl octadecyl siloxane)	1.4430
	Poly(methyl hexyl siloxane)	1.4430
	Poly(methyl octyl siloxane)	1.4450
	Poly(isobutyl methacrylate)	1.4470
	Poly(vinyl isobutyl ether)	1.4507
	Poly(methyl hexadecyl siloxane)	1.4510
PEO	Poly(ethylene oxide)	1.4539
	Poly(vinyl ethyl ether)	1.4540
	Poly(methyl tetradecyl siloxane)	1.4550
	Poly(ethylene glycol mono-methyl ether)	1.4555
	Poly(vinyl n-butyl ether)	1.4563
PPOX	Poly(propylene oxide)	1.4570
	Poly(3-butoxypropylene oxide)	1.4580
	Poly(3-hexoxypropylene oxide)	1.4590
	Poly(ethylene glycol)	1.4590
	Poly(vinyl n-pentyl ether)	1.4590
	Poly(vinyl n-hexyl ether)	1.4591
	Poly(4-fluoro-2-trifluoromethylstyrene)	1.4600

Refractive index of polymers (Continued)

Abbr.	Polymer	Refractive Index
	Poly(vinyl octyl ether)	1.4613
	Poly(vinyl n-octyl acrylate)	1.4613
	Poly(vinyl 2-ethylhexyl ether)	1.4626
	Poly(vinyl n-decyl ether)	1.4628
	Poly(2-methoxyethyl acrylate)	1.4630
	Poly(acryloxypropyl methyl siloxane)	1.4630
PMP	Poly(4-methyl-1-pentene)	1.4630
	Poly(3-methoxypropylene oxide	1.4630
PtBuMA	Poly(t-butyl methacrylate)	1.4638
	Poly(vinyl n-dodecyl ether)	1.4640
	Poly(3-ethoxypropyl acrylate)	1.4650
	Poly(vinyl propionate)	1.4664
PVAC	Poly(vinyl acetate)	1.4665
	Poly(vinyl propionate)	1.4665
	Poly(vinyl methyl ether)	1.4670
	Poly(ethyl acrylate)	1.4685
	Poly(vinyl methyl ether)(isotactic)	1.4700
	Poly(3-methoxypropyl acrylate)	1.4710
	Poly(1-octadecene)	1.4710
	Poly(2-ethoxyethyl acrylate)	1.4710
PIPA	Poly (isopropyl acrylate)	1.4728
	Poly(1-decene)	1.4730
	Poly(propylene)(atactic)	1.4735
	Poly(lauryl methacrylate)	1.4740
	Poly(vinyl sec-butyl ether) (isotactic)	1.4740
P-nBuA	Poly(n-butyl acrylate)	1.4740
	Poly(dodecyl methacrylate)	1.4740
	Poly(ethylene succinate)	1.4744
	Poly(tetradecyl methacrylate)	1.4746
	Poly(hexadecyl methacrylate)	1.4750
CAB	Cellulose acetate butyrate	1.4750
CA	Cellulose acetate	1.4750
	Poly(vinyl formate)	1.4757
EVA-40% vinyl acetate	Ethylene/vinyl acetate copolymer-40% vinyl acetate	1.4760
	Poly(2-fluoroethyl methacrylate)	1.4768
	Poly(octyl methyl silane)	1.4780
EC	Ethyl cellulose	1.4790
PMA	Poly(methyl acrylate)	1.4793
	Poly(dicyanopropyl siloxane)	1.4800

Refractive index of polymers (Continued)

Abbr.	Polymer	Refractive Index
POM	Poly(oxymethylene) or Polyformaidehyde	1.4800
	Poly(sec-butyl methacrylate)	1.4800
	Poly(dimethylsiloxane-co-alpha-methylstyrene)	1.4800
	Poly(n-hexyl methacrylate)	1.4813
EVA-33% vinyl acetate	Ethylene/vinyl acetate copolymer-33% vinyl acetate	1.4820
PnBuMA	Poly(n-butyl methacrylate)	1.4830
	Poly(ethylidene dimethacrylate)	1.4831
	Poly(2-ethoxyethyl methacrylate)	1.4833
	Poly(n-propyl methacrylate)	1.4840
	Poly(ethylene maleate)	1.4840
EVA-28% vinyl acetate	Ethylene/vinyl acetate copolymer-28% vinylacetate	1.4845
	Poly(ethyl methacrylate)	1.4850
PVB	Poly(vinyl butyral)	1.4850
PVB-11% hydroxl	Poly(vinyl butyral)-11% hydroxl	1.4850
	Poly(3,3,5-trimethylcyclohexyl methacrylate)	1.4850
	Poly(2-nitro-2-methylpropyl methacrylate)	1.4868
	Poly(dimethylsiloxane-co-diphenylsiloxane)	1.4880
	Poly(1,1-diethylpropyl methacrylate)	1.4889
	Poly(triethylcarbinyl methacrylate)	1.4889
PMMA	Poly(methyl methacrylate)	1.4893
	Poly(2-decyl-1,4-butadiene)	1.4899
PP-isotactic	Poly(propylene), isotactic	1.4900
PVB-19% hydroxyl	Poly(vinyl butyral)-19% hydroxyl	1.4900
	Poly(mercaptopropyl methyl siloxane)	1.4900
	Poly(ethyl glycolate methacrylate)	1.4903
	Poly(3-methylcyclohexyl methacrylate)	1.4947
	Poly(cyclohexyl alpha-ethoxyacrylate)	1.4969
MC	Methyl cellulose	1.4970
	Poly(4-methylcyclohexyl methacrylate)	1.4975

Refractive index of polymers (Continued)

Abbr.	Polymer	Refractive Index
	Poly(decamethylene glycol dimethacrylate)	1.4990
PVAL	Poly(vinyl alcohol)	1.5000
PVFM	Poly(vinyl formal)	1.5000
	Poly(2-decyl-1,4-butadiene)	1.4899
PVFM	Poly(vinyl formal)	1.5000
	Poly(2-bromo-4-trifluoromethyl styrene)	1.5000
	Poly(1,2-butadiene)	1.5000
	Poly(sec-butyl alpha-chloroacrylate)	1.5000
	Poly(2-heptyl-1,4-butadiene)	1.5000
	Poly(vinyl methyl ketone)	1.5000
	Poly(ethyl alpha-chloroacrylate)	1.5020
PVFM	Poly(vinyl formal)	1.5020
	Poly(2-isopropyl-1,4-butadiene	1.5020
	Poly(2-methylcyclohexyl methacrylate)	1.5028
	Poly(bornyl methacrylate)	1.5059
	Poly(2-t-butyl-1,4-butadiene)	1.5060
	Poly(ethylene glycol dimethacrylate)	1.5063
PCHMA	Poly(cyclohexyl methacrylate)	1.5065
	Poly(cyclohexanediol-1,4-dimethacrylate)	1.5067
IIR or PIBI	Butyl rubber(unvulcanized)	1.5080
	Gutta percha b	1.5090
	Poly(tetrahydrofurfuryl methacrylate)	1.5096
PIB	Poly(isobutylene)	1.5100
LDPE	Polyethylene, low density	1.5100
EMA	Ethylene/methacrylic acid ionomer, sodium ion	1.5100
PE	Polyethylene	1.5100
CN	Cellulose nitrate	1.5100
	Polyethylene lonomer	1.5100
	Polyacetal	1.5100
	Poly(1-methylcyclohexyl methacrylate)	1.5111
	Poly(2-hydroxyethyl methacrylate)	1.5119
	Poly(1-butene)(isotactic)	1.5125
	Poly(vinyl methacrylate)	1.5129

Refractive index of polymers (Continued)

Abbr.	Polymer	Refractive Index
	Poly(vinyl chloroacetate)	1.5130
	Poly(N-butyl methacrylamide)	1.5135
	Gutta percha a	1.5140
	Poly(2-chloroethyl methacrylate)	1.5170
PMCA	Poly(methyl alpha-chloroacrylate)	1.5170
	Poly(2-diethylaminoethyl methacrylate)	1.5174
	Poly(2-chlorocyclohexyl methacrylate)	1.5179
	Poly(1,4-butadiene)(35% cis; 56% trans; 7% 1,2-content)	1.5180
PAN	Poly(acrylonitrile)	1.5187
	Poly(isoprene),cis	1.5191
	Poly(allyl methacrylate)	1.5196
	Poly(methacrylonitrile)	1.5200
	Poly(methyl isopropenyl ketone)	1.5200
	Poly(butadiene-co-acrylonitrile)	1.5200
	Poly(2-ethyl-2-oxazoline)	1.5200
	Poly(1,4-butadiene)(high cis-type)	1.5200
	Poly(N-2-methoxyethyl) methacrylamide	1.5246
	Poly(2,3-dimethylbutadiene) {methyl rubber}	1.5250
	Poly(2-chloro-1-(chloromethyl) ethyl methacrylate)	1.5270
	Poly(1,3-dichloropropyl methacrylate)	1.5270
PAA	Poly(acrylic acid)	1.5270
	Poly(N-vinyl pyrrolidone)	1.5300
NYLON-6	Nylon 6{Poly(caprolactam)}	1.5300
	Poly(butadiene-co-styrene)(30%) styrene)block copolymer	1.5300
	Poly(cyclohexyl alpha-chloroacrylate)	1.5320
	Poly(methyl phenyl siloxane)	1.5330
	Poly(2-chloroethyl alpha-chloroacrylate)	1.5330
	Poly(butadiene-co-styrene) (75/25)	1.5350
	Poly(2-aminoethyl methacrylate)	1.5370
	Poly(furfuryl metacrylate)	1.5381

Refractive index of polymers (Continued)

Abbr.	Polymer	Refractive Index
PVC	Poly(vinyl chloride)	1.5390
	Poly(butylmercaptyl methacrylate)	1.5390
	Poly(1-phenyl-n-amyl methacrylate)	1.5396
	Poly(N-methyl methacrylamide)	1.5398
HDPE	Polyethylene, high density	1.5400
	Cellulose	1.5400
	Poly(cyclohexyl alpha-bromoacrylate)	1.5420
	Poly(sec-butyl alpha-bromoacrylate)	1.5420
	Poly(2-bromoethyl methacrylate)	1.5426
	Poly(dihydroabietic acid)	1.5440
	Poly(abietic acid)	1.5460
	Poly(ethylmercaptyl methacrylate)	1.5470
	Poly(N-allyl methacrylamide)	1.5476
	Poly(1-phenylethyl methacrylate)	1.5487
	Poly(2-vinyltetrahydrofuran)	1.5500
	Poly(vinylfuran)	1.5500
	Poly(methyl m-chlorophenylethyl siloxane)	1.5500
	Poly(p-methoxybenzyl methacrylate)	1.5520
	Poly(isopropyl methacrylate)	1.5520
	Poly(p-isopropyl styrene)	1.5540
	Poly(isoprene), chlorinated	1.5540
	Poly(p,p′-xylylenyl dimethacrylate)	1.5559
	Poly(cyclohexyl methyl silane)	1.5570
	Poly(1-phenylallyl methacrylate)	1.5573
	Poly(p-cyclohexylphenyl methacrylate)	1.5575
CR	Poly(chloroprene)	1.5580
	Poly(2-phenylethyl methacrylate)	1.5592
	Poly(methyl m-chlorophenyl siloxane)	1.5600
	Poly{4,4-heptane bis(4-phenyl) carbonate}	1.5602
	Poly{1-(o-chlorophenyl)ethyl methacrylate)}	1.5624

Refractive index of polymers (Continued)

Abbr.	Polymer	Refractive Index
S/MA	Styrene/maleic anhydride copolymer	1.5640
	Poly(1-phenylcyclohexyl methacrylate)	1.5645
NYLON 6,10	Nylon 6,10{Poly(hexamethylene sebacamide)}	1.5650
NYLON 6,6	Nylon 6,6{Poly(hexamethylene adipamide)}	1.5650
NYLON 6(3)	Nylon 6(3)T {Poly(trimethyl hexamethylene terephthalamide)}	1.5660
	Poly(2,2,2′-trimethylhexamethylene terephthalamide)	1.5660
	Poly(methyl alpha-bromoacrylate)	1.5672
	Poly(benzyl methacrylate)	1.5680
	Poly{2-(phenylsulfonyl)ethyl methacrylate}	1.5682
	Poly(m-cresyl methacrylate)	1.5683
SAN	Styrene/acrylonitrile copolymer	1.5700
	Poly(o-methoxyphenol methacrylate)	1.5705
PPhMA	Poly(phenyl methacrylate)	1.5706
	Poly(o-cresyl methacrylate)	1.5707
PDAP	Poly(diallyl phthalate)	1.5720
	Poly(2,3-dibromopropyl methacrylate)	1.5739
	Poly(2,6-dimethyl-p-phenylene oxide)	1.5750
PET	Poly(ethylene terephthalate)	1.5750
PVB	Poly(vinyl benozoate)	1.5775
	Poly{2,2-propane bis[4-(2-methylphenyl)]carbonate}	1.5783
	Poly{1,1-butane bis(4-phenyl) carbonate}	1.5792
	Poly(1,2-diphenylethyl methacrylate)	1.5816
	Poly(o-chlorobenzyl methacrylate)	1.5823
	Poly(m-nitrobenzyl methacrylate)	1.5845
	Poly(oxycarbonyloxy-1,4-phenyleneisopropylidene-1,4-phenylene)	1.5850

Refractive index of polymers (Continued)

Abbr.	Polymer	Refractive Index
	Poly{N-(2-phenylethyl) methacrylamide}	1.5857
	Poly{1,1-cyclohexane bis[4-(2,6-dichlorophenyl)]carbonate}	1.5858
PC	Polycarbonate resin	1.5860
BPA	Bisphenol-A polycarbonate	1.5860
	Poly(4-methoxy-2-methylstyrene)	1.5868
	Poly(o-methyl styrene)	1.5874
PS	Polystyrene	1.5894
	Poly{2,2-propane bis[4-(2-chlorophenyl)]carbonate}	1.5900
	Poly{1,1-cyclohexane bis(4-phenyl)carbonate}	1.5900
	Poly(o-methoxy styrene)	1.5932
	Poly(diphenylmethyl methacrylate)	1.5933
	Poly{1,1-ethane bis(4-phenyl) carbonate}	1.5937
	Poly(propylene sulfide)	1.5960
	Poly(p-bromophenyl methacrylate)	1.5964
	Poly(N-benzyl methacrylamide)	1.5965
	Poly(p-methoxy styrene)	1.5967
MeOS	Poly(4-methoxystyrene)	1.5967
	Poly{1,1-cyclopentane bis(4-phenyl)carbonate}	1.5993
PVDC	Poly(vinylidene chloride)	1.6000
	Poly(o-chlorodiphenylmethyl methacrylate)	1.6040
	Poly{2,2-propane bis[4-(2,6-dichlorophenyl)]carbonate}	1.6056
	Poly(pentachlorophenyl methacrylate)	1.6080
	Poly(2-chlorostyrene)	1.6098
PaMes	Poly(alpha-methylstyrene)	1.6100
	Poly(phenyl alpha-bromoacrylate)	1.6120
	Poly{2,2-propane bis[4-(2,6-dibromophenyl)cabonate]}	1.6147
	Poly(p-divinylbenzene)	1.6150
	Poly(N-vinyl phthalimide)	1.6200
	Poly(2,6-dichlorostyrene)	1.6248
	Poly(chloro-p-xylene)	1.6290

Refractive index of polymers (Continued)

Abbr.	Polymer	Refractive Index
	Poly(beta-naphthyl methacrylate)	1.6298
	Poly(alpha-naphthyl carbinyl methacrylate)	1.6300
PEI-ULTEM	Polyetherimide (880 nm wavelength)	1.630
	Polyetherimide (643.8 nm wavelength)	1.651
	Polyetherimide (587.6 nm wavelength)	1.660
	Polyetherimide (546.1 nm wavelength)	1.668
	Polyetherimide (480 nm wavelength)	1.687
	Poly(phenyl methyl silane)	1.6300
	Poly(sulfone) {Poly[4,4'-isopropylidene diphenoxy di(4-phenylene)sulfone]}	1.6330
PSU	Polysulfone resin	1.6330
	Poly(2-vinylthiophene)	1.6376
Mylar Film	Polyethylene terephthalate (boPET)	1.64–1.67
	Poly(2,6-diphenyl-1,4-phenylene oxide)	1.6400
	Poly(alpha-naphthyl methacrylate)	1.6410
	Poly(p-phenylene ether-sulphone)	1.6500
	Poly{diphenylmethane bis(4-phenyl)carbonate}	1.6539
	Poly(vinyl phenyl sulfide)	1.6568
	Poly(styrene sulfide)	1.6568
	Butylphenol formaldehyde resin	1.6600
	Poly(p-xylylene)	1.6690
PVN	Poly(2-vinylnapthalene)	1.6818
PVK	Poly(N-vinyl carbazole)	1.6830
	Naphthalene-formaldehyde rubber	1.6960
PF	Phenol-formaldehyde resin	1.7000
	Poly(pentabromophenyl methacrylate)	1.7100
MFA	Polytetrafluoroethylene-Perfluoromethylvinylether	unknown
PEEK1	(amorphous) Polyetheretherketone	1.65–1.71
PEEK2	(crystalline) Polyetheretherketone	1.68–1.77

Dielectric Properties

Electrical Properties

Polymer Dielectric Properties

Most plastics are dielectrics or insulators (poor conductors of electricity) and resist the flow of a current. This is one of the most useful properties of plastics and makes much of our modern society possible through the use of plastics as wire coatings, switches and other electrical and electronic products. Despite this, dielectric breakdown can occur at sufficiently high voltages to give current transmission and possible mechanical damage to the plastic.

The application of a potential difference (voltage) causes in the movement of electrons and when the electrons are free to move there is a flow of current. Metals can be thought of as a collection of atomic nuclei existing in a 'sea of electrons' and when a voltage is applied the electrons are free to move and to conduct a current. Polymers and the atoms that make them up have their electrons tightly bound to the central long chain and side groups through 'covalent' bonding. Covalent bonding makes it much more difficult for most conventional polymers to support the movement of electrons and therefore they act as insulators.

Polar and Non-Polar Plastics

Not all polymers behave the same when subjected to voltage and plastics can be classified as 'polar' or 'non-polar' to describe their variations in behavior.

The polar plastics do not have a fully covalent bond and there is a slight imbalance in the electronic charge of the molecule. A simple example of this type of behavior would be that of the water molecule (H_2O). The conventional representation of the molecule is that shown at right. The two hydrogen atoms are attached to the oxygen atom and the overall molecule has no charge.

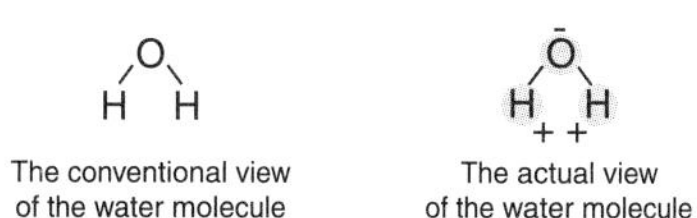

In reality, the electrons tend to be around the oxygen atom more than around the hydrogen atoms and this gives the oxygen a slightly negative charge and the hydrogen atoms a slightly positive charge. This is shown in the diagram at right where the grey areas show where the electrons are more often found. The overall water molecule is neutral and does not carry a charge but the imbalance of the electrons creates a 'polar' molecule. This 'polar dipole' will move in the presence of an electric field and attempt to line up with the electric field in much the same way as a compass needle attempts to line up with the earth's magnetic field.

In polar plastics, dipoles are created by an imbalance in the distribution of electrons and in the presence of an electric field the dipoles will attempt to move to align with the field. This will create 'dipole polarization' of the material and because movement of the dipoles is involved there is a time element to the movement. Examples of polar plastics are PMMA, PVC, PA (Nylon), PC and these materials tend to be only moderately good as insulators.

The non-polar plastics are truly covalent and generally have symmetrical molecules. In these materials there are no polar dipoles present and the application of an electric field does not try to align any dipoles. The electric field does, however, move the electrons slightly in the direction of the electric field to create 'electron polarization', in this case the only movement is that of electrons and this is effectively instantaneous. Examples of non-polar plastics are PTFE (and many other fluoropolymers), PE, PP and PS and these materials tend to have high resistivities and low dielectric constants.

The structure of the polymer determines if it is polar or non-polar and this determines many of the dielectric properties of the plastic.

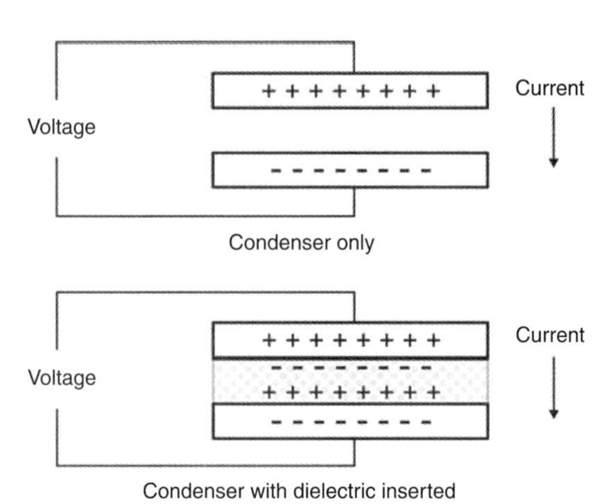

For polar plastics the alternating current frequency is an important factor because of the time taken to align the polar dipoles. At very low frequencies the dipoles have sufficient time to align with the field before it changes direction and the dielectric constant is high. At very high frequencies the dipoles do not have time to align before the field changes direction and the dielectric constant is lower. At intermediate frequencies the dipoles move but have not completed their movement before the field changes direction and they must realign with the changed field. Polar plastics at low frequencies (60 Hz) generally have dielectric constants of between 3 and 5.

For non-polar plastics the dielectric constant is independent of the alternating current frequency because the electron polarization is effectively instantaneous. Non-polar plastics always have dielectric constants of less than 3.

Measurements of Electrical Properties

Dielectric Constant (Alternating Current)

The dielectric constant is a measure of the influence of a particular dielectric on the capacitance of a condenser. It measures how well a material separates the plates in a capacitor and is defined as the ratio of the capacitance of a set of electrodes with the dielectric material between them to the capacitance of the same electrodes with a vacuum between them. The dielectric constant for a vacuum is 1 and for all other materials it is greater than 1.

Power Factor (Alternating Current)

The power factor is a measure of the energy absorbed by the material as the alternating current constantly changes direction and the dipoles try to align themselves with the field. As the dipoles try to align themselves with the external field they will always be slightly out of phase and will 'lag' behind the field. The amount of lagging is measured by the phase angle (q) and the power factor is defined as cos q. The power factor can be thought of as a measure of the internal friction created by the alternating current and will define how much a material heats up when placed in an alternating field.

For polar plastics the power factor is dependent on the alternating current frequency. At very low and at very high frequencies both the power factor and the amount of internal heating are low- the dipoles either have time to align or do not have time to align before the field changes direction. At intermediate frequencies the power factor goes through a maximum and the internal friction is high and substantial heating of the plastic can take place.

This maximum in the power factor is also the basis for microwave ovens. The microwave generator in the oven applies an alternating field (in the microwave region) to the food. The frequency of the microwave field is matched to the frequency that is the maximum for the power factor of the water dipole. The polar dipole water molecules constantly attempt to align with the alternating field and the resulting internal friction heats up the food. Non-polar materials or polar materials with a maximum in the power factor at different frequencies either do not heat up at all or gain relatively little heat. The fact that the microwaves act directly on the water molecules means that foods heat up evenly throughout their volume and cooking takes place as much internally as it does externally.

For non-polar plastics the electronic polarization is effectively in phase with the external field (i.e., q >> 0 and cos q is also approximately 0) and the power factor is generally less than 0.0003. non-polar plastics suffer from very little internal friction and minimal internal heating.

Dielectric Strength (Direct Current)

The dielectric strength is the direct current voltage between two electrodes at which dielectric breakdown occurs and is an indicator of how good an insulator the material is. The voltage is increased until the material breaks down, there is an arc across the electrodes and substantial current flows.

Most plastics have good dielectric strengths (in the order of 100 to 300 kV/cm).

Volume Resistivity (Direct Current)

The volume resistivity is a measure of the resistance of the material in terms of its volume. A voltage is applied across the plates and the current measured to allow calculation of the volume resistivity. Most plastics have very high volume resistivity (in the order of 1016 Ωm) and are therefore good insulators.

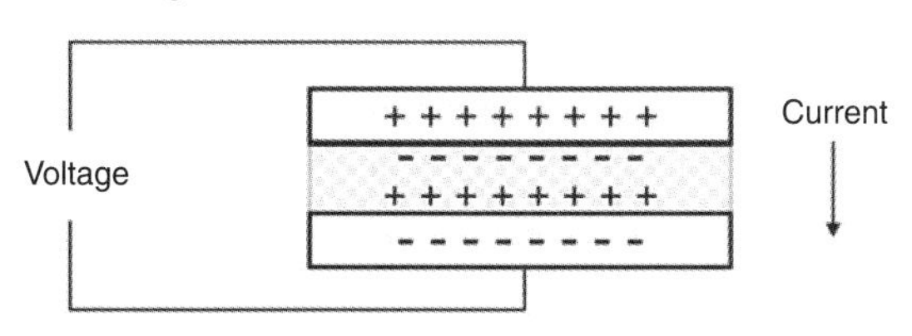

Measurement of volume resistivity

$$\rho = R\frac{A}{l}$$

where

ρ = Resistivity, Ω.m
R=Resistance, Ω
A=Cross-sectional area of specimen, and
l=Length of specimen

Material	Resistivity [Ω·m] at 20 °C
Silver	1.59×10^{-8}
Copper	1.68×10^{-8}
Gold	2.44×10^{-8}
Aluminium	2.82×10^{-8}
Calcium	3.36×10^{-8}
Tungsten	5.60×10^{-8}
Zinc	5.90×10^{-8}
Nickel	6.99×10^{-8}
Iron	1.0×10^{-7}
Platinum	1.06×10^{-7}
Tin	1.09×10^{-7}
Lead	2.2×10^{-7}
Manganin	4.82×10^{-7}
Constantan	4.9×10^{-7}
Mercury	9.8×10^{-7}
Nichrome	1.10×10^{-6}
Carbon (amorphous)	$5–8\times10^{-4}$
Carbon (graphite)[8]	$2.5–5.0\times10^{-6}$ ⊥ basal plane 3.0×10^{-3} // basal plane
Carbon (diamond)[10]	$\sim10^{12}$
Germanium[10]	4.6×10^{-1}
Seawater	2×10^{-1}
Silicon[10]	6.40×10^{2}
Glass	10^{10} to 10^{14}
Hard rubber	approx. 10^{13}
Sulfur	10^{15}
Paraffin	10^{17}
Quartz (fused)	7.5×10^{17}
PET	10^{20}
Teflon	10^{22} to 10^{24}

Surface Resistivity (Direct Current)

The surface resistivity is a measure of the resistance of the material to a surface flow of current. It is the ratio of the applied direct voltage and the resulting current along the surface of the material per unit width. Surface resistivity is measured in Ω.

Tracking and Arc Resistance (Direct Current)

These are measures of how long a material can resist forming a continuous conduction path under a high voltage/low current arc.

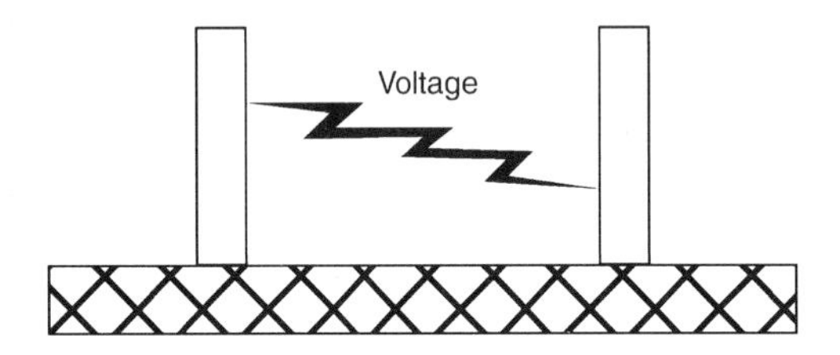

The Environment and Electrical Properties

The electrical properties of plastics may also be changed dramatically by the environmental conditions, such as moisture and/or temperature and this is particularly true for polar plastics.

The polar plastics have a tendency to absorb moisture from the atmosphere and can often contain a significant amount of water at normal room temperature. For these materials, the presence of the water generally raises the dielectric constant and lowers both the volume and surface resistivity.

Raising the temperature of a polar plastic allows faster movement of the polymer chains and faster alignment of the dipoles. This is particularly true if the temperature is raised above T_g (see Newsletter on Low Temperature Plastics) because above T_g much more molecular movement is possible. Raising the temperature inevitably raises the dielectric constant of polar plastics.

Non-polar plastics, such as the fluoropolymers, are not as affected by the water because they tend not to absorb water and temperature effects are not generally as severe because increased temperature does not affect the electronic polarization.

The Fluorocarbons

The fluorocarbon plastics family is generally non-polar and as such these plastics have very low dielectric constants (less than 3) and also the power factor is both frequency independent and low (less than 0.0003 across a wide range of frequencies). The tracking and arc resistance properties are excellent and even when arcing does occur there is little mechanical damage to the surface. Other materials will form a carbonized arc path when arced and this will act as a path for arcing in the future, PTFE does not generally leave a deep arc path and it is often possible to use the product again without repeat arcing along the same path.

Conclusions

The dielectric properties of polymers are largely predictable from the chemical structure of the polymer. The chemical structure determines the polar or non-polar nature of the final polymer and this then largely determines the behavior of the polymer under a variety of electrical situations. This exceptional electrical properties of the fluoropolymers is again no accident or coincidence. It is predictable from the non-polar structure of these polymers – their chemical structure results in their exceptional properties.

Dynamic Mechanical Properties

Properties of Polymers

Dynamic Mechanical Analysis and Principles of Polymer Structure

The complexity of polymer behavior relates ultimately to viscoelasticity. Most classical materials exhibit either elastic or viscous behavior in response to an applied stress. Elastic responses are typical in solid materials.

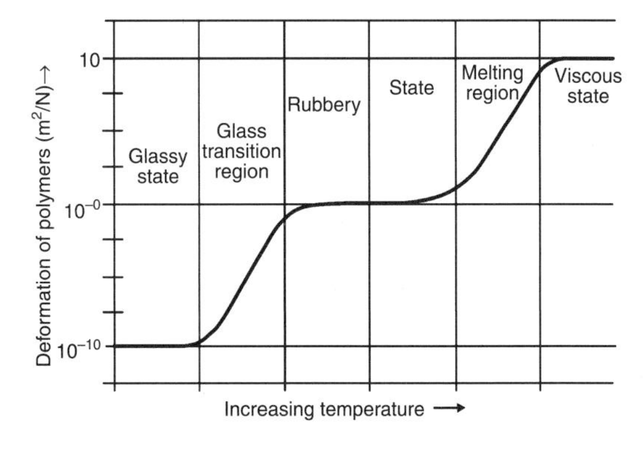

When a stress is applied to an elastic system it deforms proportionally by a quantity identified as the strain. We can quantitatively express the relationship between the applied stress and the resulting strain as:

$$\tau = G \bullet \gamma$$

where t is the stress in shear, γ is the strain, and G is the shear modulus. The same equation can be written for other modes of stress such as tension. The response of an elastic system to applied stress is instantaneous and completely recoverable. We say that the system stores the energy and can return it to the system completely when the stress is removed. The equation is familiar to us as Hooke's Law and a spring is used as the model for materials governed by this law. The plot in *Relationship of stress and strain with time for a pure elastic system* illustrates the behavior of an elastic system in time.

Viscous behavior is a characteristic of fluids, materials where the bond energies necessary for long range translational order have been overcome. In these systems an applied stress results in a strain that increases proportionally with time until the stress is removed. The strain is not recoverable; when the stress is removed the deformation is completely retained. We say that the energy has been lost to the system. The model of a dashpot is frequently used as an analogy. The plot in *Relationship of stress and strain with time for purely viscous system* his behavior graphically. Newton first defined the mathematical relationship between the applied stress and the resulting strain rate in a fluid and termed the resulting ratio the viscosity,

$$\tau = \eta \bullet \gamma$$

where t is the stress, $\gamma_\bullet$ is the strain rate, and ŋ is the viscosity.

The large size and conformational variety of polymer molecules prevent these materials from forming the fully ordered systems that we normally associate with solid materials. By the same token, in the fluid state, the high degree of chain entanglement that is possible in these systems produces behavior that departs significantly from that of classical Newtonian fluids.

In both the solid and the fluid state, these materials exhibit a combination of elastic and viscous responses when placed under stress. In a solid plastic beam we can perform classical measurements of stress versus strain that allow us to calculate a modulus. However, if we maintain a constant applied load, we find that the resulting strain is not constant; it continues to increase as a function of time. In engineering terms we refer to this as creep or cold flow and it is actually a manifestation of viscous flow in the apparently solid polymer. A counterpart to this behavior is known as stress relaxation. Here the strain is held as the constant and the stress required to maintain that strain is measured as a function of time.

In a viscoelastic system the stress decreases with time. Since modulus is defined as the ratio of stress to strain, it can be seen that the modulus calculation in viscoelastic systems must incorporate a time function and cannot be considered as an immutable property independent of the period over which the measurement is made. At a structural level, the polymer chains are slowly rearranging in response to the applied stress. Knowledge of the rate at which this occurs is critical to an accurate determination of a material's fitness-for-use in a particular application.

Similarly, we can perform viscosity determinations on a polymer in the fluid state by applying a known stress and measuring the resulting strain rate or rate of flow. In a Newtonian fluid the viscosity is a constant that is independent of the strain rate. However, if we measure the viscosity of a polymer fluid at various strain rates, we find that it changes, becoming lower at higher strain rates. At a structural level this effect is produced when the long, entangled polymer chains become oriented in the direction of flow and the entire system moves with reduced resistance. When the strain is suddenly removed the long chains re-entangle and the fluid exhibits aspects of elastic recovery.

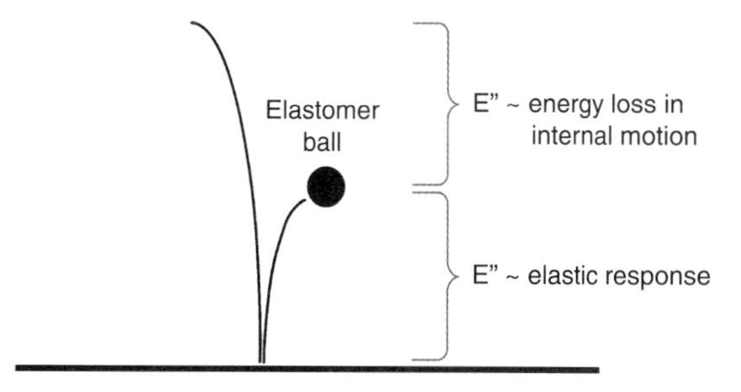

To further complicate the picture, the balance between the elastic and viscous response changes for a given material as a function of temperature. In the solid state this balance is reflected in terms of load-bearing properties – time-dependent behavior such as creep and stress relax ation, as well as impact properties. In the fluid state, viscoelasticity provides information on molecular weight, molecular weight distribution, thermal stability, and crosslinking. The equation relating stress and strain in a viscoelastic system introduces the aspect of time dependency:

$$\tau = G(t) \bullet \gamma$$

where G(t) is the stress relaxation modulus. The material initially responds in an elastic manner, then as a viscous fluid. When the stress is removed, the elastic portion recovers over an extended period of time. The plot in *Relationship of stress and strain with time for a viscoelastic* provides a generalized illustration of this compound behavior.

Determining the proportion of the elastic and viscous components in a polymer, and the factors that cause that balance to change, is crucial to understanding howa material will perform in a given application environment. It can also provide valuable information regarding structure and composition. DMA accomplishes this resolution. While it is possible to perform dynamic mechanical measurements on solids and fluids, the focus of this work improved material selection for end-use applications. Therefore, this work will concentrate solid-state properties.

While the best dynamic mechanical analyzers can be operated in a controlled stress or controlled strain mode, the primary value of the method is in the dynamic experiment. In this mode of operation, the DMA instrument applies an oscillatory stress with a controlled frequency. Dynamic modulus values using this method are a function of frequency rather than time. The stress function is sinusoidal. In a perfectly elastic system the applied stress and the resulting strain will be in phase as shown in *The behavior of an elastic system under oscillatory stress, stress and strain in phase.* For an ideal fluid the stress will lead the strain by 90° ($\pi/2$ radians) as illustrated in *The behavior of a viscous system under oscillatory stress, stress and strain are 90° out of phase.* A viscoelastic material will give some hybrid of these two responses. The stress and strain will be out of phase by some quantity known as the phase angle and commonly referred to as delta (δ). A small phase angle indicates high elasticity while a large phase angle is associated with highly viscous properties. The complex response of the material is resolved into the elastic or storage modulus (G′) and the viscous or loss modulus (G″) if the deformation is in the shear mode. If the deformation is in the tensile or flexural mode, then E′ and E″ are used. A summary of the key terms are shown in the table *Key Viscoelastic Terms.*

Key Viscoelastic Terms	
Complex Modulus	G^*or $E^* = \sigma^*/\gamma$
Elastic Modulus	G' or $E' = \sigma'/\gamma = (\sigma^*/\gamma)\cos$
Viscous Modulus	G'' or E'' $\sigma'/\gamma = (\sigma^*/\gamma)\sin$
Complex Viscosity	$n = G^*/\gamma$
Loss Tangent	$\tan\delta = G''/G'$ or E''/E'

When tensile, flexural, or shear modulus are measured by traditional methods, it is the complex modulus that is the result of the test. It is defined as the slope of the stress-strain curve in the linear region. The DMA resolves this complex modulus into the storage and loss component. The smaller the phase angle is, the closer the elastic modulus is to the complex modulus. It is convenient to think of the elastic and viscous component in the vector terms illustrated in the figure, that shows the relationship between the stress and strain vectors: *(a) Relationship of the stress and the strain vectors in a dynamic experiment; (b) Stress vectors resolved into the loss and storage components; and (c) Corresponding modulus vectors with loss vectors transposed to a right triangle.*

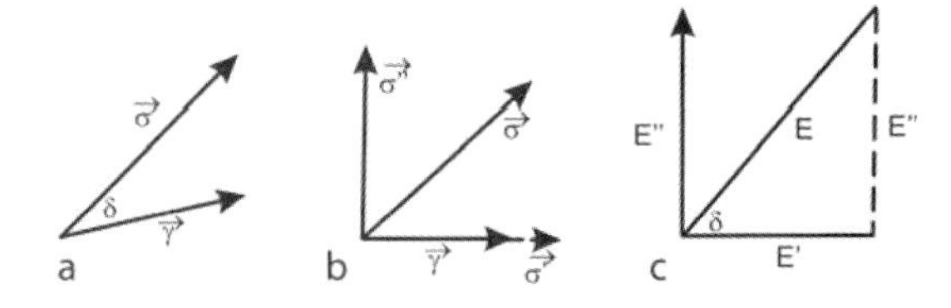

The relationship in (b) shows the stress vectors resolved into their storage and loss component. The storage component is in phase with the strain. The above stress and strain relationship in (c) expresses the vectors in terms of the modulus. The transposed loss modulus shows that the complex modulus can be thought of as the hypotenuse of a right triangle and the storage and loss components as the two shorter legs that are perpendicular to each other. The tangent of the phase angle, often referred to as tan delta, can be used to deduce the shape of the right triangle. In the solid state, tan delta for a polymeric material rarely rises above 0.1 until the material approaches the softening temperature. A tan delta of 0.1 is analogous to a right triangle with a long side of 10 units and a short side of 1 unit. A triangle of these dimensions will have a hypotenuse 10.05 units long. This quantifies the relationship between the complex modulus measured by a classical stress-strain test and the elastic modulus measured by DMA. For the vast majority of the conditions at which DMA measurements are made on

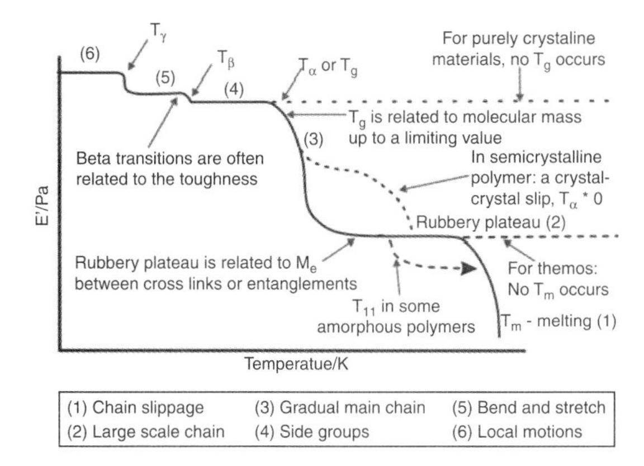

solid polymers, the complex modulus and the elastic modulus can be considered equivalent. The following table, Effect of Variance between Complex and Storage Modulus shows the relationship between tan delta and the degree of variation between the elastic and complex modulus.

A brief note about the frequency of the measurement is in order here. Many DMA instruments provide the experimenter with the option of operating the device in either the fixed frequency or the resonant frequency mode. Many older instruments offer only the resonant frequency option. In the resonant frequency method, the instrument finds the natural frequency of the material and this frequency varies with the rigidity of the sample. As the sample is heated and the modulus changes, the change is measured in terms of a reduction in the frequency, which is then converted to modulus values. In rigid systems the resonant frequency will typically fall between 15–30 Hz. While this method can be useful for making rapid and approximate determinations of transition temperatures, it is primarily designed for handling very stiff samples that are rarely encountered when working with polymer systems. In addition, there are two disadvantages to operating in the resonant frequency mode. First, subtle transitions that may appear in a multi-phase system such as a polymer blend do not resolve well at high frequencies. Second, since viscoelastic properties are time dependent and therefore frequency dependent, a method that allows the frequency to change during the scan will be inherently less accurate than a method that controls the frequency as a constant. For this reason, the ASTM method written for dynamic mechanical analysis specifies a frequency of 1 Hz. This standard is adhered to in the data contained in the appendix. This ensures that results from different experimenters will not contain discrepancies based on the frequency-dependent behavior of the materials.

Effect of Tan Delta on Variance between Complex and Storage Modulus

Tan Delta	Variance (E*/E′)
0.00	1.00000
0.01	1.00005
0.03	1.00045
0.05	1.00125
0.10	1.00499
0.20	1.01980
0.30	1.04403
0.50	1.11803
0.75	1.25000
1.00	1.41421

Data Presentation

The information from DMA tests can be configured in a variety of ways depending upon the design of the test. For solid materials, the most common experiment is a temperature sweep. A frequency and amplitude of oscillatory stress are selected and maintained as constants throughout the experiment. A heating routine is selected and the material temperature is raised from the desired starting temperature to an endpoint. Two types of heating routines are sanctioned by ASTM D-4065, the method governing dynamic mechanical analysis.

The first is a stairstep method where the sample temperature is raised in 5°C increments and allowed to equilibrate at each temperature for 3.5 minutes before performing the measurements. Since the sample thermocouple is typically 1 mm away from the face of the material, and the sample will have some thickness that may vary from 0.5–5 mm, this method is designed to overcome the problems associated with thermal lag between the measured temperature and the actual bulk temperature of the material. However, the method has the disadvantage of only providing a data point every 5°C. This may be adequate for instances where the objective of the test is an approximate storage modulus value, since interpolation is possible for applications where the temperature of interest falls between measurement points. However, for identifying exact transition temperatures, which appear as peaks in the loss modulus and tan delta curves, this method is less satisfactory than a continuous heating method.

Continuous heating routines using heating rates of 1–2°C are also permitted in the ASTM method. These typically provide 5–20 distinct data points per degree and allow for the study of materials where the temperature and the peak height of important transitions are critical. The heating rate of 2°C/minute is particularly useful since it is also the heating rate used in determining the heat deflection temperature (HDT) of plastic materials by ASTM D-648 or ISO 75. Most users of DMA data for engineering purposes come from a tradition of short-term property charts where the only attempt to address elevated temperature performance comes in the form of an HDT value. We will discuss the relationship between HDT values, and DMA data in section 3.3. In order to allow the user to readily relate HDT values to DMA data, the data provided in the appendix is generated using a heating rate of 2°C/minute. The stairstep method is useful for more advanced tests that will be discussed in section 4. These involve evaluations at multiple frequencies or stress relaxation and creep tests where multiple measurements or long-time measurements must be made at a constant temperature.

The most common graphic presentation involves plotting the elastic or storage modulus (E′ or G′), the viscous or loss modulus (E″ or G″), and tan delta as a function of temperature. The deformation mode for the data provided in the appendix is flexure and therefore E′ and E″ are used. From an engineering standpoint, these are more useful values for evaluating solid-state performance while shear results are more significant for flexible systems such as uncured crosslinkable materials, adhesives, pastes, and melts. In addition, experience has shown that tensile and flexural modulus values are nearly equivalent for a homogenous system. It is therefore possible to approximate tensile modulus values from the flexural modulus data provided. Conventionally, the y-axis data is plotted on a logarithmic scale. This can be particularly useful for amorphous polymers where the glass transition may reduce the storage modulus of the material by 2–3 orders of magnitude and obscure changes related to molecular weight that may occur above the glass transition. However, in semi-crystalline systems the changes in storage modulus are typically less than an order of magnitude until the material approaches the melting point. If the softening of the material is included in the plot, it can obscure the effects of the glass transition. In addition, logarithmic scales tend to obscure differences between materials in a comparative plot. For loss properties, logarithmic scales tend to diminish the visual impact of transitions. For data that focuses on solid-state performance, clarity is enhanced by utilizing a linear scale for all y-axis data, and this convention has been chosen for the graphs in the appendix.

Structural Characteristics of Polymers

In order to make the best use of DMA data, it is useful to relate representative plots to the structural characteristics of different polymer families. Since this initial version of the database is devoted to rigid and semi-rigid thermoplastics, this discussion will focus on the two most important polymer families within this category – amorphous and semi-crystalline materials. Thermoplastic elastomer and a rigid crosslinked systems show significant contrast.

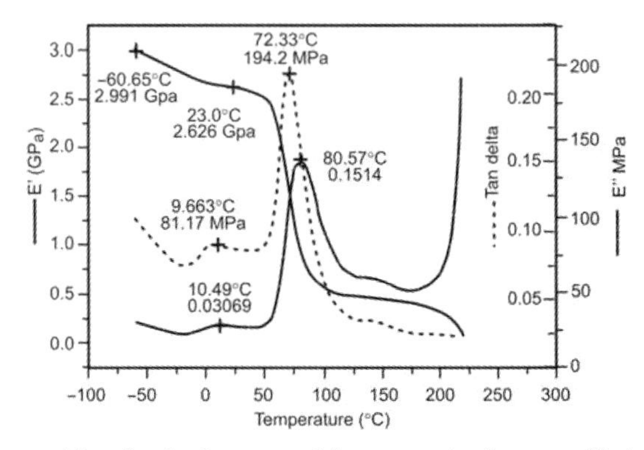

The plot in *Storage and loss properties for an unfilled polycarbonate* shows a typical DMA result for polycarbonate, an amorphous thermoplastic. The full-scale plot begins at −60°C and ends at 175°C. It can be seen that there is little change in the storage modulus between the initial temperature and 140°C. However, between 140 and 160°C the storage modulus drops by over two orders of magnitude and the material has lost its usefulness as a structural material. This abrupt change in physical properties is associated with the onset of short-range molecular motions known as the glass transition. The amorphous structure in a polymer is often likened to that of glass because there is structural rigidity without the presence of a well-organized intermolecular structure. In an amorphous polymer the glass transition can be thought of as a softening temperature.

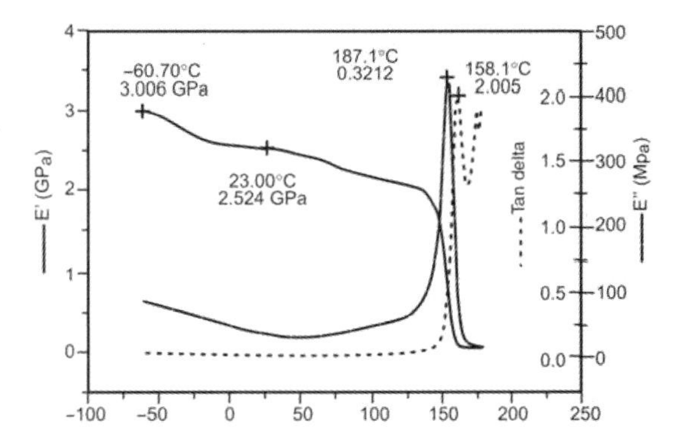

The plot in *Expanded plot of storage and loss properties for polycarbonate at* T_g expands the graph to show the glass transition in more detail. We can see that the loss modulus rises to a maximum as the storage modulus is in its most rapid rate of descent. The peak of the loss modulus is conventionally identified as the glass transition temperature (T_g), even though the DMA plot clearly shows that the transition is a process that spans a temperature range. In most amorphous polymers the temperature range is relatively narrow, 25–40°C for materials that do not contain polymeric modifiers such as elastomeric toughening agents. The tan delta curve follows the loss modulus curve closely and provides a running tally on the ration of the elastic and viscous phases in the polymer. At low temperatures leading up to the glass transition, tan delta is well below 0.1. The rapid rise in the tan delta curve coincides with the rapid decline in the storage modulus. Above 150°C the tan delta curve rises rapidly and reaches a peak above 2.0. In this region the contribution of the loss modulus to the complex modulus is equal to or greater than that of the storage modulus. Once the glass transition is complete, the loss modulus drops back to a level close to the pre-transition values. However, because of the drastic reduction in elastic properties, the tan delta values do not decline significantly. The low storage modulus indicates that the material is easily deformed by an applied load. More significantly, the high tan delta values mean that once the deformation is induced, the material will not recover its original shape. It is considered to be soft and pliable. The pattern observed here for polycarbonate is typical of all amorphous materials. The key difference lies in the glass transition temperature (T_g) and the storage modulus below T_g.

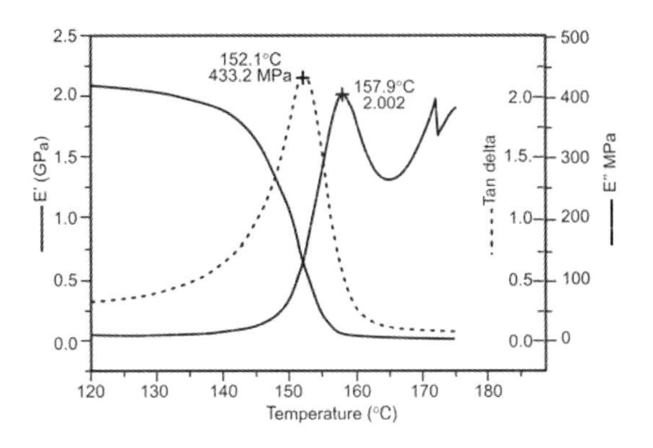

The plot in *Storage and loss properties for unfilled nylon 6* shows a DMA plot for nylon 6, a semi-crystalline polymer. Semi-crystalline polymers are so named because the large extended chain molecules are not capable of achieving the perfect lattice order that is typical of the crystalline structure in lower molecular weight materials. We speak, therefore, in terms of degree of crystallinity. If the degree of crystallinity reaches 30–35% in a polymer matrix, then there is sufficient order to produce a material with an identifiable crystalline melting point. These materials are actually a mixture of amorphous and crystalline regions. Consequently, they exhibit both a melting point and a glass transition. The glass transition can be readily identified in the DMA plot. The storage modulus declines rapidly and the loss modulus and the tan delta curve rise to maximum values. However, because of the presence of a crystalline matrix, the material does not soften above the glass transition. The new mobility of the amorphous regions causes a reduction in the storage modulus, but the material exhibits useful solid-state properties until the material approaches the melting point, some 150°C above the glass transition. The diminished effect of the glass transition on the properties of the semi-crystalline material can also be seen in the tan delta peak value. Instead of rising above 1.0 as in most amorphous materials, the peak height for this material barely exceeds 0.15. Nylon 6 gives a result that is typical for a semi-crystalline polymer. The primary differences between semi-crystalline materials are in the actual glass transition temperatures, melting points, and degree of storage modulus decline associated with the glass transition. The glass transition can be thought of as the softening point of the amorphous regions, and the melting point represents the solid-liquid transition for the semi-crystalline structure. Therefore, the reduction in the storage modulus through the glass transition can serve as a relative indicator of degree of crystallinity. We will see later that there are other modifications that increase the elastic properties of a material and decrease the effect of the glass transition on the storage modulus. Therefore, care must be taken in interpreting the structural details behind DMA data. Once the semicrystalline material approaches the melting point, the tan delta value will rapidly increase as the material changes from an elastic solid to a viscous fluid. The plot in *Storage and loss properties for an unfilled nylon 6/12 showing the rapid rise in the tan delta as the material softens* shows a DMA plot for a nylon 6/12 heated above the melting point. The tan delta value above the melting point is great enough to dwarf the glass transition event. The onset temperature for the rapid increase in tan delta will agree closely with the melting point measure by colorimetric methods.

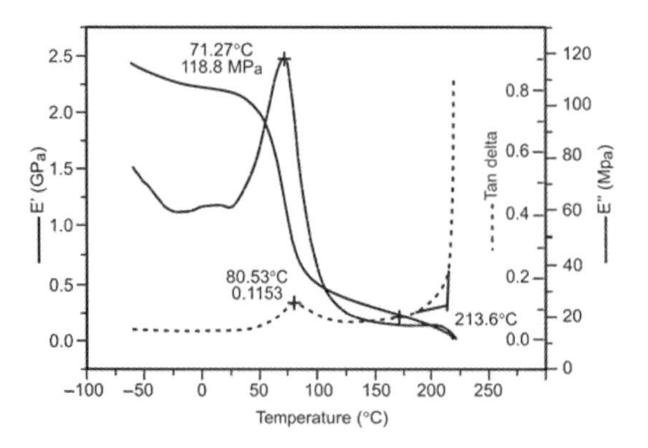

Systems such as rigid thermosets produce DMA results that are somewhat unique to the type of matrix polymer. Epoxies and phenolics, for example, have distinct temperature-dependent behaviors that make them easily distinguishable. However, in general these materials all have a well-defined glass transition that produces the typical behavior of a declining storage modulus coincident with a rising loss modulus and tan delta. The plot in *Storage and loss properties for an epoxy circuit board material shows* the storage and loss properties for an epoxy material used in printed circuit boards. Since the material is crosslinked, it has no melting point and in this respect it resembles an amorphous material. However, due to the crosslinking, the plateau modulus beyond the glass transition does not decline to near zero. Instead, the material will still exhibit useful load-bearing characteristics even 50–75°C above T_g. Note also that in crosslinked systems the tan delta values above T_g return to pre-T_g levels.

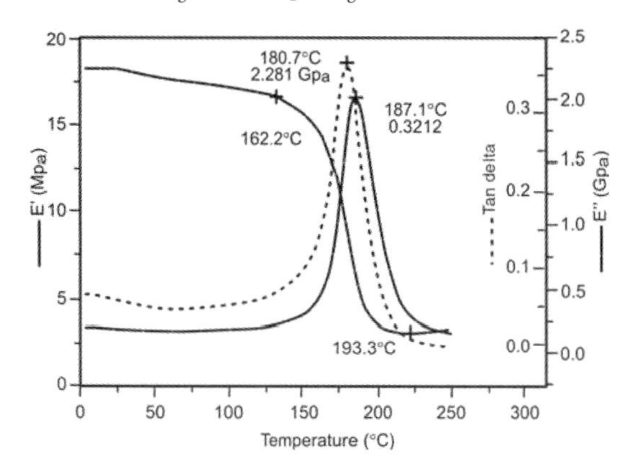

Elastomers have glass transition temperatures below room temperature and their storage modulus properties are typically very low at ambient conditions. In this respect, they resemble a rigid amorphous material that has been heated above T_g. However, unlike the amorphous materials, elastomers exhibit relatively low tan delta properties above T_g, indicating that while little force is required to deform the material, recovery will be good once the applied load is removed. Intuitively this confirms our physical experience with elastomeric compounds. When the temperature is lowered, the material passes through the glass transition and presents itself as a rigid system. If the material is a crosslinked elastomer then it will have a low but measurable modulus to very high temperatures while a thermoplastic elastomer will exhibit a second modulus decline associated with the melting point. This difference is most easily observed by plotting the storage modulus on a logarithmic scale. In addition, the tan delta values will be much higher for the melted thermoplastic system than for the crosslinked thermoset elastomer. The plot in *Storage and loss lproperties for a thermoset elastomer* shows a typical DMA result for a crosslinked elastomer.

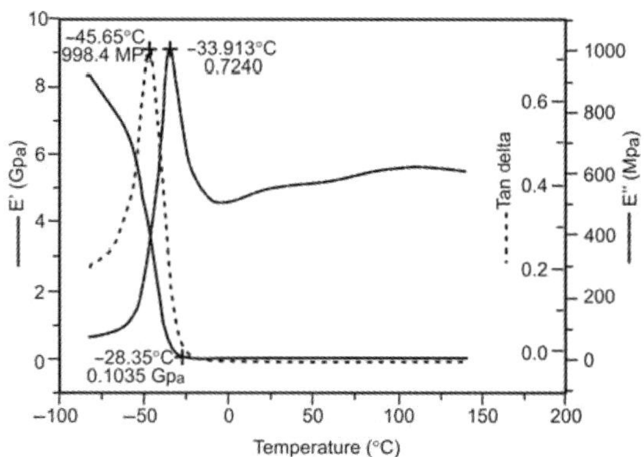

References

Menard, K., *Dynamic Mechanical Analysis*, CRC Press, Boca Raton, 1999.

Sepe, M. *Dynamic Mechanical Analysis for Plastics Engineering*, PDL Handbook Series, Plastics Design Library, Norwich, New York, 1998.

Appendix D: Microorganisms, Biochemistry and Nomenclature

Appendix D1

Nomenclature of Biochemistry and Microorganisms

Abiotic Factor A physical feature of the environment that interacts with organisms.

ABO Blood Group System One of the blood typing systems that is based on the presence or absence of blood group antigens A and B on red blood cells.

Abortive Infection Viral infection in which viruses enter a cell but are unable to express all of their genes to make infectious progeny.

Abscess An accumulation of pus in a cavity hollowed out by tissue damage.

Absorbance (A^{λ}) A dimensionless number that indicates how well a solution of a substance absorbs light of a given wavelength. It is defined as the negative logarithm of the fraction of light wavelength λ that passes through a sample of the solution; its value depends on the length of the light path, the concentration of the solution, and the extinction coefficient of the substance at that wavelength.

Absorption Process in which light rays are neither passed through nor reflected off an object but are retained and either transformed to another form of energy or used in biological processes.

Accidental Parasite A parasite that invades an organism other than its normal host.

Acetylcholinesterase An enzyme found in cholinergic synapses that breaks down acetylcholine and thus terminates its action on the postsynaptic cell.

Acid A substance that releases hydrogen ions when it is dissolved in water.

Acidic Dye See ▶ Anionic Dye.

Acidophile An acid-loving organism that grows best in an environment with a pH of 4.0 to 5.4.

Acme (sometimes referred to a fulminating). During the illness phase of the disease process, the time of most intense signs and symptoms.

Acne Skin condition caused by bacterial infection of hair follicles and the ducts of sebaceous glands.

Acquired Immune Deficiency Syndrome (AIDS) An infectious disease caused by the human immunodeficiency virus that destroys the individual's immune system.

Acquired Immunity Immunity obtained in some manner other than by heredity.

Acridine Derivative A chemical mutagen that can be inserted between bases of the DNA double helix, causing frameshift mutations.

Acrobe An organism that uses oxygen, including ones that must have oxygen.

Actinomycetes Gram-positive bacteria that tend to form filaments.

Action Potential A wave of transient depolarization that travels along the membrane of a nerve cell (or any other kind of excitable cell, such as a muscle cell) as a result of fluxes of ions across the membrane. A nerve impulse.

Activated Sludge System Procedure in which the effluent from the primary stage of sewage treatment is agitated, aerated, and added to sludge containing aerobic organisms that digest organic matter.

Activated State With respect to a chemical reaction, a transient high-energy state of a reactant molecule (such as an unfavorable electron configuration or strained conformation) that enables the molecule to undergo the reaction.

Activation Energy The energy required to start a chemical reaction.

Active Immunity Immunity created when an organism's own immune system produces antibodies or other defenses against an agent recognizes as foreign.

Active Immunization Use of vaccines to control diseases by increasing herd immunity through stimulation of the immune response.

Active Site The site on an enzyme molecule where the substrate binds and where the reaction is facilitated. It is often a cleft or pocket in the surface of the enzyme.

Active Transport (1) Movement of molecules or ions across a membrane against a concentration gradient;

requires expenditure of energy from ATP. (2) The transport of a substance across a biological membrane by a mechanism that can work against a concentration (or electrochemical) gradient. It always requires the expenditure of cellular energy. Compare ▶ Facilitated Transport, Passive Transport.

Acute Disease A disease that develops rapidly and runs it course quickly.

Acute Hemorrhagic Conjunctivitis Eye disease caused by an enterovirus.

Acute Inflammation The relatively short duration of inflammation during which time host defenses destroy invading microbes and repair tissue damage.

Acute Necrotizing Ulcerative Gingivitis (ANUG) A severe form of periodontal disease. *Also known as Trench Mouth.*

Acute Phase Protein Protein, such as C-reactive protein or mannose-binding protein, that forms a nonspecific host-defense mechanism during an acute phase response.

Acute Phase Response A response to an acute illness that produces specific blood proteins called acute phase proteins.

Acute Respiratory Disease (ARD) Viral disease that occurs in epidemics with cold symptoms as well as fever, headache, and malaise, sometimes causes viral pneumonia.

Adenovirus A medium-sized, naked DNA virus that is highly resistant to chemical agents and often causes respiratory infections or diarrhea.

Adenylylation In cells, the transfer of an adenylyl moiety from AT to another molecule. Some enzymes are regulated by reversible adenylylation.

Adherence The attachment of a microorganism to a host's cell surface.

Adhesin A protein or glycoprotein on attachment pili (fimbriae) or capsules that help a microorganism attach to a host cell.

Adipocytes Fat cells; cells that are specialized for storing triacylglycerols and for releasing them to the blood in the form of fatty acids and glycerol as required.

Adrenergic Receptors Cell-surface receptors that bind epinephrine and norepinephrine. There are several different types with somewhat different ligand specificities and effects. (The term comes from *adrenaline*, the old name for epinephrine.)

Adsorption The attachment of the virus to the host cell in the replication process.

Aerobic Respiration Processes in which aerobic organisms gain energy from the catabolism of organic molecules via the Krebs cycle and oxidative phosphorylation.

Aerosol A cloud of tiny liquid droplets suspended in air.

Aerotolerant Anaerobe A bacterium that can survive in the presence of oxygen but does not use oxygen in its metabolism.

Affinity Constant See ▶ Association Constant.

Aflatoxin Fungal toxin that is a potent carcinogen, found in food made from contaminated grain or peanuts infested with *Aspergillus flavus* and other aspergilli.

African Sleeping Sickness Disease of equatorial Africa caused by protozoan blood parasites of the genus *Trypanosoma. Also known as Trypanosomiasis.*

Agammaglobulinemia Primary immunodeficiency disease caused by failure of B cells to develop, resulting in lack of antibodies.

Agar A polysaccharide extracted from certain marine algae and used to solidify medium for the growth of microorganisms.

Agar Plate A plate of nutrient medium solidified with agar.

Agglutination Reaction A reaction of antibodies with antigens the results in agglutination, the clumping together of cells or other large particles.

Agonist In molecular biology, a substance that mimics the cellular effects of a natural compound (such as a hormone or neurotransmitter) by binding to and activating the same cellular receptor. Compare ▶ Antagonist.

Agranulocyte A leukocyte (monocyte or lymphocyte) that lacks granules in the cytoplasm and has rounded nuclei.

A Helix A right-hand helix structure of nucleic acid duplexes that has a smaller pitch and a larger diameter than the B-DNA helix. It is the structure adopted by RNA duplexes and RNA-DNA hybrid molecules.

AIDS (Acquired Immune Deficiency Syndrome) An infectious disease caused by the human immunodeficiency virus that destroys the individual's immune system.

Alcoholic Fermentation Fermentation in which pyruvic acid is reduced to ethyl alcohol by electrons from reduced NAD (NADH).

Alditols Compounds that are produced by reducing the carbonyl group on a monosaccharide (that is, reducing $R-CH=O$ to $R-CH_2-OH$).

Aldose A monosaccharide in which the carbonyl group comes at the end of the chain and thus represents an aldehyde group Compare ▶ Ketose.

Algae (singular: *alga*) Photosynthesis, eukaryotic organisms in the kingdoms Protista and Plantae.

Alkaline Condition caused by an abundance of hydroxyl ions (OH^-) resulting in a pH of greater than 7.0. *Also known as Basic.*

Alkaliphile A base- (alkaline) loving organism that grows best in an environment with a pH of 7.0 to 11.5.

Alkaloids A large group of nitrogenous basic substances found in plants. Most of them taste bitter, and many are pharmacologically active. The term can also be used for synthetic compounds.

Alkylating Agent A chemical mutagen that can add alkyl groups ($—CH^3$) to DNA bases, altering their shapes and causing errors in base pairing.

Allele The form of a gene that occupies the same place (locus) on the DNA molecule as another form but may carry different information for a trait.

Allergen An ordinarily innocuous foreign substance that can elicit an adverse immunological response in a sensitized person.

Allergy When the immune system reacts in an exaggerated or inappropriate way to a foreign substance. *Also known as hypersensitivity.*

Allograft A graft of tissue between two organisms of the same species that are not genetically identical.

Allosteric With respect to enzymes, an effect that is produced on the activity of one part of an enzyme (such as an active site) by the binding of an effector to a different part of the enzyme.

Allosteric Site The site at which a noncompetitive inhibitor binds.

Alpha (α) Hemolysin A type of enzyme that partially lyses red blood cells, leaving a greenish ring in the blood agar medium around the colonies.

Alpha (α) Hemolysis Incomplete lysis of red blood cells by bacterial enzymes.

Alternative Pathway One of the sequences of reactions in nonspecific host responses by which proteins of the complement system are activated.

Alternative Splicing The splicing of a eukaryotic RNA transcript in different ways, to include or exclude certain exons from the final mRNA.

***Alu* Elements** DNA sequences about 300 base pairs long that occur in many copies scattered throughout the genome of mammals; the human genome has hundreds of thousands of them. They may serve an unknown function, or they may be purely "parasitic," spreading as mobile elements through the genome.

Alveolus A saclike structure arranged in clusters at the ends of the respiratory bronchioles, having walls one cell layer thick, where gas exchange occurs.

Amantadine An antiviral agent that prevents penetration by influenza A virus.

Ames Test Test used to determine whether a particular substance is mutagenic, based on its ability to induce mutations in auxotrophic bacteria. A strain of the bacterium *Salmonella typhimurium* having a mutation that disables an enzyme necessary for histidine utilization is exposed to the substance in question and plated on a medium lacking histidine. A reversion mutation that activates the mutant enzyme causes the cells to grow on this medium.

Amino Acid An organic acid containing an amino group and a carboxyl group, composing the building blocks of proteins.

Aminoglycoside An antimicrobial agent that blocks bacterial protein synthesis.

Amino Terminus See ▶ N-Terminus.

Amoebic Dysentery Severe, acute form of amebiasis, caused by *Entamoeba histolytica*.

Amoeboid Movement Movement by means of pseudopodia that occurs in cells without walls, such as amoebas and some white blood cells.

Amphibolic Pathway A metabolic pathway that can yield either energy or building blocks for synthetic reactions.

Amphipathic For a molecule, the property of having both hydrophobic and hydrophilic portions. Usually one end or side of the molecule is hydrophilic and the other end or side is hydrophobic.

Amphitrichous The presence of flagella at both ends of the bacterial cell.

Ampholyte A substance whose molecules have both acidic and basic groups.

Anabolic Pathway A chain of chemical reactions in which energy is used to synthesize biologically important molecules.

Anabolism Chemical reactions in which energy is used to synthesize large molecules from simple components. *Also known as Synthesis.*

Anaerobe An organism that does not use oxygen, including some organisms that are killed by exposure to oxygen.

Anaerobic Refers to the absence of oxygen or the absence of a need for it; processes that must or can occur without oxygen are called anerobic processes.

Anaerobic Respiration Respiration in which the final electron acceptor in the electron transport chain is an inorganic molecule other than oxygen, e.g., sulfate, nitrate, etc.

Analytical Study An epidemiological study that focuses on establishing cause-and-effect relationships in the occurrence of diseases in populations.

Anamnestic Response Prompt immune response due to "recall" by memory cells. See ► Secondary Response.

Anaphylactic Shock Condition resulting from a sudden extreme drop in blood pressure caused by an allergic reaction.

Anaphylaxis An immediate, exaggerated allergic reaction to antigens, usually leading to detrimental effects.

Androgens The male sex hormones; specifically, the steroid hormones testosterone, dihydrotestosterone, and androstenedione, which act mainly to promote male sexual development and maintain male sex characteristics.

Angstrom (Å) Unit of measurement equal to 0.0000000001 m, or 10^{-10} m. No longer officially recognized.

Animalia The kingdom of organisms to which all animals belong.

Animal Passage The rapid transfer of a pathogen through animals of a species susceptible to infection by the pathogen.

Anion \ˈa-ˌn ī - ə n\ *n.* [GK, neut. of *aniōn, prp.* Of *anienai* to go up, fr. *ana-* + *ienai* to go](1834). A negatively charged ion.

Anionic \ˌa-(ˌ)n ī-ˈä-nik\ *adj.* (ca. 1920).

Anionic Dye An ionic compound, used for staining bacteria, in which the negative ion imparts the color. *Also known as Acidic Dye.*

Anomers Stereoisomes of cyclized monosaccharide molecules differing only in the configuration of the substituents on the carbonyl carbon. (This carbon is a center of Chirality in the cyclized but not in the open-chain form of the molecule.)

Antagonism The decreased effect when two antibiotics are administered together.

Antagonist In biochemistry, a substance that counteracts the cellular effects of a natural compound (such as a hormone or neurotransmitter) by binding to the cellular receptor for the compound and blocking its action. Compare ► Agonist.

Anthrax A zoonosis caused by *Bacillus anthracis* that exist in cutaneous, respiratory ("woolsorters disease"), or intestinal forms; transmitted by endospores.

Antibiosis The natural production of an antimicrobial agent by a bacterium or fungus.

Antibiotic A chemical substance produced by microorganisms that can inhibit the growth of or destroy other microorganisms.

Antibodies A set of related proteins that are produced by B lymphocytes and can bind with specificity to antigens. Some types are released into body fluids and mediate humoral immunity; other types are retained on the surface of the B cell or are taken up and displayed by some other cell types.

Antibody A protein in response to an antigen that is capable of binding specifically to that antigen. *Also known as Immunoglobulin.*

Antibody Titer The quantity of a specific antibody in an individual's blood, often measured by means of agglutination, reactions.

Anticodon A three-base sequence in tRNA that is complementary to one of the mRNA codons, forming a link between each codon and the corresponding amino acid.

Antigen A substance that the body identifies as foreign and toward which it counts an immune response. *Also known as Immunogen.*

Antigen Binding Site The site on the antibody to which the antigen ((epitope) binds).

Antigen Challenge Exposure to a foreign antigen.

Antigenic Determinant See ► Epitope.

Antigenic Drift Process of antigenic variation that results from mutations in genes coding for hemagglutinin and neuramindase.

Antigenic Mimicry Self-antigen that is similar to an antigen on a pathogen.

Antigenic Shift Process of antigenic variation probably caused by a reassortment of viral genes.

Antigenic Variation Mutations of influenza viruses that occur by antigenic drift and antigenic shift.

Antigenic Presenting Cell An immunological cell, such as a macrophage, dendritic cell, or B cell, that processes antigen fragments and presents peptide fragments from the antigen on its cell surface.

Antihistamine Drug that alleviates symptoms caused by histamines.

Antimetabolite A substance that is a structural analog of a normal metabolite or otherwise resembles it and that interferes with the utilization of the metabolite by the cell.

Antimicrobial Agent A chemotherapeutic agent used to treat diseases caused by microbes.

Antioxidant A strongly reducing compound, such as ascorbic acid, which counteracts the tendency of a metabolite to undergo oxidation to a potentially toxic or harmful species.

Antiparallel The opposite head-to-tail arrangement of the two strands in a DNA double helix.

Antiport A membrane transport process that couples the transport of a substance in one direction across a membrane to the transport of a different substance in the other direction. Compare ▶ Symport.

Antisense RNA An RNA molecule that is complementary to an mRNA; it can block translation of the mRNA by forming a duplex with it. Gene expression can be regulated by the production of antisense RNAs.

Antiseptic A chemical agent that can be safely used externally on tissues to destroy microorganisms or to inhibit their growth.

Antiserum (plural: *antisera*) Serum that contain a high concentration of antibodies against a particular antigen.

Antitoxin An antibody against a specific toxin.

Antiviral Protein A protein induced by interferon that interferes with the replication of viruses.

Apicomplexan A parasite protozoan such as *Plasmodium*, that generally has a complex life cycle. *Also known as Sporozoan.*

Aplastic Crisis A period during which erythrocyte production ceases.

Apoenzyme The protein portion of an enzyme.

Apolipoproteins The specific proteins that constitute the protein fraction of lipoproteins; they mediate the interactions of lipoproteins with tissues.

Apoptosis Programmed cell death (as distinguished from necrosis) See ▶ Autolysis.

Arachnid An anthropod with two body regions, four pairs of legs, and mount parts that are used in capturing and tearing apart prey.

Archaea One of the three Domains of living things; all members are bacterial.

Archaebacteria A group of prokaryotes that are biochemically distinct from the true bacteria (Eubacteria) and that separated from them early in the history of life. Modern archaebacteria mostly live in extreme environments, such as acid hot springs.

Archaeobacteria Prokaryotic organisms lacking peptidoglycan in their cell walls and differing from eubacteria in many ways.

Arenavirus An enveloped RNA virus that causes Lassa fever and certain other hemorrhagic fevers.

Anthropod Makes up the largest group of living organisms, characterized by a jointed chitinous exoskeleton, segmented body, and joined appendages, associated with some or all of the segments.

Arthus Reaction A local reaction seen in the skin after subcutaneous or intradermal injection of an antigenic substance, immune complex (type III) hypersensitivity.

Artificially Acquired Active Immunity When an individual is exposed to a vaccine containing live, weakened, or dead organisms or their toxins, the host's own immune system responds specifically to defend the body, e.g., by making specific antibodies.

Artificially Acquired Immunity When an individual's immune system is stimulated to react by some man-made process, e.g., given a vaccine or an immune serum.

Artificially Acquired Passive Immunity When antibodies made by other hosts are introduced into a new host, e.g., via mother's milk or shots of gamma globulin.

Ascariasis Disease caused by a large roundworm, *Ascaris lumbricodes*, acquired by ingestion of food or water contaminated with eggs.

Ascomycota See ▶ Sac Fungus.

Ascospore One of the eight sexual spores produced in each ascus of a sac fungus.

Ascus (plural: *asci*) Saclike structures produced by sac fungi during sexual reproduction.

Aseptic Technique A set of procedures used to minimize chances that cultures will be contaminated by organisms from the environment.

Asiatic Cholera Severe gastrointestinal disease caused by *Vibrio cholerae*, common in areas of poor sanitation and fecal contamination of water.

Aspergillosis Skin infection caused by various species of *Aspergillus*, which can cause severe pneumonia in immuno-suppressed patients. *Known also as Farmer's Lung Disease.*

Association Constant (*K*) An equilibrium constant that indicates the tendency of two chemical species to associate with each other; it is equal to the concentration of the associated form divided by the product of the concentrations of the free species at equilibrium. *Known also as Affinity constant.*

Asthma Respiratory anaphylaxis caused by inhaled or ingested allergies or by hypersensitivity to endogenous microorganisms.

Asymmetric Carbon A carbon molecule that carries four different substituents and therefore acts as a center of chirality, meaning that the substance can occur in two different enantiomers (stereoisomers that are nonsuperimposable mirror images of each other).

Atherosclerotic Plaques The protruding masses that form on the inner walls of arteries in atherosclerotic disease. A mature plaque consists partly of lipid, mainly cholesterol esters, which may be free or contained in lipid-engorged macrophages called foam cells, and partly of an abnormal proliferation of smooth-muscle and connective-tissue cells.

Athlete's Foot A form of ringworm in which hyphae invade the skin between the toes, causing dry, scaly lesions. *Also known as Tinea Pedis.*

Atom The smallest chemical unit of matter.

Atomic Force Microscope (AFM) Advanced member of the family of scanning tunneling microscopes, allowing 3-dimensional views of structures from atomic size to about 1 μm.

Atomic Number The number of protons in an atom of a particular element.

Atomic Weight The sum of the number of protons and neutrons in an atom.

Atopy Localized allergic reactions that occur first at the site where an allergen enters the body.

Atrichous A bacterial cell without flagella.

Attachment Pilus Type of pilus that helps bacteria adhere to surfaces. *Also known as Fimbria.*

Attenuation (1) A genetic control mechanism that terminates transcription of an operon prematurely when the gene products are not needed. (2) The weakening of the disease-producing ability of an organism.

Auditory Canal Part of the outer ear lined with skin that contains many small hairs and ceruminous glands.

Autoantibody An antibody against one's own tissue.

Autocatalytic Refers to a reaction that an enzyme catalyzes on part of its own structure, such as cleavage performed by a protease on its own polypeptide precursor.

Autoclave An instrument for sterilization by means of moist heat under pressure.

Autograft A graft of tissue from one part of the body to another.

Autoimmune Disorder An immune disorder in which individuals are hypersensitive to antigens on ells of their own body.

Autoimmunity A condition in which the body mounts an immune response against one of its own normal components.

Autoimmunization The process by which hypersensitivity to "self" develops, it occurs when the immune system responds to a body components as if it were foreign.

Autolysis Programmed cell death; the orderly self-destruction of a cell in a multicellular organism. It is the process by which unwanted cells are eliminated in the body. *Known also as Apoptosis.*

Autonomously Replicating Sequences (ARSs) Sequences in yeast chromosomes that, when incorporated into an artificial plasmid, enable the plasmid to replicate efficiently in yeast cells.

Autotroph Organisms that can synthesize their organic compounds entirely from inorganic precursors in particular needing only CO_2 as a carbon source. Compare ▶ Heterotrophs.

Autotrophy "Self-feeding" – the use of CO_2 as a source of carbon atoms for the synthesis of biomolecules.

Auxotrophic Mutant An organism that has lost the ability to synthesize one or more metabolically important enzymes through mutation, therefore requires special substances in its growth medium.

Auxotrophs Microorganism strains that requires as a nutrient a particular substance that is not required by the prototype strain. Usually the requirement results from a mutation that disables an enzyme necessary for the endogenous synthesis of the substance.

Axial Filament A subsurface filament attached near the ends of the cytoplasmic cylinder of spirochetes that causes the spirochete body to rotate like a corkscrew. *Also known as Endoflagellum.*

Axis of Symmetry An imaginary axis through a structure, such that rotating the structure around the axis through an appropriate angle leaves the appearance of the structure unchanged.

Axon A threadlike process extending from a nerve cell by which impulses are transmitted to other nerve cells or to effector cells such as muscle or gland cells. Most nerve cells have one axon; shorter processes that function ion receiving impulses form other neutrons are called dendrites.

Babesiosis A protozoan disease caused by the Apicomplexan *Babesia microti* and other species of *Babesia.*

Bacillary Angiomatosis A disease of the small blood vessels of the skin and internal organs caused by the rickettsial organism *Bartonella hensalae.*

Bacillary Dysentery See ▶ Shigellosis.

Bacillus (plural: *bacilli*) A rodlike bacterium.

Bacteremia An infection in which bacteria are transported in the blood but do not multiply in transit.

Bacteria (singular: *bacterium*) All prokaryotic organisms.

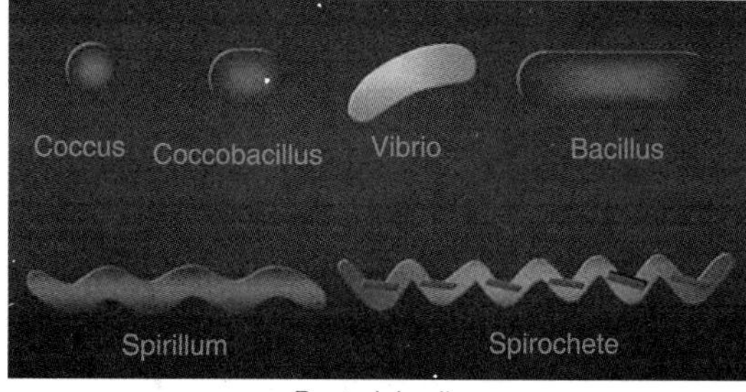

Bacterial cells

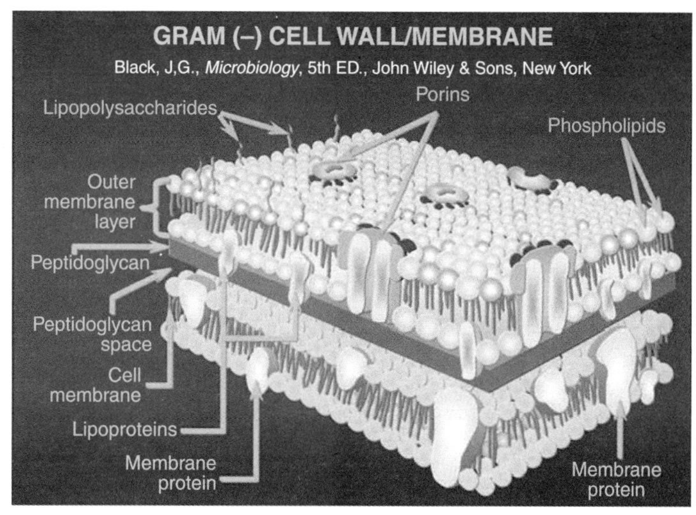

Bacterial (Gram-Neg.) cell wall

Bacteria When spelled with a capital B, it is the name of one of the three Domains of living things; all members are bacterial.

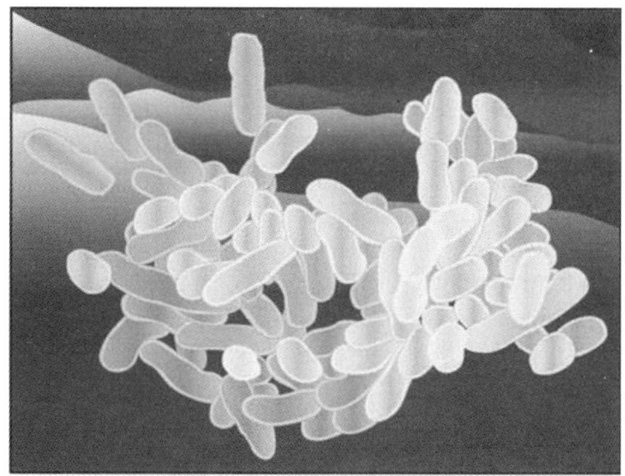

Proteus mirabilis bacteria cells

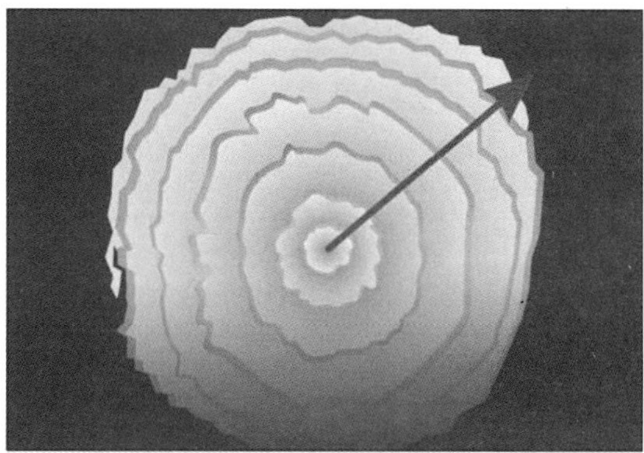

Swarming proteus mirabilis bacteria colonies

Bacterial Conjunctivitis A highly contagious inflammation of the conjunctiva caused by various bacterial species. *Also known as Pinkeye.*

Bacterial Endocarditis A life-threatening infection and inflammation of the lining and valves of the heart. *Also known as Infective Endocarditis.*

Bacterial Enteritis An intestinal infection caused by bacterial invasion of intestinal mucosa or deeper tissues.

Bacterial Lawn A uniform layer of bacteria grown on the agar surface in a Petri dish.

Bacterial Meningitis An inflammation of the meninges that cover the brain and spinal cord by any one of several bacterial species.

Bactericidal Referring to an agent that kills bacteria.

Bacteriocin A protein released by some bacteria that inhibits the growth of other strains of the same or closely related species.

Bacteriocinogen A plasmid that directs production of a bacteriocin.

Bacteriophage A virus that infects bacteria. *Also known as Phage.*

Bacteriostatic Referring to an agent that inhibits the growth of bacteria.

Bacteroid Irregularly shaped cells usually found in tight packets that develop from *Rhizobium* swarmer cells and form nodules in the roots of leguminous plants.

Balantidiasis Type of dysentery caused by the ciliated protozoan *Balantidium coli.*

Balantitis An infection of the penis.

Barophile An organism that lives under high hydrostatic pressure.

Barotholin Gland A mucus-secreting gland of the female external genitalia.

Bartonellosis Rickettsial disease, caused by *Bartonella bacilliformis*, that occurs in two forms. See also ► Oroyo Fever and ► Verruga Peruana.

Base A substance that absorbs hydrogen ions or donates hydroxyl ions.

Base Analog A chemical mutagen similar in molecular structure to one of the nitrogenous bases found in DNA that causes point mutations.

Basic Dye See ► Cationic Dye.

Basidiomycota See ► Club Fungus.

Basidiospore A sexual spore of the club fungi.

Basidium (plural: *basidia*) A clublike structure in club fungi bearing four external spores on short, slender stalks.

Basophil A leukocyte that migrates into tissues and helps initiate the inflammatory response by secreting histamine.

B Cell See ► B Lymphocyte.

B-DNA A DNA duplex with a specific right-hand helix structure. It is the usual form of DNA duplexes in vivo.

Beer's Law The equation that relates the absorbance of a solution sample at a given wavelength to the length of the light path, the concentration of the dissolved substance, and the extinction coefficient of the substance at that wavelength. See ► Extinction ► Coefficient.

Benign Not harmful.

Beta (β) Hemolysin A type of enzyme that completely lyses red blood cells, leaving a clear ring in the blood agar medium around the colonies.

Beta (β) Hemolysis Complete lysis of red blood cells by bacterial enzymes.

Beta Oxidation A metabolic pathway that breaks down fatty acids into 2-carbon pieces.

Bile Acids A family of amphipathic cholesterol derivatives that are produced in the liver and excreted in the bile; salts of the bile acids emulsify fat in the intestine.

Bilirubin A yellow substance, the product of the breakdown of hemoglobin from red blood cells.

Binary Fission Process in which a bacterial cell duplicates its components and divides into two cells.

Binocular Referring to a light microscope having two eyepieces (oculars).

Binomial Nomenclature The system of taxonomy developed by Linnaeus in which each organism is assigned a genus and specific epithet.

Biochemistry The branch of organic chemistry that studies the chemical reactions of living systems.

Bioconversion A reaction in which one compound is converted to another by enzymes in cells.

Biogenic Amines A set of low-molecular-weight amino acid derivatives that contain a basic amino group and function in the body as intercellular mediators. Examples are serotonin, histamine, and epinephrine.

Biogeochemical Cycle Mechanism by which water and elements that serve as nutrients are recycled.

Biohydrometallurgy The use of microbes to extract metals from ores.

Biological Oxygen Demand (BOD) The oxygen required to degrade organic wastes suspended in water.

Biological Vector An organism that actively transmits pathogens that complete part of their life cycle within the organism.

Bioremediation A process that uses naturally occurring or genetically engineered microorganisms to transform harmful substances into less toxic or nontoxic compounds.

Biosphere The region of the earth inhibited by living organisms.

Biotic Factor An organism in the biosphere.

Blackfly Fever Illness resulting from bites by blackflies, characterized by an inflammatory reaction, nausea, and headache.

Blackwater Fever Malaria caused by *Plasmodium falciparum* that results in jaundice and kidney damage.

Blastomycetic Dermatitis Fungal skin disease caused by *Blastomyces dermatitidis*; characterized by disfiguring, granulomatous, pus-producing lesions.

Blastomycosis Fungal skin disease caused by *Blastomyces dermatitidis* that enters the body through wounds.

Blocking Antibody IgG antibody, elicited in allergy patients by increasing doses of allergen, that complexes with allergen before it can react with IgE antibody.

Blood Agar Type of medium containing sheep blood, used to identify organisms that cause Hemolysis, or breakdown of red blood cells.

Blood-Brain Barrier Formation in the brain of special thick-walled capillaries without pores in their walls that limit entry of substances into brain cells. Physically the barrier consists of tight junctions between endothelial cells; these cells have transporters for polar substances such as glucose that need to enter the brain.

Blood Group Antigens A group of oligosaccharides that are carried in the form of glycoproteins and glycolipids on the surface of cells, including blood cells; they are encoded by a large number of polymorphic gene loci and can provoke an immune response in an individual with different blood group antigens.

B Lymphocyte A lymphocyte that is produced in and matures in bursal-equivalent tissue, it gives rise to antibody-producing plasma cells. *Known also as B Cell.*

Body Tube Microscope part that conveys an image from the objective to the eyepiece.

Bohr Effect The effect of pH on oxygen binding by hemoglobin, by which a decrease in pH causes a decrease in oxygen affinity. The effect promotes both the release of oxygen from hemoglobin in the tissues and the release of CO_2 from the blood to the air in the lungs.

Boil See ▶ Furuncle.

Bolivian Hemorrhagic Fever A multisystem disease caused by an arenavirus with insidious onset and progressive effects.

Bone Stink Putrefaction deep in the tissues of large carcasses that is caused by several species of *Clostridium.*

Bongkrek Disease Type of food poisoning caused by *Pseudomonas cocovenenans*, named for a native Polynesian coconut dish.

Botulism Disease caused by *Clostridium botulinum.* The most common form, food-borne botulism, results from ingestion of preformed toxin and is, therefore, an intoxication rather than an infection.

Bradykinin Small peptide thought to cause the pain associated with tissue injury.

Brain Abcess A pus-filled cavity caused by microorganisms reaching the brain from head wound or via blood from another site.

Branch Migration During recombination, the migration of a cross-over point (Holiday junction) by simultaneous unwinding and rewinding in both duplexes.

Bread Mold A fungus with complex mycelia composed of aseptate hyphae with chitinous cross walls. *Known also as Zygomycota or Conjugation Fungus.*

Bright-Field Illumination Illumination produced by the passage of visible light through the condenser of a light microscope.

Brill-Zinsser Disease A recurrence of an epidemic typhus infection caused by reactivation of latent organisms harbored in the lymph nodes. *Known also as Recrudescent Typhus.*

Broad Spectrum Referring to the range of activity of an antimicrobial agent that attacks a wide variety of microorganisms.

Bronchial Pneumonia Type of pneumonia that begins in the bronchi and can spread through surrounding tissue toward the alveoli.

Bronchiole A finer subdivision of the air-conveying bronchi.

Bronchitis An infection of the bronchi.

Bronchus (plural: *bronchi*) A subdivision of the trachea that conveys air to and from the lungs.

Brucellosis A zoonosis highly infection for humans, caused by any of several species of *Brucella. Also known as Undulant Fever and Malta Fever.*

Bubo Enlargement of infected lymph nodes, especially in the groin and armpit, due to accumulation of pus, characteristic of bubonic plague and other diseases.

Bubonic Plague A bacterial disease, caused by *Yersinia pestis* and transmitted by flea bites, that spread in the blood and lymphatic system.

Budding Process that occurs in yeast and a few bacteria in which a small new cell develops from the surface of an existing cell.

Buffering The ability of a mixture of an acid and its conjugate base at a pH near their pK_a to minimize pH changes caused by an influx of acid or base. The Henderson-Hasselbach equation is useful relating pH, pK and [salt]/[acid].

Bulking Phenomenon in which filamentous bacteria multiple, causing sludge to float on the surface of water rather than settling out.

Bunyavirus An enveloped RNA virus that causes some forms of respiratory distress and hemorrhagic fever.

Burkitt's Lymphoma A tumor of the jaw, seen mainly in African children, caused by the Epstein-Barr virus.

Burst Size The number of new virions released in the replication process. *Known also as Viral Yield.*

Burst Time The time from absorption to release of phages (in the replication process).

Calorie A unit of energy defined as that amount of heat energy that will raise the temperature of 1 gram of water by 1°C. 1 calorie = 4.182 joules.

Calvin Cycle The cycle of photosynthetic dark reactions by which CO_2 is fixed, reduced, and converted to glyceraldehydes-3-phosphate (the precursor of hexose monophosphates).

Cancer An uncontrolled, invasive growth of abnormal cells.

Candidiasis A yeast infection caused by *Candida albicans* that appears as thrust (in the mount) or vaginitis. *Known also as Moniliasis.*

Canine Parvovirus A parvovirus that causes severe disease in dogs.

Canning The use of moist heat under pressure to preserve food.

Capillary A blood vessel that branches from an arteriole.

Capnophile An organism that prefers carbon dioxide gas for growth.

Capsid The protein coating of a virus, which protects the nucleic acid core from the environment and usually determines the shape of the virus.

Capsomere A protein aggregate that makes up a viral capsid.

Capsule (1) A protective structure outside the cell wall, secreted by the organism. (2) A network of connective fibers covering organs such as the lymph nodes.

Carbapenem A bactericidal antibiotic that acts on bacterial cell walls.

Carbohydrate A compound composed of carbon, hydrogen, and oxygen that serves as the main course of energy for most living things.

Carbohydrates In general, substances that have the stoichiometric formula $(CH_2)_n$, where $n \geq 3$, or that are derived from such a substance by the addition of functional groups.

Carbon Cycle Process by which carbon from atmospheric carbon dioxide enters living and nonliving things and is recycled through them.

Carboxyl Terminus See ▶ C-Terminus.

Carbuncle A massive pus-filled lesion resulting from an infection, particularly of the neck and upper back.

Carcinogen A cancer-producing substance.

Cardiovascular System Body system that supplies oxygen and nutrients to all parts of the body and removes carbon dioxide and other wastes from them.

Carnitine A low-molecular-weight lysine derivative that shuttles fatty acids through the inner mitochondrial membrane to the matrix. The fatty acyl moiety is transferred from CoA to carnitine for transit through the membrane and is then transferred back to CoA; the carnitine released on the matrix side of the membrane is shuttled back for reuse.

Carrier An individual who harbors an infectious agent without having observable clinical signs or symptoms.

Cascade A set of reactions in which magnification of effect occurs, as in the complement system.

Casein Hydrolsate A substance derived from milk protein that contains many amino acids, used to enrich certain media.

Caseous Characterizing lesions with a "cheesy" appearance that form in lung tissue of patients with tuberculosis.

Caspases A family of proteases involved in apoptosis.

Catabolic Pathway A chain of chemical reactions that capture energy by breaking down large molecules into simpler components.

Catabolism The sum of all the metabolic processes by which complex molecules are broken down to simpler ones, including the processes by which molecules are broken down to yield cellular energy. Compare ▶ Anabolism.

Catabolite Activation In bacteria, a transcriptional control system that induces the synthesis of enzymes for the catabolism of energy substrates other than glucose when glucose levels are low. It involves an activator protein, CRP, that binds cyclic AMP under conditions of low glucose; this complex then binds to DNA sites and promotes transcription of the appropriate genes.

Catabolite Repression Process by which the presence of a preferred nutrient (often glucose) represses the genes coding for enzymes used to metabolize some alternative nutrient.

Catalase An enzyme that converts hydrogen peroxide to water and molecular oxygen.

Caterrhal State Stage of whooping cough characterized by fever, sneezing, vomiting, and a mild, dry persistent cough.

Cathepsins Lysosomal proteases that function in degrading proteins in lysosomes and are also released into the cell at large during cell autolysis (programmed cell death).

Cation \ˈkat¯ˌī-ən\ n. [Gk *kation*, neut. of *katiōn*, prp. of *katienai* to go down, fr. *kata-cata* + *ienai* to go] (1834) A positively charged ion.

Cationic Dye An ionic compound, used for staining bacteria in which the positive ion imparts the color. *Known also as Basic Dye.*

Cat Scratch Fever A disease caused by *Afipia felis*, or more commonly, *Bartonella (Rochalimaea) hensalae* and transmitted in cat scratches and bites.

Cavitation The formation of a cavity inside the cytoplasm of a cell.

Cell Culture A culture in the form of a monolayer from dispersed cells and continuous cultures of cell suspensions.

Cell-Mediated Immune Response The immune response to an antigen carried out at the cellular level by T cells.

Cell-Mediated Immunity The immune response involving the direction action of T cells to activate B cells or to destroy microbe-infected cells, tumor cells, or transplanted cells (organ transplants).

Cell-Mediated (Type IV) Hypersensitivity Type of allergy elicited by foreign substances from the environment, infectious agents, transplanted tissue, and the body's own malignant cells, mediated by T cells. *Known also as Delayed Hypersensitivity.*

Cell Membrane A selectively permeable lipoprotein bilayer that forms the boundary between a bacterial cell's cytoplasm and its environment.

Cell Strain Dominant cell type resulting from subculturing.

Cell Theory Theory formulated by Schleiden and Schwann that cells are the fundamental units of all living things.

Cellular Slime Mold Fungus like protest consisting of amoeboid, phagocytic cells that aggregate to form a pseudoplasmodium.

Cell Wall Outer layer of most bacterial, algal, fungal, and plant cells that maintains the shape of the cell.

Cementum The hard, bony covering of the tooth below the gumline.

Center of Chirality With respect to organic compounds, a carbon atom that has four different substituents attached to it; such a group cannot be superimposed on its own mirror image and therefore can occur in two enantiomers.

Central Nervous System The brain and spinal cord.

Centromere The region of a chromosome where the two sister chromatids are attached together. It is also the site of attachment for spindle fibers during mitosis and meiosis.

Cephalosporin An antibacterial agent that inhibits cell wall synthesis.

Cercaria A free-swimming fluke larva that emerges from the snail or mollusk host.

Cerumen Earwax.

Ceruminous Gland A modified sebaceous gland that secrets cerumen.

Cervix An opening at the narrow lower portion of the uterus.

C_4 Cycle A cycle in some plants that minimizes the wasteful effects of photorespiration by using an enzyme other than rubisco to perform the initial fixation of CO_2. This enzyme is found in mesophyll cells, where it fixes CO_2 into a four-carbon compound (hence C_4). This fixed carbon is shuttled into sheltered bundle-sheath cells, where it is released as CO_2 and enters the Calvin cycle.

Chagas' Disease Disease caused by *Trypanosoma cruzi* that occurs in the southern United States and is endemic to Mexico; transmitted by several kinds of reduviid bugs.

Chancre A hard, painless, nondischarging lesion; a symptom of primary stage syphilis.

Chancroid Sexually transmitted disease caused by *Haemphilus ducreyi* that causes soft, painful skin lesions on the genitals, which bleed easily.

Chaotropic The property of being able to disrupt the hydrogen bonding structure of water. Substances that are good hydrogen bonders, such as urea or guanidine hydrochloride, are chaotropic. Concentrated solutions of these substances tend to denature proteins because they reduce the hydrophobic effect.

Chaperonins Proteins that are involved in managing the folding of other proteins. Some of them help proteins to fold correctly; some prevent premature folding; and some prevent polypeptides from associating with other polypeptides until they have folded properly.

Chemical Bond The interaction of electrons in atoms that form a molecule.

Chemical Cross-Linking A technique for investigating the mutual arrangement of components in a complex. The complex is exposed to a reagent that can form chemical cross-links between adjacent components and is then disaggregated and analyzed. Components that are linked together can be assumed to be neighbors in the complex.

Chemical Equilibrium A steady state in which there is no net change in the concentrations of substrates or products.

Chemically Nondefined Medium See ▶ Complex Medium.

Chemical Potential ($\bar{G}$) In a system, the free energy that resides in a chemical component per mole of the component present. For example, in a system consisting of *a* moles of component A and *b* moles of component B, the total free energy *G* would be the sum of the free energy in the two components: $G = a\bar{G}_A + b\bar{G}_B$. *Known also as Partial Molar Free Energy.*

Chemiosmosis Process of energy capture in which a proton gradient is created by means of electron transport and then used to drive the synthesis of ATP.

Chemiosmotic Coupling The coupling of an enzyme-catalyzed chemical reaction to the transport of a substance across a membrane either with or against its concentration gradient. The outstanding example is the coupling of ATP synthesis to the movement of protons across a membrane in response to a proton gradient.

Chemoautotroph An autotroph that obtains energy by oxidizing simple inorganic substances such as sulfides and nitrates.

Chemoheterotroph A heterotroph that obtains energy from breaking down ready-made organic molecules.

Chemolithotroph See ▶ Chemoautotroph.

Chemostat A device for maintaining the logarithmic growth of a culture by the continuous addition of fresh medium.

Chemotaxis The process by which bacteria sense a concentration gradient of a particular substance in the medium and move either up or down the gradient.

Chemotherapeutic Agent Any chemical substance used to treat disease. *Known also as a Drug.*

Chemothrapeutic Index The maximum tolerable dose of a particular drug per kilogram body weight divided by the minimum dose per kilogram body weight that will cure the disease.

Chemotherapy The use of chemical substances to treat various aspects of disease.

Chickenpox A highly contagious disease, characterized by skin lesions, caused by the varicella-zoster herpes virus; usually occurs in children.

Chigger Dermatitis A violent allergic reaction caused by chiggers, the larvae of *Trombicula* mites.

Childbed Fever See ▶ Puerperal Fever.

Chiral With respect to a molecule or other object, the property of being nonsuperimposable on its mirror image. An atom that makes a molecule chiral, such as a carbon with four different substituents, is called a chiral atom or center of chirality.

Chitin A polysaccharide found in the cell walls of most fungi and the exoskeltons of anthropods.

Chlamydias Tiny, nonmotile, spherical bacteria; all are obligate intracellular parasites with a complex life cycle.

Chloramphenicol A Bacteriostatic agent that inhibits protein synthesis.

Chlorination The addition of chlorine to water to kill bacteria.

Chloroplast A chlorophyll-containing organelle found in eukaryotic cells that carry out photosynthesis.

Chloroquine An antiprotozoan agent effective against the malaria parasite.

Chocolate Agar Type of medium made with heated blood, so named because it turns a chocolate brown color.

Chromatin The filamentous material of eukaryotic chromosomes, consisting of DNA with associated histones and other proteins. During interphase it is dispersed and fills most of the nucleus; during nuclear division it condenses into compact chromosomes.

Chromatophore The internal membranes of photosynthetic bacteria and cyanobacteria.

Chromophore A chemical group that absorbs light at characteristic wavelengths.

Chromosomal Resistance Drug resistance of a microorganism due to a mutation in chromosomal DNA.

Chromosome A structure that contains the DNA of organisms.

Chromosome Mapping The identification of the sequence of genes in a chromosome.

Chronic Amebiasis Chronic infection caused by the protozoan *Entamoeba histolytica.*

Chronic Disease A disease that develops more slowly than an acute disease, is usually less severe, and persists for a long, interdeterminate period.

Chronic Fatigue Syndrome (previously called Chronic EBV Syndrome). Disease of uncertain origin, similar to mononucleosis, with symptoms including persistent fatigue and fever.

Chronic Inflammation A condition in which there is a persistent, indecisive standoff between an inflammatory agent and the phagocytic cells and other host defenses attempting to destroy it.

Chylomicron A type of lipoprotein that is produced in the intestinal villi and serves to transport dietary lipids in the circulation.

Cilium (plural: *cilia*) A short cellular projection used for movement that beats in coordinated waves.

Ciliate A protozoan that moves by means of cilia that cover most of its surface.

Circular Dichroism The property of absorbing right circularly polarized light and left circularly polarized light to different extents. Stereoisomers exhibit circular dichroism. Also, some types of secondary structure, such as α helices and β sheets in proteins, exhibit a predictable circular dichroism at specific wavelengths.

Circular Dichroism Spectrum (CD Spectrum) An absorption spectrum obtained using circularly polarized light; it gives the circular dichroism of the substance over a range of wavelengths.

***Cis*-Dominant** Refers to a mutation in a genetic regulatory element that affects the expression of appropriate genes *only* on the same chromosome, not on another homologous chromosome present in the same call. *Cis*-dominance demonstrates that a regulatory element does not code for a diffusible factor.

Cistron The smallest unit of DNA that must be intact to code for the amino acid sequence of a polypeptide; thus, the coding part of a gene, minus 5′ and 3′ untranslated sequences and regulatory elements.

Citric Acid Cycle A cycle of reactions that takes place in the mitochondrial matrix and results in the oxidation of acetyl units to CO_2 with the production of reducing equivalents and ATP. It is a central pathway in oxidative respiration. Other substrates besides acetyl-CoA can enter the cycle at intermediate points. *Known also as Tricarboxylic Acid Cycle and Krebs Cycle.*

Classical Pathway One of the two sequences of reactions by which proteins of the complement system are activated.

Clathrate Structure The cagelike structure of organized water molecules that forms around a hydrophobic molecule in solution. The structure has lower entropy than liquid water, which helps explain why hydrophobic substances dissolve poorly in water.

Clonal Deletion The process in which the binding of lymphocytes to self antigens triggers a genetically programmed destruction of those lymphocytes.

Clonal Selection Theory A model (proved correct) describing how the body is ale to produce specific immune response against a vast array of antigens. The B and T cells produced by the body have randomly generated antigen specificities. When a particular antigen enters the body, it induces proliferation only in B and T cells that happen to be specific for it. Thus, the antigen selects the cells that will mount an immune response against it and stimulates them to undergo Clonal proliferation.

Clone A group of genetically identical cells, organisms, or DNA sequences descending from a single parent cell.

Club Fungus A fungus, including mushrooms, toadstools, rusts, and smuts, that produces spores on basidia. *Known also as Basidiomycota.*

Cluster of Differentiation Marker An antigen found on the cell surface of B and T cells that an be used to distinguish the cells from one another.

Coagulase A bacterially produced enzyme that accelerates the coagulation (clotting) of blood.

Coarse Adjustment Focusing mechanism of a microscope that rapidly changes the distance between the objective lens and the specimen.

Coated Pit A cell membrane pit that is lined on its cytosolic side by a meshwork of the protein clathrin. Coated pits participate in the mechanism of receptor-mediated Endocytosis, in which surface receptors that have bound specific Extracellular substances are gathered into coated pits, which pinch off to become cytoplasmic vesicles.

Coccidioidomycosis Fungal respiratory disease caused by the soil fungus *Coccidioides immitis. Known also as Valley Fever.*

Coccus (plural: *cocci*). A spherical bacterium.

Codon A sequence of three bases in mRNA that specifies a particular amino acid in the translation process.

Coelom The body cavity between the digestive tract and body wall in higher animals.

Coenzyme An organic small molecule that binds to an enzyme and is essential for its activity but is not permanently altered by the reaction. Most coenzymes are derived metabolically from vitamins.

Cofactor An inorganic ion necessary for the function of an enzyme.

Colicin A protein released by some strains of *Escherichia coli* that inhibits growth of other strains of the same organism.

Coliform Bacterium Gram-negative, nonspore-forming, aerobic or facultatively anaerobic bacterium that ferments lactose and produces acid and gas, significant numbers may indicate water pollution.

Colloid A mixture formed by particles too large to form a true solution dispersed in a liquid.

Colonization Growth of microorganisms on epithelial surfaces such as skin or mucous membranes.

Colony A group of descendants of an original cell.

Colony Hybridization A technique that is used to screen bacteria for the presence of a specific recombinant DNA sequence. Colonies of the bacteria are transferred to a filter, treated to lyse the cells and denature the DNA, and then exposed to a labeled DNA probe that is complementary to part of the sequence in question. Colonies that bind the probe possess the sequence.

Colorado Tick Fever Disease caused by an orbivirus carried by dog ticks, characterized by headache, backache, and fever.

Colostrum The protein-rich fluid secreted by the mammary glands just after childbirth, prior to the appearance of breast milk.

Commensal An organism that lives in or on another organism without harming it and that benefits from the relationship.

Commensalism A symbiotic relationship in which one organism benefits and the other one neither benefits nor is harmed by the relationship.

Common-Source Outbreak An epidemic that arises from contact with contaminated substances.

Communicable Infectious Disease Infectious disease that can be spread from one host to another. *Known also as a Contagious Disease.*

Community All the kinds of organisms present in an environment.

Competence Factor A protein released into the medium that facilitates the uptake of DNA into a bacterial cell.

Competitive Inhibitor A molecule similar in structure to a substrate that competes with the substrate by binding to the active site. The inhibitor can reversibly occupy the active site but does not undergo the reaction.

Complementary Base Pairing Hydrogen bonding between adenine and thymine (or uracil) bases or between guanine and cytosine bases.

Complement A set or more than 20 large regulatory proteins that circulate in plasma and when activated form a nonspecific defense mechanism against many different microorganisms. *Known also as a Contagious Disease.*

Complement Fixation Test A complex serologic test used to detect small quantities of antibodies.

Complement System See ▶ Complement.

Completed Test The final test for coliforms in multiple-tube fermentation in which organisms from colonies grown on eosin methylene blue agar are used to inoculate broth and agar slants.

Complex Medium A growth medium that contains certain reasonable well-defined materials but that varies slightly in chemical composition from batch to batch. *(Known also as a Chemically Nondefined Medium).*

Complex Virus A virus, such as bacteriophage or poxvirus, that has an envelope or specialized structures.

Compound A chemical substance made up of atoms of two or more elements.

Compound Light Microscope A light microscope with more than one lens.

Compromised Host An individual with reduced resistance, being more susceptible to infection.

Concatemer A DNA molecule that consists of a tandem series of complete genomes. Some phage genomes form concatemers during replication as part of a strategy for replicating the full length of a linear DNA duplex.

Condenser Device in a microscope that converges light beams so that they will pass through the specimen.

Condyloma See ▶ Genital Wart.

Confirmed Test Second stage of testing for coliforms in multiple-tube fermentation in which samples from the highest dilution showing gas production are streaked into eosin methylene blue agar.

Confocal Microscopy A light-microscopy technique that allows high resolution in thick samples.

Congenial Rubella Syndrome Complication of German measles causing death or damage to a developing embryo infected by virus that crosses the placenta.

Congenital Syphilis Syphilis passed to a fetus when treponemes cross the placenta from mother to child before birth.

Conidium (plural: *conidia*) A small, asexual, aerial spore organized into chains in some bacteria and fungi.

Conjugation (1) The transfer of genetic information from one bacterial cell to another by means of conjugation pili. (2) The exchange of information between two ciliates (protests).

Conjugation Pilus A type of pilus that attaches two bacterial together and provides a means for the exchange of genetic material. *Known also as Sex Pilus.*

Conjuctiva Mucous membranes of the eye.

Consensus Sequence For a group of nucleotide or amino acid sequences that show similarity but are not identical (for example, the sequences for a family of related regulatory gene sequences), an artificial sequence that is compiled by choosing at each position the residue that is found there mot often in the sequences under study.

Consolidation Blockage of air spaces as a result of fibrin deposits in lobar pneumonia.

Constitutive With respect to gene expression, refers to proteins that are synthesized at a fairly steady rate at all times instead of being induced and repressed in response to changing conditions.

Constitutive Enzyme An enzyme that is synthesized continuously regardless of the nutrients available to the organism.

Consumer An organism that obtains nutrients by eating producers or other consumers. *Known also as Heterotroph.*

Contact Dermatitis Cell-mediated (type IV) hypersensitivity disorder that occurs in sensitized individuals on second exposure of the skin to allergens.

Contact Transmission A mode of disease transmission effected directly, indirectly, or by droplets.

Contagious Disease See ▶ Communicable Infectious Disease.

Contamination The presence of microorganisms on inanimate objects or surfaces of the skin and mucous membranes.

Continuous Cell Line Cell culture consisting of cells that can be propagated over many generations.

Continuous Reactor A device used in industrial and pharmaceutical microbiology to isolate and purify a microbial product often without killing the organism.

Control Variable A factor that is prevented from changing during an experiment.

Convalescent Stage The stage of an infectious disease during which tissues are repaired, healing takes place, and the body regains strength and recovers.

Comb's Antiglobulin Test An immunological test designed to detect anti-Rh antibodies.

Cooperative Transition A transition in a multipart structure such that the occurrence of the transition in one part of the structure makes the transition likelier to happen in other parts.

Copy Number The number of copies per cell of a particular gene or other DNA sequence.

Core The living part of an endospores.

Cori Cycle The metabolic cycle by which lactate produced by tissues engaging in anaerobic glycolysis, such as exercising muscle, is regenerated to glucose in the liver and returned to the tissue via the bloodstream.

Cornea The transparent part of the eyeball exposed to the environment.

Coronavirus Virus with clublike projections that causes colds and acute upper respiratory distress.

Cortex A laminated layer of peptidoglycan between the membranes of the endospores septum.

Corynebacteria Club-shaped, irregular, non-spore-forming, Gram-positive rods.

Coryza The common cold.

Countable Number A number of colonies on an agar plate small enough so that one can clearly distinguish and count them (30 to 300 per plate).

Counterion Atmosphere A cloud of oppositely charged small ions (*Counterions*) that collects around a macroion dissolved in a salt solution. Counterion atmospheres partly shield macroions from each other's charges and thus affect their interactions.

Covalent Bond A bond between atoms created by the sharing of pairs of electrons.

Cowpox Disease caused by the vaccinia virus and characterized by lesions, inflammation of lymph nodes, and fever, virus is used to make vaccine against smallpox and monkeypox.

Crepitant Tissue Distorted tissue caused by gas bubbles in gas gangrene.

Creutzfeldt-Jakob Disease (CJD) A transmissible spongiform encephalopathy of the human brain caused by prions.

Crista (Cristae) A fold in the inner mitochondrial membrane that project into the mitochondrial matrix. The enzymes of the electron transport chain and oxidative phosphorylation are located mainly on the Cristae.

Cross-Reaction Immune reaction of a single antibody with different antigens that are similar in structure.

Cross-Resistance Resistance against two or more similar antimicrobial agents through a common mechanism.

Croup Acute obstruction of the larynx that produces a characteristic high-pitched barking cough.

Cruiform In a DNA duplex, a structure that can be adopted by a palindromic sequence, in which each strand base-pairs with itself to form an arm that projects from the main duplex and terminates in a hairpin loop. The two arms form a "cross" with the main duplex.

Crustacean A usually aquatic anthropod that has a pair of appendages associated with each body segment.

Cryoelectron Microscopy A variation of electron microscopy in which samples are frozen in a glassy ice matrix.

Cryptococcois Fungal respiratory disease caused by a budding, encapsulated yeast, *Filobasidiella neoformans.*

Cryptosporidiosis Disease caused by protozoans of the genus *Cryptosporidium*, common in AIDS patients.

C-Terminus The end of a polypeptide chain that carries an unreacted carboxyl group. *Known also as Carboxyl Terminus.* See also ▶ N-Terminus.

Curd The solid portion of milk resulting from bacterial enzyme addition and used to make cheese.

Curie The basic unit of radioactive decay; an amount of radioactivity equivalent to that produced by 1 g of radium, mainly 2.22×10^{12} disintegrations per minute.

Cyanobacteria Photosynthetic, prokaryotic, typically unicellular organisms that are members of the kingdom Monera.

Cyanosis Blush skin characteristic of oxygen-poor blood.

Cyclic Photophosphorylation In photosynthesis, Photophosphorylation (light-dependent ATP synthesis) that is linked to a cyclic flow of electrons from photosystem II down an electron transport chain and back to photosystem II; it is not coupled to the oxidation of H_2O or to the reduction of $NADP^+$. Compare ▶ Noncyclic Photophosphorylation.

Cyclins Proteins that regulate the cell cycle by binding to and activating specific nuclear protein kinases. Cyclin-dependent kinase activations occur at three points during the cell cycle, thus providing three decision points as to whether the cycle will proceed.

Cyst A spherical, thick-walled cell that resembles an endospores, formed by certain bacteria.

Cysticercus An oval white sac with a tapeworm head invaginated into it. *Known also as a Bladder Worm.*

Cystitis Inflammation of the bladder.

Cytochrome An electron carrier functioning in the electron transport chain; heme protein.

Cytokine One of a diverse group of soluble proteins that have specific roles in host defenses.

Cytokinesis The division of a eukaryotic cell to form two cells. It usually accompanies nuclear division, although nuclear division can occur without cytokinesis.

Cytomegalovirus One of a widespread and diverse group of herpsviruses that often produces no symptoms in normal adults but can severely affect AIDS patients and congenitally infected children.

Cytopathic Effect (CPE) The visible effect viruses have on cells.

Cytoplasm The semifluid substance inside a cell, excluding, in eukaryotes, the cell nucleus.

Cytoplasmic Streaming Process by which cytoplasm flows from one part of a eukaryotic cell to another.

Cytoskelton An organized network of rodlike and fiberlike proteins that pervades a cell and helps give it its shape and motility. The cytoskelton includes action filaments, microtubules, and a diverse group of filamentous proteins collectively called intermediate filaments.

Cytosol The fluid medium that is located inside a cell but outside the nucleus and organelles (for eukaryotes) or the Nucleoid (for prokaryotes). It is a semiliquid concentrated solution or suspension.

Cytotoxic Drug A drug that interferes with DNA synthesis, used to suppress the immune system and prevent the rejection of transplants.

Cytotoxic (Type II) Hypersensitivity Type of allergy elicited by antigens on cells, especially red blood cells, that the immune system treats are foreign.

Cytotoxin Toxin produced by cytotoxic cells that kills infected host cells.

Dark-Field Illumination In light microscopy, the light that is reflected from an object rather than passing through it, resulting in a bright image on a dark background.

Dark Reactions Part of photosynthesis in which carbon dioxide gas is reduced by electrons form reduced NADP (NADPH) to form various carbohydrate molecules, chiefly glucose. These photosynthetic subprocesses do not depend *directly* on light energy; specifically, the synthesis of carbohydrate from CO_2 and H_2O. Compare ► Light Reactions.

Dark Repair Mechanism for repair of damaged DNA by several enzymes that do not require light for activation; they excise defective nucleotide sequences and replace then with DNA complementary to the unaltered DNA strand.

Daughter Cell One of the two identical products of cell division.

Deaminating Agent A chemical mutagen that can remove an amino group (—NH_2) from a nitrogenous base, causing a point mutation.

Death Phase See ► Decline Phase.

Debridement Surgical scraping to remove the thick crust or scab that forms over burnt tissue (eschar).

Decimal Reduction Time (DRT; also called D Value) The length of time needed to kill 90 percent of the organisms in a given population at a specified temperature.

Decline Phase (1) The fourth of four major phases of the bacterial growth curve in which cells lose their ability to divide (due to less supportive conditions in the medium) and thus die. *Known also as Death Phase.* (2) In the stages of a disease, the period during which the host defense finally overcome the pathogen and symptoms begin to subside.

Decomposer Organism that obtains energy by digesting dead bodies or wastes of producers and consumers.

Defined Synthetic Medium A synthetic medium that contains known specific kinds and amounts of chemical substances.

Definitive Host An organism that harbors the adult, sexually reproducing form of a parasite.

Degranulation Release of histamine and other preformed mediators of allergic reactions by sensitized mast cells and basophils after a second encounter with an allergen.

Dehydration Synthesis A chemical reaction that builds complex organic molecules.

Delayed (Type IV) Hypersensitivity See ► Cell-Mediated (Type Iv) Hypersensitivity.

Delayed Hypersensitivity (T_{DH}) Cells Those T cells (inflammatory T_H1) that produce lymphokines in cell-mediated (Type IV) hypersensitivity reactions.

Deletion The removal of one or more nitrogenous bases from DNA, usually producing a frameshift mutation.

Delta Hepatitis See ► Hepatitis D.

Denaturation For a nucleic acid or protein, the loss of tertiary and secondary structure so that the polymer becomes a random coil. For DNA, this change involves the separation of the two strands. Denaturation can be induced by heating and by certain changes in chemical environment. It can also be stated as the disruption of hydrogen bonds and other weak forces that maintain the structure of a globular protein, resulting in the loss of its biological activity.

Dengue Fever Viral systemic disease that causes severe bone and joint pain *Known also as Breakbone Fever.*

Denitrification The process by which nitrates are reduced to nitrous oxide or nitrogen gas.

Dental Caries The erosion of enamel and deeper parts of teeth. *Known also as Tooth Decay.*

Dental Plaque A continuously formed coating of microorganisms and organic matter on tooth enamel.

Deoxyribonucleic Acid (DNA) Nucleic acid that carries hereditary information from one generation to the next.

Depurination Cleavage of the glycosidic bond between C-1′ of deoxyribose and a Purine base in DNA. Used in Maxam-Gilbert sequence analysis.

Dermal Wart A fungal skin disease.

Dermatophyte A fungus that invades keratinized tissue of the skin and nails.

Dermis The thick inner layer of the skin.

Descriptive Study An epidemiologic study that notes the number of cases of a disease, which segments of the population are affected, where the cases have occurred, and over what time period.

Desensitization Treatment designed to cure allergies by means of injections with gradually increasing does of allergen.

Deuteromycota See ► Fungi Imperfecti.

Diabetes Melllitus A disease caused by a deficiency in the action of insulin in the body, resulting either from low insulin levels or from inadequate insulin levels combined with unresponsiveness of the target cells to insulin. The disease is manifested primarily by disturbances in fuel homeostasis, including hyperglycemia (abnormally high blood glucose levels).

Dialysis The process by which low-molecular-weight solutes are added to or removed from a solution by means of diffusion across a semipermeable membrane.

Diapedsis The process in which leukocytes pass out of blood into inflamed tissues by squeezing between cells of capillary walls.

Diarrhea Excessive frequency and looseness of bowel movements.

Diastereomers Molecules that are stereoisomers but not enantiomers of each other. Isomers that differ in configuration about two or more asymmetric carbon atoms and are not complete mirror images.

Diatom An alga or plantlike protest that lacks flagella and has a glasslike outer shell.

Dichotomous Key Taxonomic key used to identify organisms, composed of paired (either-or) statements describing characteristics.

Dielectric Constant A dimensionless constant that expresses the screening effect of an intervening medium on the interaction between two charged particles. Every medium (such as a water solution or an intervening portion of an organic molecule) has a characteristic dielectric constant.

Difference Spectrum With respect to absorption spectra, a spectrum obtained by loading the sample cuvette with the substances under study and a reference cuvette with an equimolar sample of the same substances in a known state (for example, fully oxidized) and recording the difference between the two spectra.

Differential Medium A growth medium with a constituent that causes an observable change (in color or pH) in the medium when a particular chemical reaction occurs, making it possible to distinguish between organisms.

Differential Stain Use of two or more dyes to differentiate among bacterial species or to distinguish various structures of an organism; for example, the Gram stain.

Diffraction Phenomenon in which light waves, as they pass through a small opening, are broken up into bands of different wavelengths.

Diffraction Pattern The pattern that is produced when electromagnetic radiation passes through a regularly repeating structure; it results because the waves scattered by the structure interact destructively in most directions (creating dark zones) but constructively in a few directions (creating bright spots). For the pattern to be sharp, the radiation wavelength must be somewhat shorter than the repeat distance in the structure. See also ► X-Ray Diffraction.

Diffusion Coefficient (D) A coefficient that indicates how quickly a particular substance will diffuse in a particular medium under the influence of a given concentration gradient.

DiGeorge Syndrome Primary immunodeficiency disease caused by failure of the thymus to develop properly, resulting in a deficiency of T cells.

Digestive System The body system that converts ingested food into material suitable for the liberation of energy or for assimilation into body tissues.

Dikaryotic Referring to fungal cells within hyphae that have two nuclei, produced by plasmogany in which the nuclei have not united.

Dilution Method A method of testing antibiotic sensitivity in which organisms are incubated in a series of tubes containing known quantities of a chemotherapeutic agent.

Dimer Two adjacent pyrimidines bonded together in a DNA strand, usually as a result of exposure to ultraviolet rays.

Dimorphism The ability of an organism to alter its structure when it changes habitats.

Dinoflagellate An alga or plantlike protest, usually with two flagella.

Diphtheria A severe upper respiratory disease caused by *Corynebacterium diphtheriae*, can produce subsequent myocarditis and polyneuritis.

Diphtheroid Organism found in normal throat cultures that fails to produce exotoxin but is otherwise indistinguishable from diphtheria-causing organisms.

Dipicolonic Acid Acid found in the core of endospores that contributes to its heat resistance.

Diploid For a cell or an organism, the possession of two homologous sets of chromosomes per nucleus (with the possible exception of sex chromosomes, which may be present in only one copy). Compare ► Haploid.

Diploid Fibroblast Strain A culture derived from fetal tissues that retains fetal capacity for rapid, repeated cell division.

Direct Contact Transmission Mode of disease transmission requiring person-to-person body contact.

Direct Fecal – Oral Transmission Direct contact transmission of disease in which pathogens from fecal matter are spread by unwashed hands to the mouth.

Direct Microscopic Count A method of measuring bacterial growth by counting cells in a known volume of medium that fills a specially calibrated counting chamber on a microscope slide.

Disaccharide A carbohydrate formed by the joining of two monosaccharides.

Disease A disturbance in the state of health wherein the body cannot carry out all its normal functions. See also ▶ Epidemiology and ▶ Infectious Disease.

Disinfectant A chemical agent used on inanimate objects to destroy microorganisms.

Disinfection Reducing the number of pathogenic organisms on objects or in materials so that they pose no threat of disease.

Dismutation A reaction in which two identical substrate molecules have different fates; particularly, a reaction in which one of the substrate molecules is oxidized and the other reduced.

Disk Diffusion Method A method used to determine microbial sensitivity to antimicrobial agents in which antibiotic disks are placed on an inoculated Petri dish, incubated, and observed for inhibition of growth.

Dispersion Forces Weak intermolecular attractive forces that arise between molecules that are close together, because the fluctuating electron distributions of the molecules become synchronized so as to produce a slight electrostatic attraction. These forces play a role in the internal packing of many biomolecules.

Disseminated Tuberculosis Type of tuberculosis spread throughout body, not seen in AIDS patients, usually caused by *Mycobacterium avium-intercellulare.*

Dissociation Constant For an acid, the equilibrium constant K_a for the dissociation of the acid into its conjugate base and a proton. For a complex of two biomolecules, the equilibrium constant K_d for dissociation into the component molecules.

Distillation The separation of alcohol and other volatile substances from solid and nonvolatile substances.

Divergent Evolution Process in which descendants of a common ancestor species undergo sufficient change to be identified as separate species.

DNA Gyrase An enzyme that is able to introduce negative superhelical turns into a circular DNA helix.

DNA Hybridization Process in which the double strands of DNA of each of two organisms are split apart and the split strands from the two organisms are allowed to combine.

DNA Polymerase An enzyme that moves along behind each replication fork, synthesizing new DNA strands complementary to the original ones.

DNA Replication Formation of new DNA molecules.

DNA Tumor Virus An animal virus capable of causing tumors.

Domain A portion of a polypeptide chain that folds on itself to form a compact unit that remains recognizably distinct within the tertiary structure of the whole protein. Large globular proteins often consist of several domains, which are connected to each other by stretches of relatively extended polypeptide. A new taxonomic category above the kingdom level, consisting of the Archaea, Bacteria, and Eukarya.

Donovan Body A large mononuclear cell found in scrapings of lesions that confirms the presence of granuloma inguinale.

DPT Vaccine Diphtheria, killed whole cell pertussis and tetanus vaccine.

Dracunculiasis Skin disease caused by a parasitic helminth, the guinea worm *Dracunculus medinensis.*

Droplet Nucleus A particle consisting of dried mucus in which microorganisms are embedded.

Droplet Transmission Contact transmission of disease through small liquid droplets.

Drug See ▶ Chemotherapeutic Agent.

Drug Resistance Factors Bacterial plasmids that carry genes coding for resistance to antibiotics.

DTaP Vaccine Diphtheria, tetanus, and acellular pertussis vaccine.

D Value See ▶ Decimal Reduction Time.

Dyad A set of paired chromosomes in eukayotic cells that are prepared to divided by mitosis or meiosis.

Dyad Axis A two-fold axis of symmetry.

Dysentery A severe diarrhea that often contains mucus and sometimes blood or pus.

Dysuria Pain and burning on urination.

Eastern Equine Encephalitis Type of viral encephalitis seen most often in the eastern United States, infects horses more frequently than human.

Ebola Virus A filovirus that causes hemorrhagic fevers.

Eclipse Period Period during which viruses have absorbed to and penetrated host cells but cannot yet be detected in cells.

Ecology The study of relationships among organisms and their environment.

Ecosystem All the biotic and Abiotic components of an environment.

Ectoparasite A parasite that lives on the surface of another organism.

Eczema Herpecticum A generalized eruption caused by entry of the herpevirus through the skin, often fatal.

Edema An accumulation of fluid in tissues that causes swelling.

Editing See ▶ RNA Editing.

Ehrlichiosis A tick-borne disease found in dogs and human and caused by *Ehrlichia canis* and *E. chaeffeenis.*

Einstein One mole of photons.

Electrolyte A substance that is ionizable in solution.

Electron A negatively charged subatomic particle that moves around the nucleus of an atom.

Electron Acceptor An oxidizing agent in a chemical reaction.

Electron Donor A reducing agent in a chemical reaction.

Electron Micrograph A "photograph" of an image taken with an electron microscope.

Electron Microscope Microscope that uses a beam of electrons rather than a beam of light and electromagnets instead of glass lenses to produce an image.

Electron Spin Resonance A form of spectroscopy that is sensitive to the environment of unpaired electrons in a sample. *Known also as Electron Paramagnetic Resonance or EPR.*

Electron Transport Processes in which pairs of electrons are transferred between cytochromes and other compounds.

Electron Transport Chain (1) A series of compounds that pass electrons to oxygen (the final electron acceptor). Known also as *Respiratory Chain.* (2) A sequence of electron carriers of progressively higher reduction potential in a cell that is linked so that electrons can pass from one carrier to the next. The chain captures some of the energy released by the flow of electrons and uses it to drive the synthesis of ATP.

Electrophoresis (1) Process used to separate large molecules such as antigens or proteins by passing an electrical current through a sample on a gel. (2) A method for separating electrically charged substances in a mixture. A sample of the mixture is placed on a supporting medium (a piece of filter paper or a gel), to which an electrical field is applied. Each charged substance migrates toward the cathode or the anode at a speed that depends on its net charge and its frictional interaction with the medium. See also ▶ Gel Electrophoresis.

Electroporation A brief electric pulse produces temporary pores in the cell membrane, allowing entrance of vectors carrying foreign DNA.

Element Matter composed of one kind of atom.

Elementary Body An infectious stage in the life cycle of chlamydias.

Elephantiasis Gross enlargement of limbs, scrotum, and sometimes other body parts from accululation of fluid due to blockage of lymph ducts by the helminth *Wuchereria bancrofti.*

Elongation Factors Nonribosomal protein factors that are necessary participants in the chain-elongation cycle of polypeptide synthesis; they interact with the ribosome-mRNA complex or with other major cycle participants.

Enamel The hard substance covering the crown of a tooth.

Enantiomers Stereoisomers that are nonsuperimposable mirror images of each other. The term *optical isomers* comes from the fact that the enantiomers of a compound rotate polarized light in opposite directions. *Known also as Optical Isomers.*

Encephalitis An inflammation of the brain caused by a variety of viruses or bacteria.

Endemic Referring to a disease that is constantly present in a specific population.

Endemic Relapsing Fever Tick-borne cases of relapsing fever caused by several species of *Borrelia.*

Endemic Typhus A flea-borne typhus caused by *Rickettsia typhi.*

Endergonic In a nonisolated system, a process that is accompanied by a positive change in free energy (positive ΔG) and therefore is thermodynamically not favored. Compare ▶ Exergonic.

Endocrine Glands Glands that synthesize hormones and release them into the circulation. The hormone-producing gland cells are called endocrine cells.

Endocytosis Process in which vesicles form by invagination of the plasma membrane to move substances into eukaryotic cells.

Endoenzyme An enzyme that acts within the cell producing it.

Endoflagellum See ▶ Axial Filament.

Endogenous Infection An infection caused by opportunistic microorganism already present in the body.

Endogenous Pyrogen Pyrogen secreted mainly by monocytes and macrophages that circulates to the hypothalamus and causes an increase in body temperature.

Endometrium The mucous membrane lining the uterus.

Endonuclease An enzyme that cleaves a nucleic acid chain at an internal phosphodiester bond.

Endoparasite A parasite that lives within the body of another organism.

Endoplasmic Reticulum A highly folded membranous compartment within the cytoplasm that is responsible for a great variety of cellular tasks, including the glycosylation and trafficking of proteins destined for secretion or for the cell membrane or some organelles. It also functions in lipid synthesis, and the enzymes of many pathways of intermediate metabolism are located on its surface.

Endorphins A class of endogenous brain peptides that exert analgesic effects in the central nervous system by binding to opiate receptors. They are produced by cleavage of the large polypeptide pro-opiomelanocortin.

Endospore A resistant, dormant structure, formed inside some bacteria, such as *Bacillus* and *Clostridium*, that can survive adverse conditions.

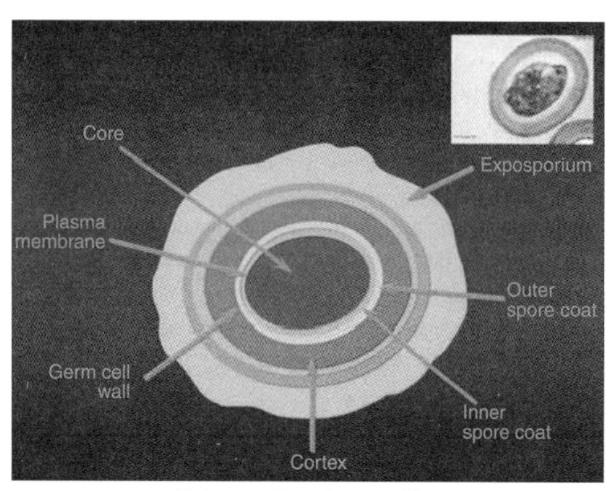

Endospore (Spore) composition
Block, S., Disfection, Sterilization, and preservation 5th Ed., Lippincott, Williams & Wilkins, New york, 2001.

Endospore Septum A cell membrane without a cell wall that grows around the core of endospores.

Endosymbiotic Theory Holds that the organelles of eukaryotic cells arose from prokaryotes that came to live, in a symbiotic relationship, inside the eukayote-to-be cell.

Endotoxin A toxin incorporated in Gram-negative bacterial cell walls and released when the bacterium dies. *Known also as lipopolysaccharide.*

End-Product Inhibition See ► Feedback Inhibition.

Energy See ► Internal Energy.

Energy Charge A quantity that indicates the state of a cell's energy reserves. It is equal to the cell's reverses of the free energy sources ATP and ADP (taking into account that ADP stores less free energy than ATP) divided by the total supply of ATP and its breakdown products ADP and AMP: ([ATP] + ½[ADP])/([ATP] + [ADP] + (AMP]).

Enhancer Sequence A DNA sequence that is distant from a gene but to which a protein factor that affects the gene's transcription can bind to exert its action. It is possible that DNA looping brings enhancer-bound proteins into proximity with the gene's promoter.

Enrichment Medium A medium that contains special nutrients that allow growth of a particular organism.

Enteric Bacteria Members of the family Enterobacteriaceae, many of which are intestinal, small facultatively anerobic Gram-negative rods with peritrichous flagella.

Enteric Fever Systemic infection, such as typhoid fever, spread throughout the body from the intestinal mucosa.

Enteritis An inflammation of the intestine.

Enterocolitis Disease caused by *Salmonella typhimurium* and 5 *paratyphi* that invade intestinal tissue and produce bacteremia.

Enterohemorrhagic Strain of *Escherichia Coli* One that causes bloody diarrhea and is often fatal; often from contaminated food.

Enteroinvasive Strain Strain of *Escherichia coli* with a plasmid-borne gene for a surface antigen (K antigen) then enables it to attach to and invade mucosal cells.

Enterotoxicosis See ► Food Poisoning.

Enterotoxigenic Strain Strain of *Escherichia coli* carrying a plasmid that enables it to make an enterotoxin.

Enterotoxin An exotoxin that acts on tissues of the gut.

Enterovirus One of the three major groups of picornaviruses that can infect nerve and muscle cells, the respiratory tract lining, and skin.

Enthalpy (*H*) A thermodynamic quantity (function of state) that is equal to the internal energy of a system plus the product of the pressure and volume: $H = E + PV$. It is equal to the heat change in constant-pressure reactions, such as most reactions in biological systems.

Entropy (*S*) A thermodynamic quantity (function of state) that expresses the degree of disorder or randomness in a system. According to the second law of thermodynamics, the entropy of an opens system tends to increase unless energy is expended to keep the system orderly.

Envelope A bilayer membrane found outside the capsid of some viruses, acquired as the virus buds through one of the host's membrane.

Enveloped Virus A virus with a bilayer membrane outside its capsid.

Enzyme A protein catalyst that controls the rate of chemical reaction in cells.

Enzyme Induction A mechanism whereby the genes coding for enzymes needed to metabolize a particular nutrient are activated by the presence of that nutrient.

Enzyme-Linked Immunosorbent Assay (ELISA) Modification of radioimmunoassay in which the anti-antibody, instead of being radioactive, is attached to an enzyme that causes a color change in its substrate.

Enzyme Repression Mechanism by which the presence of a particular metabolite represses the genes coding for enzymes used in its synthesis.

Enzyme-Substrate Complex A loose association of an enzyme with its substrate.

Eosinophil A leukocyte present in large numbers during allergic reactions and worm infections.

Epidemic Referring to a disease that has a higher than normal incidence in a population over a relatively short period of time.

Epidemic Keratoconjunctivitis Eye disease caused by an adenovirus. *Known also as Shipyard Eye.*

Epidemic Relapsing Fever Louseborne cases of relapsing fever caused by several species of *Borrelia.*

Epidemic Typhus Louseborne rickettsial disease caused by *Rickettsia prowazekii*, seen most frequently in conditions of overcrowding and poor sanitation. *Known also as Classic, European, or Louseborne Typhus.*

Epidemiologic Study A study conducted in order to learn more about the spread of a disease in a population.

Epidemiologist A scientist who studies epidemiology.

Epidemiology The study of factors and mechanisms involved in the spread of disease within a population.

Epidermis The thin outer layer of the skin.

Epiglottitis An infection of the epiglottis.

Episomes Plasmids that can undergo integration into the bacterial chromosome.

Epitope The specific portion of an antigen particle that is recognized by a given antibody or T-cell receptor. *Known also as Antigenic Determinant.*

Epstein-Barr Virus (EBV) Virus that causes infectious mononucleosis and Burkitt's lymphoma.

Ergot Toxin produced by *Claviceps, purpurea*, a parasite fungus of rye and wheat that causes ergot poisoning when ingested by humans.

Ergot Poisoning Disease caused by ingestion of ergot, the toxin produced by *Claviceps purpurea*, a fungus of rye and wheat.

Erysipelas Infection caused by hemolytic streptococci that spreads through lymphatics, resulting in septicemia and other diseases.

Erythrocyte A red blood cell.

Erythromycin An antibacterial agent that has a bacteriostatic effect on protein synthesis.

Eschar The thick crust or scab that forms over a severe burn.

Essential Amino Acids Amino acids that must be obtained in the diet because they cannot be synthesized in the body (at least not in adequate amounts).

Essential Fatty Acids Fatty acids that must be obtained in the diet because they cannot be synthesized in the body in adequate amounts. Examples are linoleic acid and linolenic acid.

Ethanbutol An antibacterial agent effective against certain strains of mycobacteria.

Etiology The assignment or study of causes and origins of a disease.

Eubacteria True bacteria.

Englenoid An alga or plantlike protest, usually with a single flagella and a pigmented eyespot (stigma).

Eukarya One of the three Domains of living things; all members are eukaryotic.

Eukaryotes Organisms whose cells are compartmentalized by internal cellular membranes to produce a nucleus and organelles. Compare ► Prokaryotes.

Eukaryote An organism composed of eukaryotic cells.

Eukaryotic Cell A cell that has a distinct cell nucleus and other membrane-bound structures.

Eutophication The nutrient enrichment of water from detergents, fertilizes, and animal manures, which cause overgrowth of algae and subsequent depletion of oxygen.

Exanthema A skin rash.

Exergonic (1) In a nonisolated system, a process that is accompanied by a negative change in free energy (negative ΔG) and therefore is thermodynamically favored. Compare ► Endergonic. (2) Releasing energy from a chemical reaction.

Exocrine Cell A cell that secretes a substance that is excreted through a duct either into the alimentary tract or to the outside of the organism. Exocrine cells are grouped together in exocrine glands.

Exocytosis Process by which vesicles inside a eukaryotic cell fuse with the plasma membrane and release their contents from the eukaryotic cell.

Exoenzyme An enzyme that is synthesized in a cell but crosses the cell membrane to act in the periplasmic space or the cell's immediate environment. *Known also as Extracellular Enzyme.*

Exogenous Infection An infection caused by microorganisms that enter the body from the environment.

Exogenous Pyrogen Exotoxins and endotoxins from infectious agents that cause fever by stimulating the release of an endogenous pyrogen.

Exon A region in the coding sequence of a gene that is translated into protein (as opposed to introns, which are not). The name comes from the fact that exons are the only parts of an RNA transcript that are seen outside the nucleus. Compare ▶ Intron.

Exonuclease An enzyme that removes segments of DNA.

Exosporium A lipid-protein membrane formed outside the coat of some endospores by the mother cell.

Exotoxin A soluble toxin secreted by microbes into their surroundings, including host tissues.

Experimental Study An epidemiological study designed to test a hypothesis about an outbreak of disease, often about the value of a particular treatment.

Experimental Variable The factor that is purposely changed in an experiment.

Exponential Rate The rate of growth in a bacterial culture characterized by doubling of the population in a fixed interval of time. *Known also as Logarithmic Rate.*

Exportins A class of proteins involved in transporting materials out of nuclei. See ▶ Importins.

Extinction Coefficient (ε_λ) A coefficient that indicates the ability of a particular substance in solution to absorb light of wavelength λ. The molar extinction coefficient, ε_M, is the absorbance that would be displayed by a 1-M solution in a 1-cm light path.

Extracellular Enzyme See ▶ Exoenzyme.

Extrachromosomal Resistance Drug resistance of a microorganism due to the presence of resistance (R) plasmids.

Extreme Thermoacidophile Organism requiring very hot and acidic environment, usually belonging to Domain Archaea.

Fab Fragment The portion of an antibody that contains an antigen binding site.

Faciliated Diffusion Diffusion (down a concentration gradient) across a membrane (from an area of higher concentration to lower concentration) with the assistance of a carrier molecule, but not requiring ATP.

Facilitated Transport The movement of a substance across a biological membrane in response to a concentration or electrochemical gradient where the movement is facilitated by membrane pores or by specific transport proteins. Compare ▶ Active Transport, ▶ Passive Transport. *Known also as Facilitated Diffusion.*

Facultative Able to tolerate the presence or absence of a particular environmental condition.

Facultative Anaerobe A bacterium that carries on aerobic metabolism when oxygen is present but shifts to anerobic metabolism when oxygen is absent.

Facultative Parasite A parasite that can live either on a host or freely.

Facultative Psychrophile An organism that grows best at temperatures below 20°C but can also grow at temperatures above 20°C.

Facultative Thermophile An organism that can grow both above and below 37°C.

FAD Flavin adenine dinucleotide, a coenzyme that carries hydrogen atoms and electrons.

Fastidious Referring to microorganisms that have special nutritional needs that are difficult to meet in the laboratory.

Fat A complex organic molecule formed from glycerol and one or more fatty acids.

Fatty Acid A long chain of carbon atoms and their associated hydrogens with a carboxyl group at one end.

Fc Fragment The tail region of an antibody that may contain sites for macrophage and complement binding.

Feces Solid waste produced in the large intestine and stored in the rectum until eliminated from the body.

Feedback Inhibition Regulation of a metabolic pathway by the concentration of one of its intermediates or, typically, its end product, which inhibits an enzyme in the pathway. *Known also as End-Product Inhibition.*

Feline Panleukopenia Virus (FPV) A parvovirus that causes severe disease in cats.

Female Reproductive System The host system consisting of the ovaries, uterine tubes, uterus, vagina, and external genitalia.

Fermentation Anaerobic metabolism of the pyruvic acid produced in glycolysis.

Fermentations Processes in which cellular energy is generated from the breakdown of nutrient molecules where there is no net change in the oxidation state of the products as compared with that of the reactants; fermentation can occur in the absence of oxygen.

Fever A body temperature that is abnormally high.

Fibroblast A new connective tissue cell that replaces fibrin as a blood clot dissolves, forming grandulation tissue.

Fibrous Proteins Proteins of elongated shape, often used as structural materials in cells and tissues. Compare ▶ Globular Proteins.

Fifth Disease A normal disease in children caused by the *Erythrovirus* called B19, characterized by a bright red rash on the cheeks and a low-grade fever. *Known also as Erythema Infectiosum.*

Filariasis Disease of the blood and lymph caused by any of several different roundworms carried by mosquitoes.

Filovirus A filamentous virus that displays unusual variability in shape. Two filoviruses, the Ebola virus and the Marburg virus, have been associated with human disease.

Filter Paper Disk Method Method of evaluating the antimicrobial properties of a chemical agent using filter paper disks placed on an inoculated agar plate.

Filtration (1) A method of estimating the size of bacterial populations in which a known volume of air or water is drawn through a filter with pores too small to allow passage of bacteria. (2) A method of sterilization that uses a membrane filter to separate bacteria from growth media. (3) The filtering of water through beds of sand to remove most of the remaining microorganisms after flocculation in water treatment plants.

Fimbria See ► Attachment Pilus.

Fine Adjustment Focusing mechanism of a microscope that very slowly changes the distance between the objective lens and the specimen.

First Law of Thermodynamics The law that states that energy cannot be created or destroyed and that it is therefore possible to account for any change in the internal energy of a system ΔE by an exchange of heat (q) and/or work (w) with the surroundings $\Delta E = Q - w$.

First-Order Reaction A reaction whose rate depends on the first power of the concentration of the reactant. Compare ► Second-Order Reaction.

Fischer Projection A convention for representing stereoisomers in a plane. The tetrahedron of bonds on a carbon is represented as a plane cross, where the bonds to the right and left are assumed to be pointing toward the viewer and the bonds to the top and bottom are assumed to be pointing away from the viewer. Fischer projections of monosaccharides are oriented with the carbonyl group at the top; the chiral carbon farthest from the carbonyl group (which is the one that determines whether the sugar is the D or the L form) is then drawn with its hydroxyl to the right for the D form and to the left for the L form.

Five-Kingdom System System of classifying organisms into one of five kingdoms: Monera (Prokaryotae), Protista, Fungi, Plantae, and Animalia.

Flagellar Staining A technique for observing flagella by coating the surfaces of flagella with a dye or a metal such as silver.

Flagellum (plural: *flagella*) A long, thin, helical appendage of certain cells that provides a means of locomotion.

Flash Pasteurization See High-Temperature Short-Time Pasteurization.

Flat Sour Spoilage Spoilage due to the growth of spores that does not cause cans to bulge with gas.

Flatworm A primitive, unsegmented, hermaphroditic often parasitic worm. *Known also as Platyhelminthes.*

Flavin Adenine Dinucleotide (FAD), Flavin Mononucleotide (FMN) Coenzymes derived from vitamin B_2 (riboflavin) that function as electron acceptors in enzymes that catalyze electron transfer reactions.

Flavivirus A small, enveloped, (+) sense RNA virus that causes a variety of encephalitides, including yellow fever.

Flavorprotein An electron carrier in oxidative phosphorylation.

Flocculation The addition of alum to cause precipitation of suspended colloids, such as clay, in the water purification process.

Flourescence-Activated Cell Sorter (FACS) A machine that collects quantities of a particular cell type under sterile conditions for study.

Fluctuation Test A test to determine that resistance to chemical substances occurs spontaneously rather than being induced.

Fluid-Mosaic Model A model describing cellular membrane structure, according to which the proteins are embedded in a phospholipid bilayer and are free to move in the plane of the membrane. This model is basically correct.

Fluke A flatworm with a complex life cycle; can be an internal or external parasite.

Fluorescence (1) Emission of light of one color when irradiated with another, shorter wavelength of light. (2) The phenomenon by which a substance that absorbs light at a given wavelength reradiates a portion of the energy as light of a longer wavelength.

Flourescence Activated Cell Sorter (FACS) Device that separates cells within a population based on whether or not they fluoresce.

Flourescence Miscroscopy Use of ultraviolet light in a microscope to excite molecules so that they release light of different colors.

Fluorescent Antibody Staining Procedure in fluorescence microscopy that uses a fluorochrome attached to antibodies to detect the presence of an antigen.

Fluoride Chemical that helps in reducing tooth decay by poisoning bacterial enzymes and hardening the surface enamel of teeth.

Flux With reference to a chemical pathway, the rate (in moles per unit time) at which reactant "flows through" the pathway to emerge as product. The term can be used for the rate at which particles undergo any process in which they either flow or can be thought of metaphorically as flowing.

Focal Infection An infection confined to a specific area from which pathogens can spread to other areas.

Folliculitis Local infection produced when hair follicles are invaded by pathogenic bacteria. *Known also as Pimple or Pustule.*

Fomite A nonliving substance capable of transmitting disease, such as clothing, dishes, or paper money.

Food Poisoning A gastrointestinal disease caused by ingestion of foods contaminated with preformed toxins or other toxic substances. *Known also as Enterotoxicosis.*

Footprinting With respect to molecular genetics, a technique used to identify the DNA segment in contact with a given DNA-binding protein. The DNA-protein complex is subjected to digestion with a nonspecific nuclease, which cleaves at the residues that are not protected by the protein.

Formed Elements Cells and cell fragments comprising about 40 percent of the blood.

F Pilus A bridge formed from an F1 cell to an F2 cell for conjugation.

F Plasmid Fertility plasmid containing genes direction synthesis of proteins that form an F pilus (sex pilus, or conjugation pilus).

F^- Cell A cell lacking the F plasmid; called recipient or female cell.

F^+ Cell A cell having an F plasmid, called donor or male cell.

F′ Plasmid An F plasmid that has been imprecisely separated from the bacterial chromosome so that it carries a fragment of the bacterial chromosome.

Frameshift Mutation A mutation that changes the reading for a gene by adding or deleting one or two nucleotides, thereby reducing the remainder of the message 3′ to the mutation to gibberish.

Frameshift Suppressor A mutant tRNA that contains either two or four bases in the Anticodon loop and can suppress the effects of a particular frameshift mutation in a gene.

Free Energy (G) thermodynamic quantity (function of state) that takes into account both enthalpy and entropy: $G = H - TS$, where H is enthalpy, S is entropy, and T is absolute temperature. The *change in free energy* (ΔG) for a process, such as a chemical reaction, takes into account the changes in enthalpy and entropy and indicates whether the process will be thermodynamically favored at a given temperature. *Known also as Gibbs Free Energy.*

Freeze-Etching Technique in which water is evaporated under vacuum from the freeze-fractured surface of a specimen before the observation with electron microscopy.

Freeze-Fracturing Technique in which a cell is first frozen and then broken with a knife so that the fracture reveals structures inside the cell when observed by electron microscopy.

Frictional Coefficient A coefficient that determines the frictional force on a particular particle (such as a molecule) in a particular medium at a given velocity. In the context of electrophoresis or centrifugation, it determines how fast a chemical species will move in a particular medium in response to a given electrical field or centrifugal force.

Fulminating See ► Acme.

Functional Group Part o a molecule that generally participates in chemical reactions as a unit and gives the molecule some of its chemical properties.

Fungi (singular: *fungus*) The kingdom of nonphotosynthetic, eukaryotic organisms that absorb nutrients from their environment.

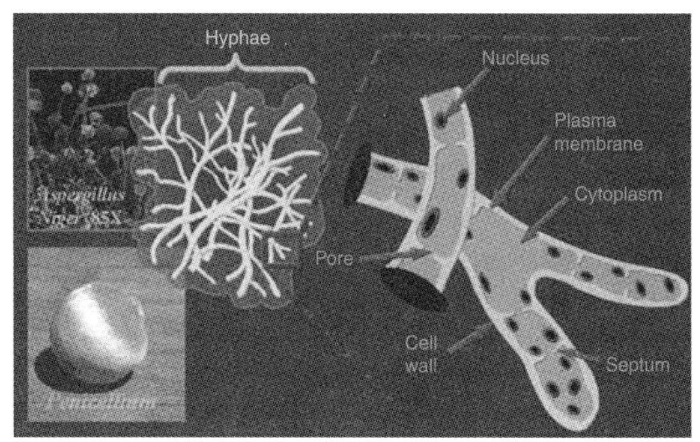

Fungi cellular structure
Black, J.G., Microbiology, 5th Ed., John Wiley & Sons, New Work

Fungi Imperfecti Group of fungi termed "imperfect" because no sexual stage has been observed in their life cycles. *Known also as Deuteromycota.*

Furuncle A large, deep pus-filled infection. *Known also as a Boil.*

Fusion Proteins Genetically engineered proteins that are made by splicing together coding sequences from two or more genes. The resulting protein thus combines portions from two different parent proteins.

F_0F_1 Complex The enzyme complex in the inner mitochondrial membrane that uses energy from the transmembrane proton gradient to catalyze ATP synthesis. The F_0 portion of the complex spans the membrane, and the F_1 portion, which performs the ATP synthase activity, projects into the mitochondrial matrix.
Gamete A male or female reproductive cell.
Gametocyte A male or female sex cell.
Gamma Globulin See ▶ Immune Serum Globulin.
Ganglion An aggregation of neuron cell bodies.
Gas Gangrene A deep would infection, destructive of tissue, often caused by a combination of two or more species of *Clostridium.*
Gated Channel A membrane ion channel that can open or close in response to signals from outside or within the cell.
Gel Electrophoresis A type of electrophoresis in which the supporting medium is a thin slab of gel held between glass plates. The technique is widely used for separating proteins and nucleic acids. See also ▶ Electrophoresis and ▶ Isoelectric Focusing.
Gene A linear sequence of DNA nucleotides that form a functional unit within a chromosome or plasmid.
Gene Amplification A technique of genetic engineering in which plasmids or bacteriophages carrying a specific gene are induced to reproduce at a rapid rate within host cells.
Generalized Anaphylaxis See ▶ Anaphylactic Shock.
Generalized Transduction Type of transduction in which a fragment of DNA from the degraded chromosome of an infected bacteria cell is accidentally incorporated into a new phage particle during viral replication and thereby transferred to another bacterial cell.
Generation Time Time required for a population of organisms to double in number.
Genetic Code The code by which the nucleotide sequence of a DNA or RNA molecule specifies the amino acid sequence of a polypeptide. It consists of three-nucleotide codons that either specify a particular amino acid or tell the ribosome to stop translating and release the polypeptide. With a few minor exceptions, all living things use the same code.
Genetic Engineering The use of various techniques to purposefully manipulate genetic material to alter the characteristics of an organism in a desired way.
Genetic Fusion A technique of genetic engineering that allows transposition of genes from one location on a chromosome to another location; the coupling of genes from two different operons.
Genetic Homology The similarity of DNA base sequences among organisms.
Genetic Immunity Inborn or innate immunity.
Genetic Recombination Any process that results in the transfer of genetic material from one DNA molecule to another. In eukaryotes, it can refer specifically to the exchange of matching segments between homologous chromosomes by the process of crossing over.
Genetics The science of heredity, including the structure and regulation of genes and how these genes are passed between generations.
Gene Transfer Movement of genetic information between organisms by transformation, transduction, or conjugation.
Genital Herpes See ▶ Herpes Simplex Virus Type 2.
Genital Wart An often malignant wart associated with sexual transmitted viral disease having a very high association rate with cervical cancer. *Known also ad Condyloma.*
Genome The total genetic information contained in a cell, an organism, or virus.
Genotype The genetic information contained in the DNA of an organism. Compare ▶ phenotype.
Genus A taxon consisting of one or more species, the first name of an organism in the binomial system of nomenclature; for example, *Escherichia in Escherichia coli.*
German Measles See ▶ Rubella.
Germination The start of the process of development of a spore or an endospore.
Germ Theory of Disease Theory that microorganisms (germs) can invade other organisms and cause disease.
Giardiasis A gastrointestinal disorder caused by the flagellated protozoan *Giardia intestinalis.*
Gibberellins A family of diterpene plant growth hormones.
Gibbs Free Energy See ▶ Free Energy.
Gingivitis The mildest form of periodontal disease, characterized by inflammation of the gums.
Gingivostomatitis Inflammation of and damage to the glomeruli of the kidneys. *Known also as Bright's Disease.*
Globular Proteins Proteins whose three-dimensional folded shape is relatively compact. compare ▶ Fibrous Proteins.
Glomerulus A coiled cluster of capillaries in the nephron.
Glucocorticoids The steroid hormones cortisol and corticosterone, which are secreted by the adrenal cortex. In addition to other functions, they promote gluconeogenesis in response to low blood sugar levels.
Glucogenic In fuel metabolism, refers to substances (such as some amino acids) that can be used as substrates for glucose synthesis.
Gluconeogenesis The processes by which glucose is synthesized from noncarbohydrate precursors such as

glycerol, lactate, some amino acids, and (in plants) acetyl-CoA.

Glucose Transporter A membrane protein that is responsible for transporting glucose across a cell membrane. Different tissues may have glucose transporters with different properties.

Glycan Another name for polysaccharide.

Glycocalyx Term used to refer to all substances containing polysaccharides found external to the cell wall.

Clycolipids Lipids that have saccharides attached to their head groups.

Glycolysis The initial pathway in the catabolism of carbohydrates, by which a molecule of glucose is broken down to two molecules of pyruvate, with a net production of ATP molecules and the reduction of two NAD^+ molecules to NADH. Under aerobic conditions, these NADH molecules are reoxidized by the electron transport chain; under anaerobic conditions, a different electron acceptor is used. An anaerobic metabolic pathway used to break down glucose into pyruvic acid while producing some ATP.

Glycoprotein A long, spikelike molecule made of carbohydrate and protein that projects beyond the surface of a cell or viral envelope; some viral glycoproteins attach the virus to receptor sites on host cells, while other said fusion of viral and cellular membranes.

Glycosaminoglycans Polysaccharides composed of repeating disaccharide units in which one sugar is either *N*-acetylgalactosamine or *N*-acetylglucosamine. Typically the disaccharide unit carries a carboxyl group and often one or more sulfates, so that most glycosaminoglycans have a high density of negative charges. Glycosaminoglycans are often combined with protein to form proteoglycans and are an important component of the Extracellular matrix of vertebrates. *Known also as Mucopolysaccharides.*

Glycosidic Bond A covalent bond between two monosaccharides.

Glyoxysome A specialized type of Peroxisome found in plant cells. It performs some of the reactions of photorespiration, and it also breaks down fatty acids to acetyl-CoA by β-oxidation and converts the acetyl-CoA to succinate via the glyoxylate cycle, thus enabling plants to convert fatty acids to carbohydrates.

Golgi Apparatus An organelle in eukaryotic cells that receives, modifies, and transports substances coming from the endoplasmic reticulum.

Golgi Complex A stack of flattened membranous vesicles in the cytoplasm. It serves as a routing center for proteins destined for secretion or for lysosomes or the cell membrane; it performs similar functions for membrane lipids, and it also modifies and finishes the oligosaccharide moieties of glycoproteins.

Gonorrhea A sexually transmitted disease caused by *Neisseria Gonorrhoeae.*

G Proteins A family of membrane-associated proteins that transduce signals received by various cell-surface receptors. They are called G proteins because binding of GTP and GDP is essential to their action.

Graft Tissue Tissue that is transplanted from one site to another.

Graft-Versus-Host (GVH) Disease Disease in which host antigens elicit an immunological response from graft cells that destroys host tissue.

Gram Molecular Weight See ► Mole.

Gram Stain A differential stain that uses crystal violet, iodine, alcohol, and safranin to differentiate bacteria. Gram-positive bacteria stain dark purple; Gram-negative ones stain pink/red.

Granulation Tissue Fragile, reddish, grainy tissue made up of capillaries and fibroblasts that appears with the healing of an injury.

Granule An inclusion that is not bounded by a membrane and contains compacted substances that do not dissolve in the cytoplasm.

Granulocyte A leukocyte (basophil, mast cell, Eosinophil, neutrophil) with granular cytoplasm and irregularly shaped, lobed nuclei.

Granuloma In a chronic inflammation, a collection of epithelial cells, macrophages, lymphocytes, and collagen fibers.

Granuloma Inguinale A sexually transmitted disease caused by *Calymmatobacterium Granulomatis. Known also as Donovanosis.*

Granulomatous Hypersensitivity Cell-mediated hypersensitivity reaction that occurs when macrophages have engulfed pathogens but have failed to kill them.

Granulomatous Inflammation A special kind of chronic inflammation characterized by the presence of granulomas.

Granzyme A Cytotoxin produced by cytotoxic T cells that help kill infected host cells.

Griseofulvin An antifungal agent that interferes with fungal growth.

Ground Itch Bacterial infection of sites of penetration by hookworms.

Group Translocation An active transport process in bacteria that chemically modifies substance so it cannot diffuse out of the cell.

Growth Curve The different growth periods of a bacterial or phage population.

Growth Factors Peptide mediators that influence the growth and/or differentiation of cells; they differ from growth hormones in being produced by many tissues and in acting locally.

Gumma A granulomatous inflammation, symptomatic of syphilis, that destroys tissue.

Gut-Associated Lymphatic Tissue (GALT) Collective name for the tissues of lymphoid nodules, especially those in the digestive, respiratory, and urogenital tracts; main site of antibody production.

Half-Life For a chemical reaction, the time at which half the substrate has been consumed and turned into product. The term can also refer to the analogous point in other processes, such as the radioactive decay of an isotope. *Known also as Half-Time.*

Halobacteria One of the groups of the Archaeobacteria that live in very concentrated salt environments.

Halophile A salt-loving organism that requires moderate to large concentrations of salt.

Hanging Drop A special type of wet mount often used with dark-field illumination to study motility of organisms.

Hansen's Disease The preferred name for leprosy, caused by *Mycobacterium leprae*, it exhibits various clinical forms ranging from tuberculoid to lepromatous.

Hantavirus Pulmonary Syndrome (HPS) The "Sin Nombre" hantavirus responsible for severe respiratory illness.

Haploid (1) A eukaryotic cell that contains a single, unpaired set of chromosomes. (2) A molecule that is too small to stimulate an immune response by itself but can do so when coupled to a larger, immunogenic carrier molecule (usually a protein).

Hapten A small molecule that can act as an antigenic determinant when combined with a larger molecule.

Haworth Projection A conventional planar representation of a cyclized monosaccharide molecule. The hydroxyls that are represented to the right of the chain in a Fischer projection are shown below the plane in a Haworth projection.

Heat Fixation Technique in which air-dried smears are passed through an open flame so that organisms are killed, adhere better to the side, and take up dye more easily.

Heat-Shock Proteins A group of Chaperonins that accumulate in a cell after it has been subjected to a sudden temperature jump or other stress. They are thought to help deal with the accumulation of improperly folded or assembled proteins in stressed cells.

Heavy Chain (H chain) Larger of the two identical pairs of chains comprising immunoglobulin molecules. Helicases Enzymes that catalyze the unwinding of duplex nucleic acids.

Helix-Loop-Helix Motif A binding motif that is found in calmodulin and some other calcium-binding proteins as well as in some DNA-binding proteins, It consists of two α helix segments connected by a loop.

Helix-Turn-Helix Motif A DNA-binding motif that is responsible for sequence-specific DNA binding in many transcription factors. It consist of two α helix segments connected by a β turn; one of the helices occupies the DNA major groove and makes specific base contacts.

Helminth A worm, with bilateral symmetry; includes the roundworms and flatworms.

Helper T Cell (T_H) (1) Lymphocytes that stimulate other immune cells, such as B cells and macrophages. (2) T lymphocytes whose role is to recognize antigens and help other defensive cells to mount an immune response. They help activate antigen-stimulated B cells (resulting in production of specific antibodies) and/or antigen-stimulated cytotoxic T cells (resulting in attack on antigenic cells), and they also produce immune mediators that stimulate nonspecific defense responses.

Hemagglutination Agglutination (clumping) of red blood cells, used in blood typing.

Hemagglutination Inhibition Test Serologic test used to diagnose measles, influenza, and other viral diseases, based on the ability of antibodies to viruses to prevent viral hemagglutination.

Heme A molecule consisting of a porphyrin ring (either protoporphyrin IX or a derivative) with a central complexed iron; it serves as a prosthetic group in proteins such as myoglobin, hemoglobin, and cytochromes.

Hemimethylated With respect to DNA, refers to the condition in which one strand of the duplex is methylated and the other is not. Newly replicated DNA is hemimethylated; normally a methylase enzyme then methylates appropriate bases in the new strand.

Hemoglobin The oxygen-binding compound found in erythocytes.

Hemolysin An enzyme that lyses red blood cells.

Hemolysis The lysis of red blood cells.

Hemolytic Disease of the Newborn Disease in which a baby is born with enlarged liver and spleen caused by efforts of these organs to destroy red blood cells damaged by maternal antibodies; mother is Rh-negative and baby is Rh-positive. *Known also as Erythroblastosis Fetalis.*

Hemorrhagic Uremic Syndrome (HUS) Infection with 0157-H7 strain of *Escherichia coli* causing kidney damage and bleeding in the urinary tract.

Hepadnavirus A small, enveloped DNA virus with circular DNA, one such virus causes hepatitis B.

Hepatitis An inflammation of the liver, usually caused by viruses but sometimes by an amoeba or various toxic chemicals.

Hepatitis A (formerly called infectious hepatitis) Common form of viral hepatitis caused by a single-stranded RNS virus transmitted by the fecal-oral route.

Hepatitis B (formerly called serum hepatitis) Type of hepatitis caused by a double-stranded DNA virus usually transmitted in blood or semen.

Hepatitis C (formerly called non-A, non-B hepatitis) Type of hepatitis distinguished by a high level of the liver enzyme alanine transferase, usually mild or inapparent infection but can be severe in compromised individuals.

Hepatitis D Severe type of hepatitis caused by presence of both hepatitis D and hepatitis B viruses, hepatitis D virus is an incomplete virus and cannot replicate without presence of hepatitis B virus as a helper. *Known also as Delta Hapatitis.*

Hepatitis E Type of hepatitis transmitted through fecally contaminated water supplies.

Hepatovirus One of three major groups of picornaviruses that can infect nerves and is responsible for causing hepatitis A.

Herd Immunity The proportion of individuals in a population who are immune to a particular disease. *Known also as Group Immunity.*

Heredity Having both male and female reproductive systems in one organism.

Herpes Gladiatorium Herpesvirus infection that occurs in skin injuries of wrestlers, transmitted by contact or on mats.

Herpes Labialis Fever blisters (cold sores) on lips.

Herpes Meningoencephalitis A serious disease caused by herpesvirus that can cause permanent neurological damage or death and that sometimes follows a generalized herpes infection or ascends from the trigeminal ganglion.

Herpes Pneumonia A rare form of herpes infection seen in burn patients, AIDS patients, and alcoholics.

Herpes Simplex Virus Type 1 (HSV-1) A virus that most frequently causes fever blisters (cold sores) and other lesions of the oral cavity, and less often causes genital lesions.

Herpes Simplex Virus Type 2 (HSV-2) A virus that typically causes genital herpes, but which can also cause oral lesions. *Known also as Herpes Hominis Virus.*

Herpesvirus A relatively large, enveloped DNA virus that can remain latent in host cells for long periods of time.

Heterogencity The ability of the immune system to produce many different kinds of antibodies, each specific for a different antigenic determinant.

Heterotroph An organism that uses compounds to produce biomolecules.

Heterotrophs Organisms that cannot synthesize their organic compounds entirely from inorganic precursors but most consume at least some organic compounds made by other organisms. In particular, these organisms cannot use CO_2 as a carbon source. compare ▶ Autotrophs.

Heterotrophy "Other-feeding" the use of carbon atoms from organic compounds for the synthesis of biomolecules.

Heterozygous In a diploid organism, the possession of two different alleles for a given gene (as opposed to two copies of the same allele). Compare ▶ Homozygous.

Hib Vaccine Vaccine against *Haemophilus influenzae b.*

High-Density Lipoprotein (HDL) A type of lipoprotein particle that functions mainly to scavenge excess cholesterol from tissue cells and transport it to the liver, where it can be excreted in the form of bile acids.

High-Energy Bond A chemical bond that releases energy when hydrolyzed; the energy can be used to transfer the hydrolyzed product to another compound.

High Frequency of Recombination (Hfr) Strain A strain of F^+ bacteria in which the F plasmid is incorporated into the bacterial chromosome.

High-Temperature Short-Time (HTST) Pasteurization Process in which milk is heated to 71.6°C for at least 15 seconds. *Known also as Flash Pasteurization.*

Hill Coefficient (n_H) A coefficient that indicates the degree of cooperativity of a cooperative transition; it is the maximum slope of a Hill plot of the transition.

Histamine Amine release by basophils and tissues in allergic reactions.

Histocompatibility Antigen An antigen found in the membranes of all human cells that is unique in all individuals except identical twins.

Histone A protein that contributes directly to the structure of eukaryotic chromosomes.

Histones The proteins that participate in forming the nucleosomal structure of chromatin. Four of the five kinds of histones make up the core particle of the nucleosome; the fifty is associated with the linker DNA between nucleosomes. All histones are small, very basic proteins.

Histoplasmosis Fungal respiratory disease endemic to the central and eastern United States, caused by the soil fungus *Histoplasma capsulatium. Known also as Darling's Disease.*

Holding Method See ▶ Low-Temperature Long-Time Pasteurization.

Holliday Junction An intermediate during homologous recombination; a four-armed structure in which each of the participating DNA duplexes has exchanged one strand with the other duplex.

Holoenzyme A functional enzyme consisting of an apoenzyme and a coenzyme or cofactor.

Homeo Box A common sequence element of about 180 base pairs that is found in homeotic genes. It codes for a sequence-specific DNA-binding element of the helix-loop-helix class. See also ▶ Homeotic Genes.

Homeotic Genes Genes that contain homeo box elements and typically are involved in controlling the pattern of organismal development. Homeotic mutations, which scramble portions of this pattern, affect homeotic genes. The nuclear DNA-binding proteins encoded by these genes presumably serve as transcriptional regulators for the coordinated expression of groups of genes. See also ▶ Homeo Box.

Homolactic Acid Fermentation A pathway in which pyruvic acid is directly converted to lactic acid using electrons from reduced NAS (NADH).

Homologous Recombination Genetic recombination that requires extensive sequence homology between the recombining DNA molecules. Meiotic recombination by crossing over in eukaryotes is an example.

Homopolymer (biological) A polymer that is made of only one kind of monomer. Starch, made only of glucosyl units, is an example. Polymers that include more than one kind of monomer, like polypeptides and nucleic acids, are called heteropolymers.

Homozygous In a diploid organism, the possession of two identical alleles for a given gene. Compare ▶ Heterozygous.

Hookworm A disease caused by two species of small roundworms, *Ancylostoma duodenale* and *Necator americanus*, whose larvae burrow through skin and feet, enter the blood vessels, and penetrate lung and intestinal tissue.

Horizontal Transmission Direct contact transmission of disease in which pathogens are usually passed by handshaking, kissing, contact with sores, or sexual contact.

Hormone A substance that is synthesized and secreted by specialized cells and carried via the circulation to target cells, where it elicits specific changes in the metabolic behavior of the cell by interacting with a hormone-specific receptor.

Hormone-Responsive Element A DNA site that binds an intracellular hormone-receptor complex; binding of the complex to a hormone-responsive element affects the transcription of specific genes.

Host Any organism that harbors another organism.

Host-Induced Restriction and Modification A genetic system found in bacteria whereby a genetic element (often a plasmid) encodes both an enzyme for the methylation of DNA at a specific base sequence and an Endonuclease that cleaves unmethylated DNA at that sequence. The system thus *restricts* the DNA that can survive in the cell to DNA that is *modified* by methylation at the correct sequences.

Host Range The different types of organisms that a microbe can infect.

Host Specificity The range of different ghosts in which a parasite can mature.

Human Immunodeficiency Virus (HIV) One of the retroviruses that is responsible for AIDS.

Human Leukocyte Antigen (HLA) A lymphocyte antigen used in laboratory tests to determine compatibility of donor and recipient tissues for transplants.

Human Papillomavirus (HPV) Virus that attacks skin and mucous membranes, causing papillomas or warts.

Humoral Immune Response A response to foreign antigens carried out by antibodies circulating in the blood.

Humoral Immunity The immune response most effective in defending the body against bacteria, bacterial toxins, and viruses that have not entered cells.

Humus The nonliving organic components of soil.

Hyaluronidase A bacterially produced enzyme that digests hyaluronic acid, which helps hold the cells of certain tissues together, thereby making tissues more accessible to microbes. *Known also as Spreading Factor.*

Hybridoma A hybrid cell resulting from the fusion of a cancer cell with another cell, usually an antibody-producing white blood cell.

Hybridomas Cultured cell lines that are made by fusing antibody-producing B lymphocytes with cells derived from a mouse myeloma (a type of lymphocyte cancer). Like B cells, they produce specific antibodies, and like myeloma cells, they can proliferate indefinitely in culture.

Hydatid Cyst An enlarged cyst containing many tapeworm heads.

Hydrogen Bond An attractive interaction between the hydrogen atom of a donor group, such as —OH or ==NH, and a pair of nonbonding electrons on an acceptor group, such as O==C. The donor group atom that carries the hydrogen must be fairly electronegative for the attraction to be significant.

Hydrologic Cycle See ▶ Water Cycle.

Hydrolysis A chemical reaction that produces simpler products from more complex organic molecules.

Hydrophilic Refers to the ability of an atom or a molecule to engage in attractive interactions with water molecules. Substances that are ionic or can engage in hydrogen bonding are hydrophilic. Hydrophilic substances are either soluble in water or, at least, wettable. Compare ▶ Hydrophobic.

Hydrophobic The molecular property of being unable to engage in attractive interactions with water molecules. Hydrophobic substances are nonionic and nonpolar; they are nonwettable and do not readily dissolve in water. Compare ▶ Hydrophilic.

Hydrophobic Effect With respect to globular proteins, the stabilization of tertiary structure that results from the packing of hydrophobic side chains in the interior of the protein.

Hydrostatic Pressure Pressure exerted by standing water.

Hyperimmune Serum A preparation of immune serum globulins having high titers of specific kinds of antibodies. *Known also as Convalescent Serum.*

Hyperparasitism The phenomenon of a parasite itself having parasites.

Hypersensitivity Disorder in which the immune system reacts inappropriately, usually by responding to an antigen it normally ignores. *Known also as an Allergy.*

Hypertonic Solution A solution containing a concentration of dissolved material greater than that within a cell.

Hypha (plural *hyphae*) A long, threadlike structure of cells in fungi or actinomycetes.

Hypochromism With respect to DNA, a reduction in the absorbance of ultraviolet light of wavelength of about 260 nm that accompanies the transition from random-coil denatured strands to a double-strand helix. It can be used to track the process of Denaturation or renaturation.

Hypothesis A tentative explanation for an observed condition or event.

Hypotonic Solution A solution containing a concentration of dissolved material lower than that within a cell.

IgA Class of antibody found in the blood and secretions.

IgD Class of antibody found on the surface of B cells and rarely secreted.

IgE Class of antibody that binds to receptors on basophils in the blood or mast cells in the tissues, responsible for allergic or immediate (type I) hypersensitivity reactions.

IgG The main class of antibodies found in the blood; produced in largest quantities during secondary response.

IgM The first class of antibody secreted into the blood during the early stages of a primary immune response (a rosette of five immunoglobulin molecules) or found on the surface of B cells (a single immunoglobulin molecule).

Illness Phase In an infectious disease, the period during which the individual experiences the typical signs and symptoms of the disease.

Imidazole An antifungal agent that disrupts fungal plasma membranes.

Immediate (Type I) Hypersensitivity Response to a foreign substance (allergen) resulting from prior exposure to the allergen. *Known also as Anaphylactic Hypersensitivity.*

Immersion Oil Substance used to avoid refraction at a glass-air interface when examining objects through a microscope.

Immune Complex An antigen-antibody complex that is normally eliminated by phagocytic cells.

Immune Complex Disorder A disorder caused by antigen-antibody complexes that precipitate in the blood and injure tissues; elicited by antigens in vaccines, on microorganisms, or on a person's own cells. *Known also as Immune Complex (Type III) Hypersensitivity.*

Immune Cytolysis Process in which the membrane attack complex of complement produces lesions on cell membranes through which the contents of the bacterial cells leak out.

Immune Serum Globulin A pooled sample of antibody-containing fractions of serum from many individuals. *Known also as Gamma Globulin.*

Immune System Body system that provides the host organism with specific immunity to infectious agents.

Immunity The ability of an organism to defend itself against infectious agents.

Immunocompromised Referring to an individual whose immune defenses are weakened due to fighting another infectious disease, or because of an immunodeficiency disease or an immunosuppressive agent.

Immunodeficiency Inborn or acquired defects in lymphocytes (B or T cells).

Immunodeficiency Disease A disease of impaired immunity caused by lack of lymphocytes, defective lymphocytes, or destructive lymphocytes.

Immunodiffusion Test serologic test similar to the precipitin test but carried out in agar gel medium.

Immunoelectrophoresis Serologic test in which antigens are first separated by gel electrophoresis and then allowed to react with antibody placed in a trough in the gel.

Immunofluorescence Referring to the use of antibodies to which a fluorescent substance is bound and used to detect antigens, other antibodies, or complement within tissue.

Immunogen See ▶ Antigen.

Immunogenic Something that is a potent stimulator of antibody production and defense cell activity.

Immunoglobulin (Ig) The class of protective proteins produced by the immune system in response to a particular epitope. *Known also as an Antibody.* See ▶ Antibodies.

Immunological Disorder Disorder that results from an inappropriate or inadequate immune system.

Immunological Memory The ability of the immune system to recognize substances it has previously encountered.

Immunology The study of specific immunity and how the immune system responds to specific infectious agents.

Immunosuppression Minimizing of immune reactions using radiation or cytotoxic drugs.

Impetigo A highly contagious pyoderma caused by staphylococci, streptococci, or both.

Importins A class of proteins involved in importing molecules into the nucleus. See ▶ Exportins.

Inapparent Infection An infection that fails to produce symptoms, either because too few organisms are present or because host defenses effectively combat the pathogens. *Known also as Subclinical Infection.*

Inborn Errors of Metabolism Human mutations that result in specific derangements of intermediary metabolism. Usually the problem is an enzyme that is inactive, overactive, too scarce, or too abundant; symptoms may result from the insufficient production of a necessary metabolite and/or from the accumulation of another metabolite to toxic levels.

Incidence Rate The number of new cases of a particular disease per 100,000 population seen in a specific period of time.

Inclusion A granule or vesicle found in the cytoplasm of a bacterial cell.

Inclusion Blennorrhea A mild chlamydial infection of the eyes in infants.

Inclusion Body (1) An aggregation of reticulate bodies within chlamydias. (2) A form of Cytopathic effect consisting of viral components, masses of viruses, or remnants of viruses.

Inclusion Conjunctivitis A chlamydial infection that can result from self-inoculation with *Chlamydia trachomatis.*

Incubation Period In the stages of infectious disease, the time between infection and the appearance of signs and symptoms.

Index Case The first case of a disease to be identified.

Index of Refraction A measure of the amount that light rays bend when passing from one medium to another.

Indicator organism An organism such as *Escherichia coli* whose presence indicates the contamination of water by fecal matter.

Indigenous Organism An organism native to a given environment. *Known also as a Native Organism.*

Indirect Contact Transmission Transmission of disease through fomites.

Indirect Fecal-Oral Transmission Transmission of disease in which pathogens from feces of one organism infect another organism.

Induced Dipole A molecule has an induced dipole if an external electric field induces an asymmetric distribution of charge within it.

Induced Fit Model A model for how enzymes interact with substrates to achieve catalysis. According to this model, the empty active site of the enzyme only roughly fits the substrate(s), and the entry of substrate causes the enzyme to change its shape so as to both tighten the fit and causes the substrate to adopt an intermediate state that resembles the transition state of the uncatalyzed reaction. This is currently the dominant model for enzymatic catalysis.

Induced Mutation A mutation produced by agents called mutagen that increase the mutation rate.

Inducer A substance that binds to and inactivates a repressor protein.

Inducible Enzyme An enzyme coded for by a gene that is sometimes active and sometimes inactive.

Induction (1) The stimulation of a temperature phase (prophage) to excise itself from the host chromosome and initiate a lytic cycle of replication. (2) In cellular metabolism, the synthesis of a particular protein in response to a signal; for example, the synthesis of an enzyme in response to the appearance of its substrate.

Induration A raised, hard, red region on the skin resulting from tuberculin hypersensitivity.

Industrial Microbiology Branch of microbiology concerned with the use of microorganisms to assist in the manufacture of useful products or disposal of waste products.

Infant Botulism Form of botulism in infants associated with ingestion of honey. *Known also as "Floppy Baby" Syndrome.*

Infection The multiplication of a parasite organism, usually microscopic, within or upon the host's body.

Infectious Disease Disease caused by infectious agents (bacteria viruses, fungi, protozoa, and helminths).

Infectious Hepatitis See ▶ Hepatitis A.

Infectious Mononucleosis An acute disease that affects many systems, caused by the Epstein-Barr virus.

Infestation The presence of heminths (worms) or anthropods in or on a living host.

Inflammation The body's defensive response to tissue damage caused by microbial infection.

Influenza Viral respiratory infection caused by orthomyxoviruses that appears as epidemics.

Initiating Segment That part of the F plasmid that is transferred to the recipient cell in conjugation with an Hfr bacterium.

Innate Immunity Immunity to infection that exists in an organism because of genetically determined characteristics.

Insect An anthropod with three body regions, three pairs of legs, and highly specialized mouthparts.

Insertion The addition of one or more bases to DNA, usually producing a frameshift mutation.

In Situ Hybridization A technique for finding the chromosomal location of a particular DNA sequence by probing the chromosomes with a radiolabeled sequence that will hybridize with the sequence in question. The location of the probe is then visualized with radioautography.

Intercalation With respect to DNA, refers to the fitting (intercalation) of a small molecule between adjacent bases in a DNA helix.

Interferon (1) A small protein often released from virus-infected cells that binds to adjacent uninfected cells, causing them to produce antiviral proteins that interfere with viral replication. (2) All of the reactions in an organism that are concerned with storing and generating metabolic energy and with the biosynthesis of low-molecular-weight compounds and energy-storage compounds. It does not include nucleic acid and protein synthesis.

Interleukin A cytokine produced by leukocytes.

Intermediate Host An organism that harbors a sexually immature stage of a parasite.

Internal Energy (*E*) The energy contained in a system. For the purposes of biochemistry, the term encompasses all the types of energy that might be changed by chemical or nonnuclear physical processes, including the kinetic energy of motion and vibration of atoms and molecules and the energy stored in bonds and noncovalent interactions.

Intoxication The ingestion of a microbial toxin that leads to a disease.

Intron (1) Region of a gene (or mRNA) in eukaryotic cells that does not code for a protein. *Known also as the Intervening Region.* (2) A region in the coding sequence of a gene that is not translated into protein. Introns are common in eukaryotic genes but are rarely found in prokaryotes. They are excised from the RNA transcript before translation. Compare ► Exon.

Invasiveness The ability of a microorganism to take up residence in a host.

Invasive Stage Disease spreads into body from site of energy causing symptoms to appear.

Ion An electrically charged atom produced when an atom gains or loses one or more electrons.

Ion-Exchange Resins Polycationic or polyanionic polymers that are used in ion-exchange column chromatography to separate substances on the basis of electrical charge.

Ionic Bond A chemical bond between atoms resulting from attraction of ions with opposite charges.

Ionic Strength (I) A quantity that reflects the total concentration of ions in a solution and the stoichiometric charge (charge per atom or molecule) of each ion. It is defined as $I = {}^1\!/_2 \sum M_i Z_i^2$ where M_i and Z_i are respectively the molarity and stoichiometeric charge of ion i. It is used, for example, in calculating the effective radius of a counterion atmosphere.

Ion Pore A pore in a cellular membrane through which ions can diffuse. It is formed by a transmembrane protein and can discriminate among ions to some degree on the basis of size and charge. Many ions pores are gated, meaning that they can open and close in response to signals.

Iris Diaphragm Adjustable device in a microscope that controls the amount of light passing through the specimen.

Ischemia Reduced blood flow to tissues with oxygen and nutrient deficiency and waste accululation.

Isoelectric Focusing A version of gel electrophoresis that allows ampholytes to be separated almost purely on the basis of their isoelectric points. The ampholytes are added to a gel that contains a pH gradient and are subjected to an electric field, Each Ampholyte migrates until it reaches the pH that represents its isoelectric point, at which point it ceases to have a net electric charge and therefore comes to a halt and accumulates. See also ► Gel Electrophoresis, Isoelectric Point.

Isoelectric Point (pI) The pH at which the net charge on an ampholyte is, on average, zero.

Isoenzymes Different but related forms of an enzyme that catalyze the same reaction. Often differ in only a few amino acid substitutions. *Known also as Isozymes.*

Isograft A graft of tissue between genetically identical individuals.

Isolation Situation in which a patient with a communicable disease is prevented from contact with the general population.

Isomer An alternative form of a molecule having the same molecular formula but different structure.

Isomorphorous Replacement The replacement of one atom in a macromolecule with a heavy metal atom in such a way that the structure of the macromolecule does not change. It is used in the determination of molecular structure by x-ray crystal diffraction.

Isoniazid An Antimetabolite that is bacteriostatic against the tuberculosis-causing mycobacterium.

Isotonic Fluid containing the same concentration of dissolved materials as is in a cell; causes no change in cell volume.

Isotope An atom of a particular element that contains a different number of neutrons.

Isozymes See ► Isoenzymes.

Joule (J) A unit for energy or work, defined as the work done by a force of 1 newton when its point of application moves 1 meter in the direction of the force. It is the unit of energy used in the Système Internationale (SI).

Kala Azar Visceral leishmaniasis caused by *Leishmania donovani.*

Kaposi's Sarcoma A malignancy often found in AIDS patients in which blood vessels grow into tangled masses that are filled with blood and easily ruptured.

Karyogamy Process by which nuclei fuse to produce a diploid cell.

Keratin A waterproofing protein found in epidermal cells.

α-Keratins A class of keratins that are the major proteins of hair. They consist of long α-helical polypeptides, which are wound around each other to form triplet helices.

Keratitis An inflammation of the cornea.

Keratoconjunctivitis Condition in which vesicles appear in the cornea and eyelids.

Ketone Bodies The substances acetoacetate, β-hydroxybutyrate, and acetone, which are produced from excess acetyl-CoA in the liver when the rate of fatty acid β-oxidation in liver mitochondria exceeds the rate at which acetyl-CoA is used for energy generation or fatty acid synthesis.

Ketose A monosaccharide in which the carbonyl group occurs within the chain and hence represents a ketone group. Compare ► Aldose.

Kidney One of a pair of organs responsible for the formation of urine.

Kirby-Bauer Method See ► Disk Diffusion Method.

Koch's Postulates Four postulated formulated by Robert Koch in the 19th century, used to prove that a particular organism causes a particular disease.

Koplik's Spots Red spots with central bluish specks that appear on the upper lip mucosa in early stages of measles.

Krebs Cycle A sequence of enzyme-catalyzed chemical reactions that metabolizes 2-carbon units called acetyl groups to CO_2 and H_2O. *Known also as Tricarboxylic Acid Cycle and the Citric Acid Cycle.*

Kupffer Cells Phagocytic cells that remove foreign matter from the blood as it passes through sinusoids.

Kura Transmissible spongiform encephalopathy disease of the human brain, caused by prions, associated with cannibalism and tissue/organ transplants.

Lacrimal Gland Tear-producing gland of the eye.

Lactobacilli Type of regular, nonsporing, Gram-positive rod found in many foods; used in production of cheeses, yogurt, sourdough, and other fermented foods.

Lagging Strand During DNA replication, the strand that is synthesized in the opposite direction to the direction of movement of the replication fork; it is synthesized as a series of fragments that are subsequently joined. Compare ► Leading Strand.

Lag Phase First of four major phases of the bacterial growth curve, in which organisms grow in size but do not increase in number.

Large Intestine The lower area of the intestine that absorbs water and converts undigested food into feces.

Laryngeal Papilloma Benign growth caused by herpesviruses that can be dangerous is such papillomas block the airway, infants are often infected during birth by mothers having genital warts.

Laryngitis An infection of the larynx, often with loss of voice.

Larynx The voicebox.

Lassa Fever Hemorrhagic fever, caused by arenaviruses, that begins with pharyngeal lesions and proceeds to severe liver damage.

Latency The ability of a virus to remain the host cells for long periods of time while retaining the ability to replicate.

Latent Disease A disease characterized by periods of inactivity either before symptoms appear or between attacks.

Latent Period Period of a bacteriophage growth curve that spans the time from penetration through biosynthesis.

Latent Viral Infection An infection typical of herpesviruses in which an infection in childhood that is brought under control later in life is reactivated.

Lateral Gene Transfer Genes pass from one organism to another within the same generation.

Leader Sequence For an mRNA, the nontranslated sequence at the 5′ end of the molecule that precedes the initiation codon. For a protein, a short N-terminal hydrophobic sequence that causes the protein to be translocated into or through a cellular membrane. *Known also as Signal Sequence.*

Leading Strand During DNA replication, the strand that is synthesized in the same direction as the direction of movement of the replication fork; it is synthesized continuously rather than in fragments. Compare ▶ Lagging Strand.

Leavening Agent An agent, such as yeast, that produces gas to make dough rise.

Legionellas The causative bacterial agent in Legionnaires' disease, *Legionella pneumophila.*

Leishmaniasis A parasitic systemic disease caused by three species of protozoa of the genus *Leishmania* and transmitted by sandflies.

Leproma An enlarged, disfiguring skin lesion that occurs in the lepromatous form of Hansen's disease.

Lepromatous Referring to the nodular form of Hansen's disease (leprosy) in which a granulomatous response causes enlarged, disfiguring skin lesions called leptromas.

Lepromin Skin Test Test used to detect Hansen's disease (leprosy), similar to the tuberculin test.

Leprosy See ▶ Hansen's Disease.

Leptospirosis A zoonosis caused by the spirochete *Leptospira interrogans*, which enters the body through mucous membranes or skin abrasions.

Leukocidin An exotoxin produced by many bacteria, including the streptococci and staphylococci, that kills phagocytes.

Leukocyte A white blood cell.

Leukocyte-Endogenous Mediator A substance that helps raise the body temperature while decreasing iron absorption (increasing iron storage).

Leukocytosis An increase in the number of white blood cells (leukocytes) circulating in the blood.

Leukostatin An exotoxin that interferes with the ability of leukocytes to engulf microorganisms that release the toxin.

Leukotriene A reaction mediator released from mast cells after Degranulation that causes prolonged airway construction, dilation, and increased permeability of capillaries, increased thick mucous secretion, and stimulation of nerve endings that cause pain and itching.

Leukotrienes A family of molecules that are synthesized from arachidonic acid by the lipoxygenase pathway and function as local hormones, primarily to promote inflammatory and allergic reactions (such as the bronchial construction of asthma).

L Forms Irregularly shaped naturally occurring bacteria with defective cell walls.

Library With respect to molecular genetics, a large collection of random cloned DNA fragments from a given organism, sometimes representing the entire nuclear genome.

Ligand In general, a small molecule that binds specifically to a larger one – for example, a hormone that binds to a receptor, the term can also be used to mean a chemical species that forms a coordination complex with a central atom, which is usually a metal atom.

Ligase An enzyme that joins together DNA segments.

Light Chain (L chain) Smaller of the two identical pairs of chains constituting immunoglobulin molecules.

Light Microscopy The use of any type of microscope that uses visible light to make specimens observable.

Light Reaction The part of photosynthesis in which light energy is used to excite electrons from chlorophyll, which are then used to generate ATP and NADPH.

Light Reactions The photosynthetic subprocesses that depend *directly* on light energy; specifically, the synthesis of ATP by Photophosphorylation and the reduction of $NADP^+$ to NADPH via the oxidation of water. Compare ▶ Dark Reactions.

Light Repair Repair of DNA dimers by a light-activated enzyme. *Known also as Photoreactivation.*

Lineweaver-Burk Plot A plot that allows one to derive the rate constant k_{cat} and the Michaelis constant K_M for an enzyme-catalyzed reaction. It is constructed by measuring the initial reaction rate V at various substrate concentration [S] and plotting the values on a graph of $1/V$ versus $1/[S]$.

Linkage Map A map showing the arrangement of genes on a chromosome; it is constructed by measuring the frequency of recombination between pairs of genes.

Linking Number (*L*) The total number of times the two strands of a closed, circular DNA helix cross each other by means of either twist or writhe; this equals the number of times the two strands are interlinked. It reflects both the winding of the native DNA helix and the presence of any supercoiling. See also ▶ Twist and ▶ Writhe.

Lipid One of a group of complex, water-insoluble compounds.

Lipid A Toxic substance found in the cell wall of a Gram-negative bacteria.

Lipid Bilayer A membrane structure that can be formed by amphipathic molecules in an aqueous environment; it consists of two back-to-back layers of molecules, in each of which the polar head groups face the water and the nonpolar tails face the center of the membrane. The fabric of cellular membranes is a lipid bilayer.

Lipids A chemically diverse group of biological compounds that are classified together on the basis of their generally apolar structure and resulting poor solubility in water.

Lipopolysaccharide Part of the outer layer of the cell wall in Gram-negative bacteria. *Known also as an Endoxtoxin.*

Lipoproteins Any lipid-protein conjugate. Specifically refers to lipid-protein associations that transport lipids in the circulation, Each consists of a core of hydrophobic lipids surrounded by a skin of amphipathic lipids with embedded apolipoproteins. Different kinds of lipoproteins play different roles in lipid transport.

Listeriosis A type of meningitis caused by *Listeria monocytogenes* that is especially threatening to those with impaired immune systems.

Loaiasis Tropical eye disease caused by the filarial worm *Loa loa.*

Lobar Pneumonia Type of pneumonia that affects one or more of the five major lobes of the lungs.

Local Infection An infection confirmed to a specific area of the body.

Localized Anaphylaxis An immediate (Type I) hypersensitivity restricted to only some tissue/organs resulting in e.g., reddening of the skin, watery eyes, hives, etc.

Locus The location of a gene on a chromosome.

Logarithmic Rate See ► Exponential Rate.

Log Phase Second of four major phases of the bacterial growth curve, in which cells divide at an exponential or logarithmic rate.

Long Terminal Repeats (LTRs) A pair of direct repeats several hundred base pairs long that are found at either end of a retroviral genome. They are involved in integration into the host genome and in viral gene expression.

Lophotrichous Having two or more flagella at one or both ends of a bacterial cell.

Low-Angle Neutron Scattering A set of techniques that can be used to find the size of a particle in solution or to find the size or spacing of internal regions that can be distinguished by different neutron scattering power, such as the protein and nucleic acid components of a nucleoprotein particle or labeled proteins within a multisubunit complex.

Low-Density Lipoprotein (LDL) A type of lipoprotein particle that functions mainly to distribute cholesterol from the liver to other tissues. Its protein component consists of a single molecule of apoprotein B-100.

Lower Respiratory Tract Thin-walled bronchioles and alveoli where gas exchange occurs.

Low-Temperature Long-Time (LTLT) Pasteurization Procedure in which milk is heated to 62.0°C for at least 30 minutes. *Known also as Holding Method.*

Luminescence Process in which absorbed light rays are reemitted at longer wavelengths.

Lyme Disease Disease caused by *Borrelia burgdorferi*, carried by the deer tick.

Lymph The excess fluid and plasma proteins lost through capillary walls that is found in the lymphatic capillaries.

Lymphangitis Symptom of septicemia in which red streaks due to inflamed lymphatics appear beneath the skin.

Lymphatic System Body system, closely associated with the cardiovascular system, that transports lymph in lymphatic vessels through body tissues and organs, performs important functions in host defenses and specific immunity.

Lymphatic Vessel Vessel that returns lymph to the blood circulatory system.

Lymph Node An encapsulated globular structure located along the routes of the lymphatic vessels that helps clear the lymph of microorganisms.

Lymphocyte A leukocyte (white blood cell) found in large numbers in lymphoid tissues that contribute to specific immunity.

Lymphogranuloma Venereum A sexually transmitted disease, caused by *Chlamydia trachomatis*, that attacks the lymphatic system.

Lymphoid Nodule A small, unencapsulated aggregation of lymphatic tissue that develops in many tissues, especially the digestive, respiratory, and unogenital tracts, collectively called gut-associated lymphatic tissue (GALT); they are the body's main sites of antibody production.

Lymphoid Stem Cell A cell in the bone marrow from which lymphocytes develop.

Lyophokine A cytokine secreted by T cells when they encounter an antigen.

Lymphilization The drying of a material from the frozen state, freeze-drying.

Lysis The destruction of a cell by the rupture of a cell or plasma membrane, resulting in the loss of cytoplasm.

Lysogen The combination of a bacterium and a temperate phage.

Lysogenic Pertaining to a bacterial cell in the state of lysogeny.

Lysogenic Conversion The ability of a prophage to prevent additional infections of the same cell by the same type of phage; also the conversion of a non-toxin–producing bacterium into a toxin-producing one by a temperate phage.

Lysogeny The ability of temperature bacteriophages to persist in a bacterium by the integration of the viral DNA into the host chromosome and without the replication of new viruses or cell lysis.

Lysosome A small membrane-bound organelle in animal cells that contains digestive enzymes.

Lytic Cycle The sequence of events in which a bacteriophage infects a bacterial cell, replicates, and eventually causes lysis of the cell.

Lytic Phage See ▶ Virulent Phage.

Macrolide 2w?>A large-ring compound, such as erythromycin, that is antibacterial by affecting protein synthesis.

Macrophage Ravenously phagocytic leukocytes found in tissues.

Mad Cow Disease Transmissible spongiform encephalopathy disease of the brain of cattle, caused by prions.

Madura Foot Tropical disease caused by a variety of soil organisms (fungi and actinomycetes) that often enter the skin through bare feet. *Known also as Maduromycosis.*

Maduromycosis See ▶ Madura Foot.

Major Histocompatibility Complex (MHC) A group of cell surface proteins that are essential to immune recognition reactions.

Malaria A severe parasitic disease caused by several species of the protozoan *Plasmodium* and transmitted by mosquitoes.

Male Reproductive System The host system consisting of the testes, ducts, specific glands, and the penis.

Malignant Relating to a tumor that is cancerous.

Malta Fever See ▶ Brucellosis.

Malted Referring to cereal grains that are partially germinated to increase the concentration of starch-digesting enzymes.

Mammary Gland A modified sweat gland that produces milk and ducts that carry milk to the nipple.

Mash Malted grain that is crushed and mixed with hot water.

Mass Spectrometry A method for determining the molecular mass from the velocity of motion of ions in a vacuum.

Mast Cell A leukocyte that releases histamine during an allergic response.

Mastigophoran A flagellate protozoan such as *Giardia.*

Mastoid Area Portion of the temporal bone prominent behind the ear opening.

Matrix Fluid-filled inner portion of a mitochondrion.

Maturation The process by which complete virions are assembled from newly synthesized components in the replication process.

Measles A febrile disease with rash caused by the rubeola virus, which invades lymphatic tissue and blood. *Known also as Rubeola.*

Measles Encephalitis A serious complication of measles that leaves many survivors with permanent brain damage.

Mebendazole An antihelminthic agent that blocks glucose uptake by parasitic roundworms.

Mechanical Stage Attachment to a microscope stage that holds the slide and allows precise control in moving the slide.

Mechanical Vector A vector in which the parasite does not complete any part of its life cycle during transit.

Mechanism-Based Inhibitor An enzyme inhibitor whose action depends on the enzyme's catalytic mechanism. Typically it is a substrate analog that irreversibly modifies the enzyme at a particular step in the catalytic cycle.

Medium A mixture of nutritional substances on or in which microorganisms grow.

Megakaryocyte Large cell normally present in bone marrow that gives rise to platelets.

Meiosis Division process in eukaryotic cells that reduces the chromosome number in half.

Membrane Attack Complex A set of proteins in the complement system that lyses invading bacteria by producing lesions in their cell membranes.

Membrane Electrical Potential With respect to biological membranes, a voltage difference that exists across a membrane owing to differences in the concentrations of ions on either side of the membrane.

Membrane Filter Method Method of testing for coliform bacteria in water in which bacteria are filtered through a membrane and then incubated on the membrane surface in growth a medium.

Memory Cell Long-lived B or T lymphocyte that can carry out an anamestic or secondary response.

Meninges Three layers of membrane that protect the brain and spinal cord.

Merozoite A malaria trophozoite found in infected red blood or liver cells.

Mesophile An organism that grows best at temperatures between 25 and 45°C, including most bacteria.

Mesophilic Spoilage Spoilage due to improper canning procedures or because the seal has been broken.

Messenger RNA (mRNA) (1) A type of RNA that carries the information from DNA to dictate the arrangement of amino acids in a protein. (2) RNA molecules that act as templates for the synthesis of polypeptides by ribosomes.

Metabolic Pathway A chain of chemical reactions in which the product of one reaction serves as the substrate for the next.

Metabolism The totality of the chemical reactions that occur in an organism. compare ▶ Anabolism and ▶ Catabolism.

Metacercaria The postcercarial encysted stage in the development of a fluke, prior to transfer to the final host.

Metachromasia Property of exhibiting a variety of colors when stained with a simple stain.

Metachromatic Granule A polyphosphate granule that exhibits metachromasia. *Known also as Volutin.*

Metastability For a system, the condition of being in a state that does not represent thermodynamic equilibrium but is nearly stable at the time scale of interest because progress toward equilibrium is slow.

Metastasize Relating to the spread of malignant tumors to other body tissues.

Methanogens One of the groups of the archaeobacteria that produce methane gas.

Metronidazole An antiprotozoan agent effective against *Trichomonas* infections.

Micelles Tiny droplets that form when an amphipathic substance that has a polar head group and a nonpolar tail region (such as a fatty acid) is added to an aqueous medium and shaken. Each droplet consists of a spherical cluster of amphipathic molecules arranged with their polar head groups facing out toward the water and their nonpolar tails facing in toward the center.

Michaelis-Menten Equation An equation that gives the rate of an enzyme-catalyzed reaction in terms of the concentrations of substrate and enzyme as well as two constants that are specific for a particular combination of enzyme and substrate: a rate constant, k_{cat}, for the catalytic production of product when the enzyme is saturated, and the Michaelis constant, K_M.

Microaerophile A bacterium that grows best in the presence of a small amount of free oxygen.

Microbe See ▶ Microorganism.

Microbial Antagonism The ability of normal microbiota to compete with pathogenic organisms and in some instances to effectively combat their growth.

Microbial Antagonism The ability of normal microbiota to compete with pathogenic organisms and in some instances to effectively combat their growth.

Microbial Growth Increase in the number of cells, due to cell division.

Microbiology The study of microorganisms.

Micrococci Aerobes or facultative anerobes that form irregular clusters by dividing in two or more planes.

Microenvironment A habitat in which the oxygen, nutrients, and light are stable, including the environment immediately surrounding the microbe.

Microfilament A protein fiber that makes up part of the cytoskeleton in eukaryotic cells.

Microfilaria An immature microscopic roundworm larva.

Micrometer (μm) Unit of measure equal to 0.000001 m or 10^{-6} m; formerly called a micron (μ).

Microorganism Organism studied with a microscope; includes the viruses. *Known also as a Microbe.*

Microscopy The technology for making very small things visible to the unaided eye.

Microtubule A protein tubule that forms the structure of cilia, flagella, and part of the cytoskeleton in eukaryotic cells.

Microtubule-Associated Proteins (MAPs) A class of proteins associated with microtubules that assist in dynamic processes.

Microtubules Fiberlike cytoplasmic structures that consist of units of the protein tubulin arranged helically to form a hollow tube. They are involved in various kinds of cellular motility, including the beating of cilia and flagella and the movement of organelles from one part of the cell to another.

Microvillus (plural: *Microvilli*) A minute projection from the surface of an animal cell.

Miliary Tuberculosis Type of tuberculosis that invades all tissues producing tiny lesions.

Minimum Bactericidal Concentration (MBC) The lowest concentration of an antimicrobial agent that kills microorganisms, as indicated by absence of growth following subculturing in the dilution method.

Minimum Inhibitory Concentration (MIC) The lowest concentration of an antimicrobial agent that prevents growth in the dilution method of determining antibiotic sensitivity.

Minus Strand In viral genomes, a nucleic acid strand that is complementary to the RNA strand that serves as mRNA. Compare ▶ Plus Strand.

Miracidium Ciliated, free-swimming first-stage fluke larva that emerges from an egg.

Mismatch Repair A system for the correction of mismatched nucleotides or single-base insertions or deletions produced during DNA replication; it scans the nearly replicated DNA, and when it finds an error, it removes and replaces a stretch of the strand containing the error.

Missense Mutation A mutation that alters a DNA codon so as to cause one amino acid in a protein to be replaced by a different one.

Mitochondria The organelles whose chief task it is to supply the cell with ATP via oxidative phosphorylation. They contain the enzymes for pyruvate oxidation, the citric acid cycle, the β-oxidation of fatty acids, and oxidative phosphorylation, as well as the electron transport chain.

Mitochondrion An organelle in eukaryotic cells that carries out oxidative reactions that capture energy.

Mitosis Process by which the cell nucleus in a eukaryotic cell divides to form identical daughter nuclei.

Mixed-Functional Oxidase An oxygenase enzyme that analyzes a reaction in which two different substrates are oxidized, one by the addition of an oxygen atom from O_2 and the other by supplying two hydrogen atoms to reduce the remaining oxygen atom to H_2O.

Mixed Infection An infection caused by several species of organisms present at the same time.

Mixture Two or more substances combined in any proportion and not chemically bound.

MMR Vaccine Measles, mumps, and rubella vaccine.

Mole The weight of a substance in grams equal to the sum of the atomic weights of the atoms in a molecule of the substance. *Known also as Gram Molecular Weight.*

Molecular Mimicry Imitation of the behavior of a normal molecule by an antimetabolite.

Molecule Two or more atoms chemically bonded together.

Molluscum Contagiosum A viral infection characterized by flesh-colored, painless lesions.

Molten Globule A hypothetical intermediate state in the folding of a globular protein, in which the overall tertiary framework has been established but internal side chains (especially hydrophobic ones) are still free to move about.

Monera The kingdom of prokaryotic organisms that are unicellular and lack a true cell nucleus. *Known also as Prokaryotae.*

Moniliasis See ▶ Candidiasis.

Monoclonal Antibody A single, pure antibody produced in the laboratory by a clone of cultured hybridoma cells.

Monocular Refers to a light microscope having one eyepiece (ocular).

Monocyte A ravenously phagocytic leukocyte, called a macrophage after it migrates into tissues.

Monolayer A suspension of cells that attach to plastic or glass surfaces as a sheet one cell layer thick.

Monosaccharide A simple carbohydrate, consisting of a carbon chain or ring with several alcohol groups and either an aldehyde or ketone group.

Monotrichous A bacterial cell with a single flagellum.

Morbidity Rate The number of persons contracting a specific disease in relation to the total population (cases per 100,000).

Mordant A chemical that helps a stain adhere to the cell or cell structure.

Mortality Rate The number of deaths from a specific disease in relation to the total population.

Most Probable Number (MPN) A statistical method of measuring bacterial growth, used when samples contain too few organisms to give reliable measures by the plate-count method.

Mother Cell A cell that has approximately doubled in size and is about to divide into two daughter cells. *Known also as a Parent Cell.*

Mucin A glycoprotein in mucus that coats bacteria and prevents their attaching to surfaces.

Mucociliary Escalator Mechanism involving ciliated cells that allows materials in the bronchi, trapped in mucus, to be lifted to the pharynx and spit or swallowed.

Mucopolysaccharides See ▶ Glycosaminoglycans.

Mucous Membrane A covering over those tissues and organs of the body cavity that are exposed to the exterior. *Known also as Mucosa.*

Mucus A thick but watery secretion of glycoproteins and electrolytes secreted by the mucous membranes.

Multicatalytic Proteinase Complex (MPC) A massive complex of proteolytic enzymes that is found in the cytosol of many eukaryotic cells and seems to function in the programmed destruction of cellular proteins.

Multiple-Tube Fermentation Method Three-step method of testing for coliform bacteria in drinking water.

Mumps Disease caused by a paramyxovirus that is transmitted by saliva and invades cells of the oropharynx.

Murine Typhus See ▶ Endemic Typhus.

Muscarinic Acetylcholine Receptors A class of receptors for the neurotransmitter acetylcholine that are characterized by an ability to bind the toadstool toxin muscarine. Synapses that have these receptors may be either excitatory or inhibitory. Compare ▶ Nicotinic Acetylcholine Receptors.

Mutagen An agent that increases the rate of mutations.

Mutation Any inheritable change in the nucleotide sequence of genomic DNA (or genomic RNA, in the case of an RNA virus).

Mutualism A form of symbiosis in which two organisms of different species live in a relationship that benefits both of them.

Myasthenia Gravis Autoimmune disease specific to skeletal muscle, especially muscles of the limbs and those involved in eye movements, speech, and swallowing.

Mycelium (plural: *mycelia*). In fungi, a mass of long, threadlike structures (hyphae) that branch and intertwine.

Mycobacteria Slender, acid-fast rods, often filamentous, include organisms that cause tuberculosis, leprosy, and chronic infections.

Mycology The study of fungi.

Mycoplasmas Very small bacteria with cell membranes, RNA and DNA, but no cell walls.

Mycosis (plural: *Mycoses*) A disease caused by a fungus.

Myiasis An infestation caused by maggots (fly larvae).

Mycarditis An inflammation of the heart muscle.

Nacardioforms Gram-positive, nonmotile, pheomorphic, aerobic bacteria, often filamentous and acid-fast, include some skin and respiratory pathogens.

NAD Nicotinamide dinucleotide, a coenzyme that carries hydrogen atoms and electrons.

Naked Virus A virus that lacks an envelope.

Nanometer (▶ nm) Unit of measure equal to 0.000000001 m or 10^{-9}; formerly called a millimicron (nμ).

Narrow Spectrum The range of activity of an antimicrobial agent that attacks only a few kinds of microorganisms.

Nasal Cavity Part of the upper respiratory tract where air is warmed and particles are removed by hairs as they pass through.

Nasal Sinus A hollow cavity within the skull that is lined with mucous membrane.

Naturally Acquired Active Immunity When an individual is exposed to an infectious agent, often having the disease, and their own immune system responds in a protective way.

Naturally Acquired Immunity Defense against a specific disease is acquired sometime after birth, without the intervention or use of man-made products such as vaccines or gamma globulin.

Naturally Acquired Passive Immunity When antibodies made by another individual are given to a host, e.g., in mother's milk, without intervention by man.

Natural Killer (NK) Cell A lymphocyte that can destroy virus-infected cells, malignant tumor cells, and cells of transplanted tissues.

Negative (−) Sense RNA An RNA strand made up of bases complementary to those of a positive (+) sense RNA.

Negative Staining Technique of staining the background around a specimen, leaving the specimen clear and unstained.

Nematode See ▶ Roundworm.

Neonatal Herpes Infection in infants usually with HSV-2, most often acquired during passage through a birth canal contaminated with the virus.

Neoplasm A localized tumor.

Neoplastic Transformation The uncontrollable division of host cells caused by infection with a DNA tumor virus.

Nephron A functional unit of the kidney in which fluid from the blood is filtered.

Nernst Equation An equation that relates the electrical potential across a membrane to the concentrations of ions on either side of the membrane.

Nerve A bundle of neuron fibers that relays sensory and motor signals throughout the body.

Nervous System The body system, comprising the brain, spinal cord, and nerves, that coordinates the body's activities in relation to the environment.

Neurohormones Substances that are released from neurons and modulate the behavior of target cells, which are often other neurons. Unlike neurotransmitters, they do not act strictly across a synapse. Most neurohormones are peptides.

Neuron A conducting nerve cell.

Neurotoxin A toxin that acts by disrupting nerve cell function. Fast-acting neurotoxins often act by blocking the action of an ion gate necessary for the development of an action potential.

Neurosyphilis Neurological damage, including thickening of the meninges, ataxia, paralysis, and insanity, that results from syphilis.

Neurotoxin A toxin that acts on nervous system tissues.

Neurotransmitter A low-molecular-weight substance that is released from an axon terminal in response to the arrival of an action potential and then diffuses across the synapse to influence the post-synaptic cell, which may be either another neuron or a muscle or gland cell.

Neutral Referring to a solution with a pH of 7.

Neutralization Inactivation of microbes or their toxins through the formation of antigen-antibody complexes.

Neutralization Reaction An immunological test used to detect bacterial toxins and antibodies to viruses.

Neutron An uncharged subatomic particle in the nucleus of an atom.

Neutrophil A phagocytic leukocyte. *Known also as Polymorphonuclear leukocyte, PMNL.*

Neutrophile An organism that grows best in an environment with a pH of 5.4 to 8.5.

Nicotinic Acetylcholine Receptors A class of receptors for the neurotransmitter acetylcholine that are characterized by their ability t bind nicotine. Synapses with this king of receptor are excitatory. Compare ▶ Muscarinic Acetylcholine Receptors.

Nick Translation A process in which nucleotides in one strand of a nucleic acid duplex are replaced one by one with nucleotides complementary to the other strand. The process starts at a nick in the strand and causes the location of the nick to migrate (hence the origin of the term).

Niclosamide An antihelminthic agent that interferes with carbohydrate metabolism.

Nitrification The process by which ammonia or ammonium ions are oxidized to nitrites or nitrates.

Nitrofuran An antibacterial drug that damages cellular respiratory systems.

Nitrogenase Enzyme in nitrogen-fixing bacteroids that catalyzes the reaction of nitrogen gas and hydrogen gas to form ammonia.

Nitrogen Cycle Process by which nitrogen moves from the atmospheric through various organisms and back into the atmosphere.

Nitrogen Fixation The reduction of atmospheric nitrogen gas to ammonia.

Nocardiosis Respiratory disease characterized by tissue lesions and abscesses, caused by the filamentous bacterium *Nocardia asteroids.*

Nocturia Nightime urination, often a result of urinary tract infections.

Nomarski Microscopy Differential interference contrast microscopy; utilizes differences in refractive index to visualize structures producing a nearly three-dimensional image.

Noncommunicable Infectious Disease Disease caused by infectious agents but not spread from one host to another.

Noncompetitive Inhibitor A molecule that attaches to an enzyme at an allosteric site (a site other than the active site), distorting the shape of the active site so that the enzyme can no longer function.

Noncovalent Interactions All the kinds of interactions between atoms and molecules that do not involve the actual sharing of electrons in a covalent bond; they include electrostatic interactions, permanent and induced dipole interactions, and hydrogen bonding.

Noncyclic Photophosphorylation In photosynthesis, Photophosphorylation (light-dependent ATP synthesis) that is linked to a one-way flow of electrons from water through photosystems II and I and finally to NADPH; it is thus coupled to the oxidation of H_2O and the reduction of $NADP^+$. Compare ▶ Cyclic Photophosphorylation.

Noncyclic Photoreduction The photosynthetic pathway in which excited electrons from chlorophyll are used to generate ATP and reduce NADP with the splitting of water molecules.

Nongonococcal Urethritis (NGU) A gonorrhealike sexually transmitted disease most often caused by *Chlamydia trachomatis* and mycoplasmas.

Nonindigenous Organism An organism temporarily found in a given environment.

Noninfection Disease Disease caused by any factor other than infectious agents.

Nonself Antigens recognized as foreign by an organism.

Nonsense Codon A set of three bases in a gene (or mRNA) that does not code for an amino acid. *Known also as Terminator Codon.*

Nonsense Mutation A mutation that creates an abnormal stop codon and thus causes translation to terminate prematurely; the resulting truncated protein is usually nonfunctional.

Nonspecific Defenses Those host defenses against pathogens that operate regardless of the invading agent.

Nonspecific Immunity Product by general defenses, such as skin, lysozyme and complement, that protect against many different kinds of organisms rather than a specific one or two.

Nonsynchronous Growth Natural pattern of growth during the log phase in which every cell in a culture divides at some point during the generation time, but not simultaneously.

Normal Microflora Microorganisms that live on or in the body but do not usually cause disease. *Known also as Normal Flora.*

Northern Blotting A technique for detecting the presence of a specific RNA sequence in a cell and determining its size. The total RNA of the cell is extracted, resolved by gel electrophoresis, and blotted onto a filter. There it is incubated under annealing conditions with a radiolabeled probe for the sequence in question, and heteroduplexes of the probe with RNA are detected by radioautography.

Nosocomial Infection An inflection acquired in a hospital or other medical facility.

Notifiable Disease A disease that a physician is required to report to public health officials.

N-Terminus The end of a polypeptide chain that carries an unreacted amino group. A ribosome synthesizes a polypeptide in the direction from the N-terminus to the C-terminus. *Known also as Amino Terminus.* See also ► C-Terminus.

Nuclear Envelope The double membrane surrounding the cell nucleus in a eukaryotic cell. It is pierced by nuclear pores that allow even quite large molecules, such as mRNAs and nuclear proteins, to enter or leave the nucleus.

Nuclear Magnetic Resonance (NMR) Spectroscopy A type of spectroscopy that depends on the fact that isotope nuclei having the property of spin will resonate with specific frequencies of microwave radiation when placed in a magnetic field of given strength. The resonance energy is sensitive to the local molecular environment, so NMR spectroscopy can be used to explore molecular structure. Also, different living tissues have characteristic overall NMR spectra, which are sensitive to changes in the tissue environment. NMR can thus be used in the study of tissue metabolism and the diagnosis of disease.

Nuclear Matrix A protein web that is left in the nucleus when histones and other weakly bound proteins are removed and most of the DNA is digested away. It is presumed to act as an organizing scaffold for the chromatin.

Nuclear Pore An opening in the nuclear envelope that allows for the transport of materials between nucleus and cytoplasm.

Nuclear Region Central location of DNA, RNA, and some proteins in bacteria; not a true nucleus. *Known also as Nucleoid.*

Nuclease An enzyme that cleaves nucleic acids.

Nucleic Acids Long polymers of nucleotides that encode genetic information and direct protein synthesis.

Nucleocapsid The nucleic acid and capsid of a virus.

Nucleoid The large, circular DNA molecule of a prokaryotic cell, along with its associated proteins; also sometimes called the bacterial chromosome. It is supercoiled and forms a dense mass within the cell, and the term Nucleoid is often used for the cell region occupied by this mass. See ► Nuclear Region.

Nucleolus (plural: *Nucleoli*) Area in the nucleus of a eukaryotic cell that contains RNA and serves as the site for the assembly of ribosomes.

Nucleoplasm The semifluid portion of the cell nucleus in eukaryotic cells that is surrounded by the nuclear envelope.

Nucleoside A molecule that, upon complete hydrolysis, yields 1 mole per mole of a purine or pyrimidine base and a sugar.

Nucleotide An organic compound consisting of a nitrogenous base, a five-carbon sugar, and one or more phosphate groups.

Nucleosome The first-order structural unit for the packing of DNA in chromatin, consisting of 146 bp of DNA wrapped 1.75 times around a core octamer of histone proteins. Successive nucleosomes are connected by stretches of "linker" DNA.

Nucleotide A molecule that, upon complete hydrolysis, yields at least 1 mole per mole of a Purine or pyrimidine base, a sugar, and inorganic phosphate.

Nucleus The membrane-bound structure in a eukaryotic cell that contains the chromosomal genetic material and associated components. It is also the place where RNA molecules are processed and ribosomes are assembled.

Null Cells Undifferentiated cells that cannot be identified as either B cells or T cells; include the natural killer (NK) cells.

Numerical Aperature The widest cone of light that can enter a lens.

Numerical Taxonomy Comparison of organisms based on quantitative assessment of a large number of characteristics.

Nutritional Complexity The number of nutrients an organism must obtain to grow.

Nutritional Factor One factor that influences both the kind of organisms found in an environment and their growth.

Objective Lens Lens in a microscope closest to the specimen that creates an enlarged image of the object viewed.

Obligate Requiring a particular environmental condition.

Obligate Aerobe A bacterium that must have free oxygen to grow.

Obligate Anaerobe A bacterium that is killed by free oxygen.

Obligate Intracellular Parasite An organism or virus that can live or multiply only inside a living host cell.

Obligate Parasite A parasite that must spend some or all of its life cycle in or on a host.

Obligate Psychrophile An organism that cannot grow at temperatures about 20°C.

Obligate Thermophile An organism that can grow only at temperatures above 37°C.

Ocular Lens Lens in the microscope that further magnifies the image creased by the objective lens.

Ocular Micrometer A glass disk with an inscribed scale that is placed inside the eyepiece of a microscope; used to measure the actual size of an object being viewed.

Okazaki Fragment One of the short, discontinuous DNA segments formed on the lagging strand during DNA replication.

Okazaki Fragments The discontinuous stretches in which the lagging strand is initially synthesized during DNA replication; these fragments are later joined to form a continuous strand.

Onchocerciasis Ana eye disease caused by the filarial larvae of the nematode *Onchocerca volvulus*, transmitted by blackflies, common in Africa and Central America. *Known also as River Blindness.*

Oncogene A gene that, in a mutated version, can help transform a normal cell to a cancer cell. Many oncogene codes for mutant proteins that are involved in the reception and transduction of growth factor signals. A cancer-causing gene.

Oncoprotein The protein product of an Oncogene.

ONPG and MUG Test Water purity test that relies on the ability of coliform bacteria to secrete enzymes that convert a substrate into a product that can be detected by a color change.

Oomycota See ► Water Mold.

Open-Promoter Complex A complex between RNA polymerase Holoenzyme and a promoter that has undergone initial unwinding (has "opened") preparatory to the start of transcription. It is preceded by a much less stable *closed-promoter complex*, in which the promoter has not unwound, that may either fall apart or proceed to an open-promoter complex.

Open Reading Frame A sequence within a messenger RNA that is bounded by start and stop codons and can be continuously translated. It represents the coding sequence for a polypeptide.

Operator A DNA site where a repressor protein binds to block the initiation of transcription from an adjacent promoter.

Operon (1) A sequence of a closely associated genes that includes both structural genes and regulatory sites that control transcription. (2) A set of contiguous prokaryotic structural genes that are transcribed as a unit, along with the adjacent regulatory elements that control their expressions.

Opthalmia Neonatorium Pyrogenic infection of the eyes caused by *Neisseria Gonorrhoeae. Known also as Conjunctivitis of the Newborn.*

Opportunist A species of resident or transient microbiota that does not ordinarily cause disease but can do so under certain conditions.

Opsonin An antibody that promotes phagocytosis when bound to the surface of a microorganism.

Opsonization The process by which microorganisms are rendered more attractive to phagocytes by being coated with antibodies (opsonins) and C3b complement protein. *Known also as Immune Adherence.*

Optical Isomers See ► Enantiomers.

Optical Microscope See ► Compound Light Microscope.

Optimum pH The pH at which microorganisms grow best.

Orbivirus Type of virus that causes Colorado tick fever.

Orchitis Inflammation of the testes; a symptom of mumps in postpubertal males.

Organelle An internal membrane-enclosed structure found in eukaryotic cells.

Organelles Membrane-bound compartments in the cytoplasm of eukaryotic cells. Each kind of organelle carries out a specific set of functions. Examples are mitochondria, chloroplasts, and nuclei.

Organic Chemistry The study of compounds that contain carbon.

Ornithosis Disease with pneumonialike symptoms, caused by *Chlamydia psittaci* and acquired from birds (previously called psittacosis and parrot fever).

Oroyo Fever One form of bartonellosis; an acute fatal fever with severe anemia. *Known also as Carrion's Disease.*

Orthomyxovirus A medium-sized, enveloped RNA virus that varies in shape from spherical to filamentous and has an affinity for mucus.

Osmosis A special type of diffusion in which water molecules move from an area of higher concentration to one of lower concentration across a selectively permeable membrane.

Osmotic Pressure The pressure required to prevent the net flow of water molecules by osmosis.

Otitis Externa Infection of the external ear canal.

Otitis Media Infection of the middle ear.

Outer Membrane A bilayer membrane, forming part of the cell wall of Gram-negative bacteria.

Ovarian Follicle An aggregation of cells in the ovary containing an ovum.

Ovary In the female, one of a pair of glands that produce ovarian follicles, which contain an ovum and hormone-secreting cells.

Oxidase An enzyme that catalyzes the oxidation of a substrate with oxygen as the electron acceptor.

Oxidation The loss of electrons and hydrogen atoms.

Oxidative Phosphorylation (1) Process in which the energy of electrons is captured in high-energy bonds as phosphate groups combine with ADP to form ATP. (2) The phosphorylation of ADP to ATP that occurs in conjunction with the transit of electrons down the electron transport chain in the inner mitochondrial membrane.

Oxygenase An enzyme that catalyzes the incorporation of oxygen into a substrate.

Palindrome With respect to DNA, a segment in which the sequence is the same on one strand read right to left as on the other strand read left to right; thus, a back-to-back pair of inverted repeats.
Pandemic An epidemic that has become worldwide.
Papilloma See ▸ Wart.
Papovavirus A small, naked DNA virus that causes both benign and malignant warts in humans, some types cause cervical cancer.
Parainfluenza Viral disease characterized by nasal inflammation, pharyngitis, Bronchitis, and sometimes pneumonia, mainly in children.
Parainfluenza Virus Virus that initially attacks the mucous membranes of the nose and throat.
Paramyxovirus A medium-sized, enveloped RNA virus that has an affinity for mucus.
Parasite An organism that lives in or on, and at the expense of, another organism, the host.
Parasitism A symbiotic relationship in which one organism, the parasite, benefits from the relationship, whereas the other organism, the host, is harmed by it.
Parasitology The study of parasites.
Parfocal For a microscope, remaining in approximate focus when minor focus adjustments are made.
Paroxysmal Stage State of whooping cough in which mucus and masses of bacterial fill the airway, causing violet coughing.
Partial Molar Free Energy See ▸ Chemical Potential.
Partition Coefficient (*K*) A coefficient that indicates how a particular substance will distribute itself between two media if allowed to diffuse to equilibrium between them; it is equal to the ratio of the solubilities of the substance in the two media.
Parvovirus A small, naked DNA virus.
Passive Immunity Immunity created when ready-made antibodies are introduced into, rather than created by, an organism.
Passive Immunization The process of inducing immunity by introducing ready-made antibodies into a host.
Passive Transport With respect to membrane transport, the movement of a substance across a biological membrane by molecular diffusion through the lipid bilayer. Compare ▸ Active Transport, ▸ Facilitated Transport. *Known also as Passive Diffusion.*
Pasteur Effect The inhibition of glycolysis by oxygen; discovered by Pasteur when he found that aerobic yeast cultures metabolize glucose relatively slowly.
Pasteurella-Haemophilus Group Very small Gram-negative bacilli and coccobacilli that lack flagella and are nutritionally fastidious.
Pasteurization Mild heating to destroy pathogens and other organisms that cause spoilage.
Pathogen An organism capable of causing disease in its host.
Pathogenicity The capacity to produce disease.
Pediculosis Lice infestation, resulting in reddened areas at bites, dermatitis, and itch.
Pellicle (1) A thin layer of bacteria adhering to the air-water interface of a broth culture by their attachment pili. (2) A strengthened plasma membrane of a protozoan cell. (3) Film over the surface of a tooth at the beginning of plaque formation.
Pelvic Inflammatory Disease (PID) An infection of the pelvic cavity in females, caused by any of several organisms including *Neisseria Gonorrhoeae* and *Chlamydia.*
Penetration The entry of the virus (or its nucleic acid) into the host cell in the replication process.
Penicillin An antibacterial agent that inhibits cell wall synthesis.
Penis Part of the male reproductive system used to deliver semen to the female reproductive tract during sexual intercourse.
Peptide Bond A covalent bond joining the amino group of one amino acid and the carboxyl group of another amino acid. It consists of an amide bond between the α-carboxyl group of one amino acid and the α-amino group of the next.
Peptidoglycan A structural polymer in the bacterial cell wall that forms a supporting net. *Known also as Murein.*
Peptidyltransferase During ribosomal polypeptide synthesis, the enzyme complex that transfers the polypeptide chain from the tRNA in the P site to the amino acid carried by the tRNA in the A site, thereby adding another amino acid to the chain, the complex is an integral part of the large ribosomal subunit.
Peptococci Anerobes that form pairs, tetrads, or irregular clusters, they lac both Catalase and the enzyme to ferment lactic acid.
Peptone A product of enzyme digestion of proteins that contains many small peptides; a common ingredient of a complex medium.
Perforin A Cytotoxin produced by cytotoxic T cells that bores holes in the plasma membrane of infection host cells.
Pericarditis An inflammation of the protective membrane around the heart.
Periodontal Disease A combination of gum inflammation, decay of cementum, and erosion of periodontal ligaments and bone that support teeth.

Periodontitis A chronic periodontal disease that affects the bone and tissue that supports the teeth and gums.

Peripheral Nervous System All nerves outside the central nervous system.

Periplasm Those substances (enzymes, transport proteins) located in the periplasmic space of Gram-negative bacteria or in the older cell wall of Gram-positive bacteria.

Periplasmic Enzyme An Exoenzyme produced by Gram-negative organisms, which acts in the periplasmic space.

Periplasmic Space The space between the cell membrane and the outer membrane in Gram-negative bacteria that is filled with periplasm.

Peritrichous Having flagella distributed all over the surface of a bacterial cell.

Permanent Dipole In chemistry, a molecule that has a permanent, asymmetric distribution of charge such that one end is negative and the other end positive, The water molecule is an example: The oxygen end has a partial negative charge, and the hydrogen end has a partial positive charge.

Permanent Parasite A parasite that remains in or on a host once it has invaded the host.

Permease An enzyme complex involved in active transport through the cell membrane.

Peroxisome (1) An organelle filled with enzymes that in animal cells oxidate amino acids and in plant cells oxidize fats. (2) A small, vesicular organelle that specializes in carrying out cellular reactions involving the transfer of hydrogen from a substrate to O_2. These reactions produce the by-product H_2O_2, which is split to H_2O and O_2 by the peroxisomal enzyme catalase.

Persistent Viral Infection The continued production of viruses within the host over many months or years.

Pertussis See ► Whooping Cough.

PEST Sequences A family of amino acid sequences that have been found on cellular proteins that undergo rapid turnover; they may target proteins for rapid proteolysis. They consist of a region about 12 to 60 residues long that is rich in praline, glutamate, serine, and threonine (P, E, E, and T in the one-letter abbreviation system).

Petechia (plural: *petechiae*) A pinpoint-size hemorrhage, most common in skin folds, that often occurs in rickettsial diseases.

pH A means of expressing the hydrogen-ion concentration, and thus the acidity, of a solution.

Phage See ► Bacteriophage.

Phage Typing Use of bacteriophages to determine similarities or differences among different bacteria.

Phagocyte A cell that ingests and digests foreign particles.

Phagocytosis Ingestion of solids into cells by means of the formation of vacuoles.

Phagolysosome A structure resulting from the fusion of lysosomes and a phagosome.

Phagosome A vacuole that forms around a microbe within the phagocyte that engulfed it.

Pharmaceutical Microbiology A special branch of industrial microbiology concerned with the manufacture of products used in treating or preventing disease.

Pharyngitis An infection of the pharynx, usually caused by a virus but sometimes bacterial in origin, a sore throat.

Pharnyx The throat, a common passage-way for the respiratory and digestive systems with tubes connecting to the middle ear.

Phase-Contrast Microscopy Use of microscope having a condenser that accentuates small differences in the refractive index of various structures within the cell.

Phenol Coefficient A numerical expression for the effectiveness of a disinfectant relative to that of phenol.

Phenotype The specific observable characteristics displayed by an organism. It results from the interaction of the organism's genetic makeup with the environment. Compare ► Genotype.

Pheromones Intercellular mediator compounds that are released from one organism and influence the metabolism or behavior of another organism, usually of the same species. Sex attractants, which elicit reproductive behavior in suitable recipients, are an example.

Phlebovirus Bunyavirus that is carried by the sandfly *Phlebotomus papatsii.*

Phorbol Esters A group of natural substances that resemble *sn*-1,2-diacyglycerol (DAG) in part of their structure and can act as tumor promoters. This effect suggests that the DAG second-messenger system may be involved in growth factor action.

Phosphodiester Link The linkage that connects the nucleotide monomers in a nucleic acid. It consists of a phosphate residue that links the sugar moieties of successive monomers by forming an ester bond with the 5′ carbon of one sugar and the 3′ carbon of the next.

Phospholipid A lipid composed of glycerol, two fatty acids, and a polar head group; found in all membranes.

Phosphorescence Continued emission of light by an object when light rays no longer strike it.

Phosphorus Cycle The cyclic movement of phosphorous between inorganic and organic forms.

Phosphorylation The addition of a phosphate group to a molecule, often from ATP, generally increasing the molecule's energy.

Phosphotransferase System A mechanism that uses energy from phosphorenolyruvate to move sugar molecules into cells by active transport.

Photoautotroph An autotroph that obtains energy from light.

Photoheterrotroph A heterotroph that obtains energy from light.

Photolysis Process in which light energy is used to split water molecules into protons, electrons, and oxygen molecule.

Photophosphorylation Phosphorylation of ADP to ATP that depends directly on energy from sunlight. The light energy is captured by a pigment such as chlorophyll and is passed in the form of excited electrons to an electron transport chain; the electron transport chain uses energy from the electrons to create a proton gradient across a membrane, which drives the synthesis of ATP.

Photoproducts The products that result when light energy causes a chemical reaction to occur in a substance. With respect to DNA, the term refers to the types of damaged DNA that can be caused by UV irradiation.

Photoreactivation A DNA repair process in which an enzyme uses light energy to break cyclobutane pyrimidine dimers created by UV irradiation and to restore the correct bonding. See ▶ Light Repair.

Photorespiration The cycle of reactions that occurs in place of the Calvin cycle when the photosynthetic enzyme rubisco adds O_2 rather than CO_2 to ribulose bisphosphate carboxylase. It takes place partly in chloroplasts, partly in peroxisomes, and partly in mitochondria; it expends ATP energy and loses a previously fixed CO_2 molecule in the process of regenerating the Calvin cycle intermediate 3-phosphoglycerate.

Photosynthesis The capture of energy from light and use of this energy to manufacture carbohydrates from carbon dioxide.

Photosystem A structural unit in a cellular membrane that captures light energy and converts a portion of it to chemical energy. The photosynthesis practiced by plants, algae and cyanobacteria involves two types of photosystem, both of which capture energy in the form of high-energy electrons and transduce it via an electron transport chain.

Phototaxis A non random movement of an organism toward or away from light.

Phylogenetic Pertaining to evolutionary relationships.

Physical Factor Factor in the environment, such as temperature, moisture, pressure, or radiation, that influences the kinds of organisms found and their growth.

Picornavirus A small, naked RNA virus; different genera are responsible for polio, the common cold, and hepatitis.

Pilus (plural: *pili*) A tiny hollow projection used to attach bacteria to surfaces (attachment pilus) or for conjugation (conjugation pilus).

Pimple See ▶ Folliculitis.

Pinna Flaplike external structure of the ear.

Pinworm A small roundworm, *Enterobius vermicularis*, that causes gastrointestinal disease.

Placebo An unmedicated, usually harmless substance given to a recipient as a substitute for or to test the efficacy of a medication or treatment.

Plantae The kingdom of organisms to which all plants belong.

Plaque (1) A clean area in a bacterial lawn culture where viruses have lysed cells. (2) A clear area that is formed by a local phage infection in a lawn of cultured bacteria in a Petri dish; for purposes of experimentation, it is the phage equivalent of a bacterial colony.

Plaque Assay A viral assay used to determine viral yield by culturing viruses on a bacterial lawn and counting plaques.

Plaque-Forming Unit A plaque counted on a bacterial lawn that gives only an approximate number of phages present, because a given plaque may have been due to more than one phage.

Plasma Liquid portion of the blood, excluding the formed elements.

Plasma Cell A large lymphocyte differentiated from a B cell that synthesizes and releases antibodies like those on the B cell surface.

Plasma Membrane A selectively permeable lipoprotein bilayer that forms the boundary between the cytoplasm of a eukaryotic cell and its environment. *Known also as Cell Membrane.*

Plasmid A small, circular, independent replicating piece of DNA in a cell that is not part of its chromosome and can be transferred to another cell. *Known also as Extrachromosomal DNA.*

Plasmids Small, extrachromosomal circular DNA molecules found in many bacteria, they replicate independently of the main chromosome and may occur in multiple copies per cell.

Plasmodial Slime Mold Funguslike protest consisting of a multinucleate amoeboid mass, or plasmodium, that moves about slowly and phagocytizes dead matter.

Plasmodium A multinucleate mass of cytoplasm that forms one of the stages in the life cycle of a plasmodial slime mold.

Plasmogamy Sexual reproduction in fungi in which haploid gametes unite and their cytoplasm mingles.

Plasmolysis Shrinking of a cell, with separation of the cell membrane from the cell wall, resulting from loss of water in a hypertonic solution.

Platelet A short-lived fragment of large cells called megakaryocytes, important component of the blood-clotting mechanism.

Pleomorphism Phenomenon in which bacteria vary widely in form, even within a single culture under optimal conditions.

Pleura Serious membrane covering the surfaces of the lungs and the cavities they occupy.

Pleurisy Inflammation of pleural membranes that causes painful breathing often accompanies lobar pneumonia.

Plus Strand In viral genomes, a nucleic acid strand that can serve as mRNA or (for DNA strand) that is homologous to one that can; as distinct from the complementary (minus) strand. Most viruses with single-strand genomes package only the plus or minus strand in virions; the other strand is made transiently during replication. Compare ▶ Minus Strand.

Pneumocystis Pneumonia A fungal respiratory disease caused by *Pneumocystis carinii.*

Pneumonia An inflammation of lung tissue caused by bacteria, viruses, or fungi.

Pneumonic Plague Usually fatal form of plague transmitted by aerosol droplets from a coughing patient.

Point Mutation Mutation in which one base is substituted for another at a specific location in a gene.

Polar Compound A molecule with an unequal distribution of charge due to an unequal sharing of electrons from between atoms.

Poliomyelitis Disease caused by any of several strains of polioviruses that attack motor neutrons of the spinal cord and brain.

Polyacrylamide Gel Electrophoresis (PAGE) A technique for separating proteins from a cell based in their molecular size.

Polyene An antifungal agent that increases membrane permeability.

Polymer (1) A large molecule that is made by linking together prefabricated molecular units (monomers) that are similar or identical to each other. The number of monomers in a polymer may range up to millions. (2) A long chain of repeating subunits.

Polymerase Chain Reaction (PCR) A technique that rapidly produces a billion or more identical copies of a DNA fragment without needing a cell.

Polymyxin An antibacterial agent that disrupts the cell membrane.

Polynucleotide A chain of many nucleotides.

Polypeptide A chain of many amino acids.

Polyribosome A long chain of ribosomes attached at different points along an mRNA molecule. *Known also as Polysome.*

Polysaccharide A carbohydrate formed when many monosaccharides are linked together by glycosidic bonds.

Polytene Chromosome An extra-thick chromosome that includes many parallel copies of the original DNA molecule; it is produced by repeated rounds of DNA replication without separation of the resulting copies. Polytene chromosomes are found in various cell types, notably *Drosophilia* salivary gland cells; they are useful in chromosome mapping because they are large and because the genes on the strands are arranged in strict register.

Pontiac Fever A mild variety of legionellosis.

Porin A protein in the outer membrane of Gram-negative bacteria that nonselectively transports polar molecules into the periplasmic space.

Portal of Entry A site at which microorganisms an gain access to body tissues.

Portal of Exit A site at which microorganisms can leave the body.

Positive Chemotaxis Movement of an organism toward a chemical.

Positive (+) Sense RNA An RNA strand that encodes information for making proteins needed by a virus.

Potable Water Water that is fit for human consumption.

Pour Plate A plate containing separate colonies and used to prepare a pure culture.

Pour Plate Method Method used to prepare pure cultures using serial dilutions, each of which is mixed with melted agar and poured into a sterile Petri plate.

Poxvirus DNA virus that is the largest and most complex of all viruses.

Precipitation Reaction Immunological test in which antibodies called precipitans react with antigens to form latticelike networks of molecules that precipitate from solution.

Precipitin Test Immunological test used to detect antibodies that is based on the precipitation reaction.

Prediction The expected outcome if a hypothesis is correct.

Preserved Culture A culture in which organisms are maintained in a dormant state.

Presumptive Test First stage of testing in multiple-tube fermentation in which gas production in lactose broth provides presumptive evidence that coliform bacteria are present.

Prevalence Rate The number of people infected with a particular disease at any one time.

Primaquine An antiprotozoan agent that interferes with a protein synthesis.

Primary Atypical Pneumonia A mild form of pneumonia with insidious onset. *Known also as Mycoplasma Pneumonia and Walking Pneumonia.*

Primary Cell Culture A culture that comes directly from an animal and is not subcultured.

Primary Immunodeficiency Disease A genetic or developmental defect in which T cells or B cells are lacking or nonfunctional.

Primary Infection An initial infection in a previously healthy person.

Primary Response Humoral immune response that occurs when an antigen is first recognized by host B cells.

Primary Structure For a nucleic acid or a protein, the sequence of the bases or amino acids in the polynucleotide or polypeptide. Compare ► Quaternary Structure, ► Secondary Structure, and ► Tertiary Structure.

Primary Treatment Physical treatment to remove solid wastes from sewage.

Primer A short piece of DNA or RNA that is base-paired with a DNA template strand and provide a free 3′—OH end from which a DNA polymerase can extend a DNA strand. Also refers to DNA oligomers used in the polymerase chain reaction.

Primosome An enzyme complex that is located in the replication fork during DNA replication; it synthesizes the RNA primers on the lagging strand and also participates in unwinding the parental DNA helix.

Prion An exceedingly small infectious particle consisting of protein without any nucleic acid.

Probe A single-stranded DNA fragment that has a sequence of bases that can be used to identify complementary DNA base sequences.

Processivity For a DNA or an RNA polymerase, the average number of nucleotides incorporated per event of binding between the polymerase and a 3′ primer terminus. It describes the tendency of a polymerase to remain bound to a template.

Prodromal Phase In an infectious disease, the short period during which nonspecific symptoms such as malaise and headache sometimes appear.

Prodrome A symptom indicating the onset of a disease.

Producer Organism that captures energy from the sun and synthesizes food. *Known also as Autotroph.*

Product The material resulting from an enzymatic reaction.

Productive Infection Viral infection in which viruses enter a cell and produce infectious progeny.

Proglottid One of the segments of a tapeworm, containing the reproductive organs.

Progressive Multifocal Leukoencephalopathy Disease caused by the JC polyomavirus with symptoms including mental deterioration, limb paralysis, and blindness.

Prokaryote Microorganism that lacks a cell nucleus and membrane-enclosed internal structures, all bacteria in the kingdom Monera (Prokaryotae) are prokaryotes.

Prokaryotes Primitive single-celled organism that are not compartmentalized by internal cellular membranes; the eubacteria and archaebacteria. Compare ► Eukaryotes.

Prokaryotic Cell A cell that lacks a cell nucleus; includes all bacteria.

Promoter A DNA sequence that can bind RNA polymerase, resulting in the initiation of transcription.

Propagated Epidemic An epidemic that arises from person-to-person contacts.

Prophage An inactive phage genome that is present in a bacterial cell and its progeny. It is integrated into the host chromosome.

Propionibacteria Pleomorphic, irregular, nonsporing, Gram-positive rods.

Prostaglandin A reaction mediator that acts as a cellular regulator, often intensifying pain.

Prostaglandins A family of compounds that are derived from certain long-chain unsaturated fatty acids (particularly arachidonic acid) by a cyclooxygenase pathway and that function as local hormones.

Prostate Gland The gland located at the beginning of the male urethra whose milky fluid discharge forms a component of semen.

Prostatitis Inflammation of the prostate gland.

Prosthetic Group A metal ion or small molecule (other than an amino acid) that forms part of a protein in the protein's native state and is essential to the protein's functioning; its attachment to the protein may be either covalent or noncovalent.

Proteases Enzymes that cleave peptide bonds in a polypeptide. Many show specificity for a particular amino acid sequence.

Proteasome A large, ATP-dependent protease complex that is found in the cytosol of cells and is involved in the selective degradation of short-lived cytoplasmic proteins.

Protein A polymer of amino acids joined by peptide bonds.

Protein Profile A technique for visualizing the proteins contained in a cell; obtained by the use of polyacrylamide gel electrophoresis.

Proteoglycans Glycoproteins in which carbohydrate is the dominant element. The carbohydrate is in the form of glycosamminoglycan polysaccharides, which are connected to extended core polypeptides to form huge, feathery molecules. Proteoglycans are important components of the intercellular matrix.

Protista The kingdom of organisms that are unicellular but contain internal organelles typical of the eukaryotes.

Protist A unicellular eukaryotic organism that is a member of the kingdom Protista.

Protofilaments The 13 linear columns of tubulin units that can be visualized in the structure of a microtubule; they result because each turn of the microtubule helix contains exactly 13 tubulin units. Each protofilament consists of alternating α and β tubulin subunits.

Proton A positively charged subatomic particle located in the nucleus of an atom.

Proton Motive Force (pmf) An electrochemical H^+ gradient that is set up across a cellular membrane by membrane-bound proton pumps, such as the ones in the inner mitochondrial membrane or thylakoid membrane. As the protons flow back down their gradient across the membrane, they can drive processes such as ATP synthesis.

Proton Pumping The active pumping of protons across a cellular membrane to form a proton gradient. For example, the electron transport chains of the inner mitochondrial and Thylakoid membranes incorporate proton pumps, which create the proton gradient that powers the ATP synthases of these membranes.

Proto-Oncogene A normal gene that can cause cancer in uncontrolled situations; often the normal gene comes under the control of a virus.

Protoplast A Gram-positive bacterium from which the cell wall has been removed.

Protoplast Fusion A technique of genetic engineering in which genetic material is combined by removing the cell walls of two different types of cells and allowing the resulting protoplasts to fuse.

Prototroph A normal, nonmutant organism. *Known also as Wild Type.*

Protozoa (singular: *protozoan*) Single-celled, microscopic, animallike protests in the kingdom Protista.

Provirus Viral DNA that is incorporated into a host-cell chromosome.

Pseudocoelom A primitive body cavity, typical of nematodes, that lacks the complete lining found in higher animals.

Pseudocyst An aggregate of trypanosome protozoa that forms in lymph nodes in Chagas' disease.

Pseudogenes Nontranscribed stretches of DNA that bear a strong sequence similarity to functioning genes and obviously arose from them during evolution. Many gene families contain pseudogene members.

Pseudomembrane A combination of bacilli, damaged epithelial cells, fibrin, and blood cells resulting from infection with diphtheria that can block the airway, causing suffocation.

Pseudomonads Aerobic motile rods with polar flagella.

Paeudoplasmodium A multicellular mass composed of individual cellular slime mold cells that have aggregated.

Pseudopodium A temporary footlike projection of cytoplasm associated with amoeboid movement.

Psittacosis See ► Ornithosis.

Psychrophile A cold-loving organism that grows best at temperatures of 15 to 20°C.

Puerperal Fever Disease caused by β-hemolytic streptococci, which are normal vaginal and respiratory microbiota that can be introduced during child delivery by medical personnel. *Known also as Childbed Fever or Puerperal Sepsis.*

Pulsed Field Gel Electrophoresis A type of gel electrophoresis in which the orientation of the electric field is charged periodically. This technique makes it possible to separate very large DNA molecules, up to the size of whole chromosomes.

Pure Culture A culture that contains only a single species of organism.

Purine The nucleic acid bases adenine and guanine.

Pus Fluid formed by the accumulation of dead phagocytes, the material they have ingested, and tissue debris.

Pustule See ► Folliculitis.

Pyelonephritis Inflammation of the kidneys.

Pyoderma A pus-producing skin infection caused by staphylococci, streptococci, and Corynebacteria, singly or in combination.

Pyrimidine Any of the nucleic acid bases thymine, cytosine, and uracil.

Pyrogen A substance that acts on the hypothalamus to set the body's "thermostat" to a higher-than-normal temperature.

Q Fever Pneumonialike disease caused by *Coxiella burnetii*, a rickettsia that survives long periods outside cells and can be transmitted aerially as well as by ticks.

Quantum Efficiency (*Q*) With respect to photosynthesis, the ratio of oxygen molecules released to photons absorbed.

Quarantine The separation of human or animals from the general population when they have a communicable disease or have been exposed to one.

Quaternary Ammonium Compound (quat) A cationic detergent that has four organic groups attached to a nitrogen atom.

Quaternary Structure (1) The three-dimensional structure of a protein molecule formed by the association of two or more polypeptide chains. (2) For a protein, the level of structure that results when separate, folded polypeptide chains (subunits) associate in a specific way to produce a complete protein. Compare ▶ Primary Structure, ▶ Secondary Structure, and ▶ Tertiary Structure.

Quinine An antiprotozoan agent used to treat malaria.

Quinolone A bactericidal agent that inhibits DNA replication.

Quinone A nonprotein, lipid-soluble electron carrier in oxidative phosphorylation. *Known also as Coenzyme Q.*

Rabies A viral disease that affects the brain and nervous system with symptoms including hydrophobia and aerophobia; transmitted by animal bites.

Rabies Virus An RNA-containing rhabdovirus that is transmitted through animal bites.

Rad A unit of radiation energy absorbed per gram of tissue.

Radial Immunodiffusion Serological test used to provide a quantitative measure of antigen or antibody concentration by measuring the diameter of the ring of precipitation around an antigen.

Radiation Light rays, such as X-rays and ultraviolet rays, that can act as mutagens.

Radioautography A technique in which an item containing radioactively labeled elements (for example, a tissue slice or a chromatography gel) is laid against a photographic film; the radioactivity exposes the film to form an image of the labeled elements. Also called autoradiography.

Radioimmunoassay (RIA) Technique that uses a radioactive anti-antibody to detect very small quantities of antigens or antibodies.

Radioisotope Isotope with unstable nuclei that tends to emit subatomic particles and radiation.

Ramachandran Plot A plot that constitutes a map of all possible backbone configurations for an amino acid in a polypeptide. The axes of the plot consist of the rotation angles of the two backbone bonds that are free to rotate (ϕ and ψ, respectively); each point ϕ,ψ on the plot thus represents a conceivable amino acid backbone configuration.

Random Coil Refers to a linear polymer that has no secondary or tertiary structure but instead is wholly flexible with a randomly varying geometry. This is the state of a denatured protein or nucleic acid.

Rat Bite Fever A disease caused by *Streptobacillus moniliformis* transmitted by bites from wild and laboratory rats.

Rate Constant With respect to chemical reactions, a constant that relates the reaction rate for a particular reaction to substrate concentrations.

Rate Equation An equation, such as the Michaelis-Menten equation that relates velocity of an enzyme-catalyzed reaction to measurable parameters.

Reactant Substance that takes part in a chemical (enzymatic) reaction.

Reaction Center In photosynthesis, a specific pair of chlorophyll molecules in a photosystem that collect light energy absorbed by other chlorophyll molecules and pass it to an electron acceptor, normally the first compound of an electron transport chain.

Reactive Oxygen Species (ROS) Oxygen species intermediate in oxidation level between O_2 and H_2O, which are more reactive than O_2; ROS include superoxide, peroxide, peroxynitrite, and hydroxyl radical.

Reagin Older name for immunoglobulin E (IgE); very important in allergies.

Receptor A protein that binds selectively to a specific molecule (such as an intercellular mediator or antigen) and initiates a biological response.

Recognition Helix In a helix-turn-helix DNA binding motif, the α-3 helix, which fits deep in the major groove and is responsible for the sequence specificity of binding.

Recombinant DNA DNA combined from two different species by restriction enzymes and ligases.

Recombinant DNA Molecule A DNA molecule that includes segments from two or more precursor DNA molecules.

Recombination (1) The combining of DNA from two different cells, resulting in a recombinant cell. (2) A process in an organism in which two parent DNA molecules give rise to daughter DNA that combines segments from both parent molecules. It may involve the integration of one DNA molecule into another, the substitution of a DNA segment for a homologous segment on another DNA molecule, or the exchange of homologous segments between two DNA molecules.

Redia The development stage of the fluke immediately following the sporocyst stage.

Reducing Equivalent An amount of a reducing compound that donates the equivalent of 1 mole of electrons in an oxidation-reduction reaction. The electrons may be expressed in the form of hydrogen atoms.

Reduction The gain of electrons and hydrogen atoms.

Reference Culture A preserved culture used to maintain an organism with its characteristics ad originally defined.

Reflection The bouncing of light off an object.

Refraction The bending of light as it passes from one medium to another medium of different density.

Regulator Gene Gene that controls the expression of structural genes of an operon through the synthesis of a repressor protein.

Regulatory Site The promotor and operator regions of an operon.

Regulon A group of unlinked (nonadjacent) genes that are all regulated by a common mechanism.

Relapsing Fever Disease caused by various species of *Borrelia,* most commonly by *B recurrentis*; transmitted by lice.

Release The exit from the host cell of new virions, which usually kills the host cell.

Release Factors Independent protein factors that are necessary participants in the release of a finished polypeptide chain from a ribosome.

Rennin An enzyme from calves' stomachs used in cheese manufacture.

Reovirus A medium-sized RNA virus that has a double-capsid with no envelope; causes upper respiratory and gastrointestinal infections in humans.

Replica Plating A technique used to transfer colonies from one medium to another.

Replication Process by which an organism or structure (especially a DNA molecule) duplicates itself.

Replication Cycle The series of steps of virus replication in a host cell.

Replication Fork A site at which the two strands of the DNA double helix separate during replication and new complementary DNA strands form.

Replicon A unit in the genome that consists of an origin of replication and all the DNA that is replicated from that origin.

Repressor In an operon it is the protein that binds to the operator, thereby preventing transcription of adjacent genes.

Repressor Protein Substance produced by host cells that keeps a virus in an inactive state and prevents the infections of the cell by another phage.

Reservoir Host An infected organism that makes parasites available for transmission to other hosts.

Reservoir of Infection Site where microorganisms can persist and maintain their ability to infect.

Resident Microflora Species of microorganisms that are always present on or in an organism.

Resistance The ability of a microorganism to remain unharmed by an antimicrobial agent.

Resistance (R) Gene A component of a resistance plasmid that confers resistance to a specific antibiotic or to a toxic metal.

Resistance (R) Plasmid A plasmid that carries genes that provide resistance to various antibiotics or toxic metals. *Known also as R Factor.*

Resistance Transfer Factor (RTF) A component of a resistance plasmid that implements transfer by conjugation of the plasmid.

Resolution The ability of an optical device to show two items as separate and discrete entities rather than a fuzzily overlapping image.

Resolving Power A numerical measure of the resolution of an optical instrument.

Respiration With respect to energy metabolism, the process in which cellular energy is generated through the oxidation of nutrient molecules with O_2 as the ultimate electron acceptor. This type of respiration is also called *cellular respiration* to distinguish it from respiration in the sense of breathing.

Respiratory Anaphylaxis Life-threatening allergy in which airways become constructed and filled with mucous secretions.

Respiratory Bronchiole Microscopic channel in the lower respiratory system that ends in a series of alveoli.

Respiratory Chain The electron transport chain that is employed during cellular respiration and has O_2 as the ultimate electron acceptor.

Respiratory Syncytial Virus (RSV) Cause of lower respiratory infections affecting children under 1 year old, causes cells in culture to fuse their plasma membranes and become multinucleate masses (syncytia).

Respiratory System Body system that moves oxygen from the atmosphere to the blood and removes carbon dioxide and other wastes from the blood.

Resting Potential The voltage difference that exists across the membrane of an excitable cell, such as a nerve cell, except in places when an action potential is in progress. It is a consequence of the ion gradients that are maintained across the membrane.

Restriction Endonuclease Enzymes that catalyze the double-strand cleavage of DNA at specific base sequences. Many restriction endonucleases with different sequence specificities have been found in bacteria; they are used extensively in molecular genetics.

Restriction Enzyme Another term for RESTRICTION ENDONUCLEASE.

Restriction Fragment Length Polymophism (RFLP) (1) A short piece of DNA snipped out by restriction enzymes. (2) A type of genetic polymorphism that is readily detected by Southern blotting and can be used to screen for genetic diseases. It is based on the fact that alleles often have different restriction Endonuclease cleavage sites and therefore produce different arrays of fragments upon cleavage with appropriate endonucleases.

Reticulate Body An intracellular stage in the life cycle of chlamydias.

Retinoids Substances that are derived from retinoic acid (a form of vitamin A) and act as intercellular mediators; they are particularly important in regulating development.

Retrovirus An enveloped RNA virus that uses its own reverse transcriptase to transcribe its RNA into DNA in the cytoplasm of the host cell.

Retroviruses A family of RNA viruses that possess reverse transcriptase. After the virus infects a cell, this enzyme transcribes the RNA genome into a double-strand DNA version, which integrates into a host chromosome. Human immunodeficiency virus (HIV) is a retrovirus.

Reverse Transcriptase An enzyme found in retroviruses that synthesizes a double-strand DNA molecule from a single-strand RNA template. It is an important tool in molecular genetics.

Reverse Transcription An enzyme found in retroviruses that copies RNA into DNA.

R Factor See ▶ Resistance (R) Plasmid.

R Group An organic chemical group attached to the central carbon atom in an amino acid.

Rhabdovirus A rod-shaped, enveloped RNA virus that infects insects, fish, various other animals, and some plants.

Rh Antigen An antigen found on some red blood cells, discovered in the cells of Rhesus monkeys.

Rheumatic Fever A multisystem disorder following infection by β-hemolytic *Streptococcus pyogenes* that can cause heart damage.

Rheumatoid Arthritis Autoimmune disease that affects mainly the joints but can extend to other tissues.

Rheumatoid Factor IgM found in the blood of patients with rheumatoid arthritis, and their relatives.

Rhinovirus A virus that replicates in cells of the upper respiratory tract and causes the common cold.

Ribonucleic Acid (RNA) Nucleic acid that carries information from DNA to sites where proteins are manufactured in cells and that directs and participates in the assembly of proteins.

Ribosomal RNA (rRNA) A type of RNA that, together with specific proteins, makes up the ribosomes.

Ribosome Site for protein synthesis consisting of RNA and protein, located in the cytoplasm.

Ribosomes Large protein – RNA complexes that are responsible for synthesizing polypeptides under the direction of mRNA templates.

Rickettsialpox Mild rickettsial disease with symptoms resembling those of chickenpox; caused by *Rickettsia akari* and carried by mites found on house mice.

Rickettsias Small, nonmotile, Gram-negative organisms, obligate intercellular parasites of mammalian and anthropod cells.

Rifamycin An antibacterial agent that inhibits ribonucleic and (RNA) synthesis.

Rift Valley Fever Disease caused by bunyaviruses that occurs in epidemics.

Ringworm A highly contagious fungal skin disease that can cause ringlike lesions.

River Blindness See ▶ Onchocerciasis.

RNA Editing A type of RNA processing that has been found in the mitochondrial mRNAs of certain eukaryotes, in which the RNA sequence is altered by the insertion of uridine residues at specific sites.

RNA Polymerase An enzyme that binds to one strand of exposed DNA during transcription and catalyzes the synthesis of RNA from the DNA template.

RNA Primer During DNA replication, the short stretch of RNA nucleotides that is laid down at the beginning of each Okazaki fragment; it provides a 3′ –OH end from which DNA polymerase can extend the fragment, It is later replaced with DNA.

RNA Tumor Virus Any retrovirus that causes tumors and cancer.

Rocky Mountain Spotted Fever Disease caused by *Rickettsia rickettsia* and transmitted by ticks.

Rotavirus Virus transmitted by the fecal-oral route that replicates in the intestine, causing diarrhea and enteritis.

Roundworm A worm with a long, cylindrical, unsegmented body and a heavy cuticle. *Known also as a Nematode.*

Rubella Viral disease characterized by a skin rash; can cause severe congenital damage. *Known also as German Measles.*

Rubeola See ▶ Measles.

Rubisco (Ribulose Bisphosphate Carboxylase-Oxygenase) The enzyme that accomplishes carbon fixation in photosynthesis by adding CO_2 to ribulose-1,5-bisphosphate. It can also add O_2 in place of CO_2, initiating photorespiration.

Rule of Octets Principle that an element is chemically stable if it contains eight electrons in its outer shell.

Sac Fungus A member of a diverse group of fungi that produces saclike asci during sexual reproduction. *Known also as Ascomycota.*

St. Anthony's Fire See ► Erysipelas.

St. Louis Encephalitis Type of viral encephalitis most often seen in humans in the central United States.

Salmonellosis A common enteritis characterized by abdominal pain, fever, and diarrhea with blood and mucus; caused by *Salmonella* species.

Sapremia A condition caused when saprophytes release metabolic products into the blood.

Saprophyte An organism that feeds on dead or decaying organic matter.

Sarcina A group of eight cocci in a cubicle packet.

Sarcodine An amoeboid protozoan.

Sarcoplasmic Reticulum A network of membranous tubules that surrounds each myofibril in a skeletal muscle cell. It is a specialized region of endoplasmic reticulum; its main function is to sequester and then release the Ca^{2+} that triggers myofibril contraction.

Sarcoptic Manage See ► Scabies.

Satellite DNA DNA consisting of multiple tandem repeats of very short, simple nucleotide sequences. It typically makes up 10% to 20% of the genome of higher eukaryotes; at least some of it may play a role in chromosome structure.

Saturated Fatty Acid A fatty acid containing only carbon-hydrogen single bonds.

Scabies Highly contagious skin disease caused by the itch mite *Sarcoptes scabiei. Known also as Sarcoptic Mange.*

Scalded Skin Syndrome Infection caused by staphylococci consisting of large, soft vesicles over the whole body.

Scanning Electron Microscope (SEM) A type of electron microscopy in which a beam of electrons is scanned across an object, and the pattern of reflected electrons is analyzed to crease an image of the object's surface. This type of microscope is used to study the surfaces of specimens. Compare ► Transmission Electron Microscopy.

Scanning Tunneling Microscope (STM) Also called scanning probe microscope; type of microscope in which electron tunnel into each other's clouds, can show individual molecules, live specimens, and work underwater.

Scarlet Fever (sometimes called scarlatina) Infection caused by *Streptococcus pyogenes* that produces an erythrogenic toxin.

Schaeffer-Fulton Spore Staining A differential stain used to make endospores easier to visualize.

Schick Test Test to determine immunity to diphtheria.

Schistosomiasis Disease of the blood and lymph caused by blood fluke of the genus *Schistosoma. Known also as Bilharzia.*

Schizogony Multiple fission, in which one cell gives rise to many cells.

Scolex Head end of a tapeworm, with suckers and sometimes hooks that attach to the intestinal wall.

Scrapie Transmissible spongiform encephalopathy disease of the brain of sheep, causing extreme itching so that the sheep repeatedly scrape themselves against posts, trees, etc.

Scrub Typhus A typhus caused by *Rickettsia tsutsugamushi*, transmitted by mites that feed on rats. *Known also as Tsutsugamushi.*

Sebaceous Gland Epidermal structure, associated with hair follicles, that secretes an oily substance called sebum.

Sebum Oily substance secreted by the sebaceous glands.

Secondary Immunodeficiency Disease Result of damage to T cells or B cells after they have developed normally.

Secondary Infection Infection that follows a primary infection, especially in patients weakened by the primary infection.

Secondary Response The folding or coiling of a polypeptide chain into a particular pattern, such as a helix or pleated sheet.

Secondary Structure (1) Local folding of the backbone of a linear polymer to form a regular, repeating structure. The B- and Z-forms of the DNA helix and the α-helix and β-sheet structures of polypeptides are examples. Compare ► Primary Structure, ► Quaternary Structure, and ► Tertiary Structure. (2) The folding or coiling of a polypeptide chain into a particular pattern such as a helix or pleated sheet.

Secondary Treatment Treatment of sewage by biological means to remove remaining solid wastes after primary treatment.

Second Law of Thermodynamics The law that states that the entropy in a closed system never decreases. An alternative statement is that processes that are thermodynamically favored at constant temperature and pressure involve a decrease in free energy.

Second Messenger An intercellular substance that relays an Extracellular signal (such as a hormonal signal) from the cell membrane to intracellular effector proteins.

Second-Order Reaction A reaction in which two reactant molecules must come together for the reaction to occur. The reaction is called second-order because the reaction rate depends on the square of reactant concentration (for two molecules of the same reactant) or on the product of two reactant concentrations (for two different reactants). Compare ▶ First-Order Reaction.

Secretory Piece A part of the IgA antibody that protects the immunoglobulin from degradation and helps in the secretion of the antibody.

Secretory Vesicle Small membrane-enclosed structure that stores substances coming from the Golgi apparatus.

Sedimentation Coefficient (*S*) A coefficient that determines the velocity at which a particular particle will sediment during centrifugation; it depends on the density of the medium, the specific density of the particle, and the size, shape, and mass of the particle.

Sedimentation Equilibrium A technique for using centrifugation to measure the mass of a large molecule such as a protein. A solution of the substance is centrifuged at low speed until the tendency of the substance to sediment is balanced by its tendency to diffuse to uniform concentration; the resulting concentration gradient is used to measure the molecular mass.

Selectively Permeable Able to prevent the passage of certain specific molecules and ions while allowing others through.

Selective Medium A medium that encourages growth of some organisms and suppresses growth of others.

Selective Toxicity The ability of an antimicrobial agent to harm microbes without causing significant damage to the host.

Self Molecules that are not recognized as antigenic or foreign by an organism.

Semen The male fluid discharge at the time of ejaculation, containing sperm and various glandular and other secretions.

Semiconservative Replication Replication in which a new DNA double helix is synthesized from one strand of parent DNA and one strand of new DNA.

Seminal Vesicle A saclike structure whose secretions form a component of semen.

Semisynthetic Drug An antimicrobial agent made partly by laboratory synthesis and partly by microorganisms.

Sense Codon A set of three DNA (or mRNA) bases that code for an amino acid.

Sense Strand For a gene, the DNA strand that is homologous to an RNA transcript of the gene – that is, it carries the same sequence as the transcript, except with T in place of U. It is thus complementary to the strand that served as a template for the RNA.

Sensitization Initial exposure to an antigen, which causes the host to mount an immune response against it.

Septicemia An infection caused by rapid multiplication of pathogens in the blood. *Known also as Blood Poisoning.*

Septicemic Plague Fatal form of plague that occurs when bubonic plague bacteria move from the lymphatics to the circulatory system.

Septic Shock A life-threatening septicemia with low blood pressure and blood-vessel collapse, caused by endotoxins.

Septic Tank An underground tank for receiving sewage, where solid material settle out as sludge, which must be pumped periodically.

Septum (plural: *septa*) A cross-wall separating two fungal cells.

Sequela (plural: *sequelae*) The aftereffect of a disease; after recovery from it.

Serial Dilution A method of measurement in which successive 1:10 dilutions are made from the original sample.

Seroconversion The identification of a specific antibody in serum as a result of an infection.

Serology The branch of immunology dealing with laboratory tests to detect antigens and antibodies.

Serovar Strain; a subspecies category.

Serum The liquid part of blood after cells and clotting factors have been removed.

Serum Hepatitis See ▶ Hepatitis B.

Serum Killing Power Test used to determine effectiveness of an antimicrobial agent in which a bacterial suspension is added to the serum of a patient whose is receiving an antibiotic; and incubated.

Serum Sickness Immune complex disorder that occurs when foreign antigens in seta cause immune complexes to be deposited in tissues.

Severe Combined Immunodeficiency (SCID) Primary immunodeficiency disease caused by failure of stem cells to develop properly, resulting in deficiency of both B and T cells.

Sewage Used water and the wastes it contains.

Sex Factors Plasmids that specify gene products that enable bacteria to engage in conjugation (bacterial mating).

Sexually Transmitted Disease (STD) An infectious disease spread by sexual activities.

Shadow Casting The coating of electron microscopy specimens with a heavy metal, such as gold or palladium, to create a three-dimensional effect.

Shigellosis Gastrointestinal disease caused by several strains of *Shigella* that invade intestinal lining cells. *Known also as Bacillary Dysentery.*

Shinbone Fever See ▶ Trench Fever.

Shingles Sporadic disease caused by reactivation of varicella-zoster herpesvirus that appears most frequently in older and Immunocompromised individuals.

Shrub of Life A diagram that represents our current understanding of the early evolution of life. There are many roots rather than a single ancestral line, and the branches criss-cross and merge again and again.

Sickle-Cell Disease A genetic disease resulting from a hemoglobin mutation. It produces fragile erythrocytes, leading to anemia.

Sign A disease characteristic that can be observed by examining the patient, such as swelling or redness.

Signal Recognition Particles (SRPs) Cytoplasmic particles that dock ribosomes on the surface of the endoplasmic reticulum (ER) if the nascent polypeptide is destined to be processed by the ER. The SRP recognizes and binds to a specific N-terminal signal sequence on the nascent polypeptide.

Simple Diffusion The net movement of particles from a region of higher to one of lower concentration; does not require energy from a cell.

Signal Sequence See ▶ Leader Sequence.

Simple Stain A single dye used to reveal basic cell shapes and arrangements.

Single-Cell Protein (SCP) Animal feed consisting of microorganisms.

Sinus A large passageway in tissues, lined with phagocytic cells.

Sinusitis An infection of the sinus cavities.

Sinusoid An enlarged capillary.

Site-Directed Mutagenesis A technique by which a specific mutation is introduced at a specific site in a cloned gene. The gene can then be introduced into an organism and expressed.

6–4 Photoproduct A type of DNA damage caused by UV irradiation in which a bond forms between carbon-6 of one pyrimidine base and carbon-4 of an adjacent pyrimidine base. This type of photoproduct appears to be the chief cause of UV-induced mutations.

Skin The largest single organ of the body that presents a physical barrier to infection by microorganisms.

Slime Layer A thin protective structure loosely bound to the cell wall that protects the cell against drying, helps trap nutrients, and sometimes binds cells together.

Slime Mold A funguslike protist.

Sludge Solid matter remaining from water treatment that contains aerobic organisms that digest organic matter.

Sludge Digester Large fermentation tank in which sludge is digested by anerobic bacteria into simple organic molecules, carbon dioxide, and methane gas.

Small Intestine The upper area of the intestine where digestion is completed.

Smallpox A formerly worldwide and serious viral disease that has now been eradicated.

Smear A thin layer of liquid specimen spread out on a microscopic slide.

Solute The substance dissolved in a solvent to form a solution.

Solution A mixture of two or more substances in which the molecules are evenly distributed and will not separate out on standing.

Solvent The medium in which substances are dissolved to form a solution.

Somatic Mutation A mutation that occurs in a cell of an organism other than a germ-like cell; it may affect the organism in which it occurs, but it cannot be passed on to progeny.

Sonication The disruption of cells by sound waves.

SOS Response A bacterial response to various potentially lethal stresses, including severe UV irradiation. It involves the coordinated expression of a set of proteins that carry out survival maneuvers, including an error-prone type of repair for thymine dimers in DNA.

Southern Blotting A technique for detecting the presence of a specific DNA sequence in a genome. The DNA is extracted, cleaved into fragments, separated by gel electrophoresis, denatured, and blotted onto a nitrocellulose filter. There it is incubated under annealing conditions with a radiolabeled probe for the sequence in question, and heteroduplexes of the probe with genomic DNA are detected by radioautography.

Specialized Transduction Type of transduction in which the bacterial DNA transduced is limited to one or a few genes lying adjacent to a prophage that are accidentally included when the prophage is excised from the bacterial chromosome.

Species A group of organisms wit many common characteristics; the narrowest taxon.

Species Immunity Innate or inborn genetic immunity.

Specific Defense A host defense that operates in response to a particular invading pathogen.

Specific Epithet The second name of an organism in the binomial system of nomenclature, following that of the genus – for example, *coli* in *Escherichia coli.*

Specific Immunity Defense against a particular microbe.

Specificity (1) The property of an enzyme that allows it to accept only certain substrates and catalyze only one particular reaction. (2) The property of a virus that restricts it to certain specific types of host cells. (3) The ability of the immune system to mount a unique immune response to each antigen it encounters.

Spectrophotometer An instrument that exposes a sample to light of defined wavelengths and measures the absorbance. Different types of spectrophotometers operate in different wavelength ranges, such as ultraviolet, visible, and infrared.

Spectrum of Activity Refers to the range of different microbes against which are antimicrobial agent is effective.

Spheroplast A Gram-negative bacterium that lacks the cell wall but had not lysed.

Spike A glycoprotein projection that extend to form the viral capsid or envelope and is used to attach to or fuse with host cells.

Spindle Apparatus A system of microtubules in the cytoplasm of a eukaryotic cell that guides the movement of chromosomes during mitosis and meiosis.

Spin Label A substance that has an unpaired electron detectable by electron spin resonance and that is used as a chemical label.

Spirillar Fever A form of rat bite fever, caused by *Spirillum minor*, first described as sodoku in Japan.

Spirillum (plural: *spirilla*) A flexible, wavy-shaped bacterium.

Spirochetes Corkscrew-shaped motile bacteria.

Spleen The largest lymphatic organ; acts as a blood filter.

Spliceosome A protein-RNA complex in the nucleus that is responsible for splicing introns out of RNA transcripts.

Spontaneous Generation The theory that living organisms can arise from nonliving things.

Spontaneous Mutation A mutation that occurs in the absence of any agent known to cause changes in DNA; usually caused by errors during DNA replication.

Sporadic Disease A disease that is limited to a small number of isolated case posing no great threat to a large population.

Spore A resistant reproductive structure formed by fungi and actinomycetes; different from a bacterial endospore.

Spore Coat A keratinlike protein material that is lad down around the cortex of an endospore by the mother cell.

Sporocyst Larval form of a fluke that develops in the body of its snail or mollusk host.

Sporotrichosis Fungal skin disease caused by *Sporothrix schenckii* that often enters the body from plants.

Sporozoite A malaria trophozoite present in the salivary glands of infected mosquitoes.

Sporulation The formation of spores such as endospores.

Spread Plate Method A technique used to prepare pure cultures by placing a diluted sample of cells on the surface of an agar plate and then spreading the sample evenly over the surface.

Stain A molecule that can bind to a structure and give it color. *Known also as a Dye.*

Standard Bacterial Growth Curve A graph plotting the number of bacteria versus time and showing the phases of bacterial growth.

Standard Reduction Potential (E_0) For a given pair consisting of an electron donor and its conjugate acceptor, the reduction potential under standard conditions (25°C; donor and acceptor both at 1 m concentration).

Standard State A reference state, with respect to which thermodynamic quantities (such as chemical potentials) are defined. For substances in solution, standard state indicates 1 m concentration at 1 atm pressure and 25°C.

Start Codon The first codon in a molecule of mRNA which begins the sequence of amino acids in protein synthesis; in bacteria it always codes for methionine.

Stationary Phase The third of four major phases of the bacterial growth curve in which new cells are produced at the same rate that old cells die, leaving the number of live cells constant.

Sterility The state in which there are no living organisms in or on a material.

Sterilization The killing or removal of all microorganisms in a material or on an object.

Steroid A lip having a four-ring structure, includes cholesterol, steroid hormones, and vitamin D.

Stock Culture A reserve culture used to store an isolated organism in pure condition for use in the laboratory.

Stop Codon (1) The last codon to be translated in a molecule of mRNA, causing the ribosome to release from the mRNA. (2) RNA codons that signal a ribosome to stop translating an mRNA and to release the polypeptide. In the normal genetic code, they are UAG, UGA, and UAA.

Strain A subgroup of a species with one or more characteristics that distinguish it from other subgroups of that species.

Streak Plate Method Method used to prepare pure cultures in which bacterial are lightly spread over the surface of agar plates, resulting in isolated colonies.

Streptococci Aerotolerant anerobes that form pairs, tetrads, or chains by dividing in one or two planes; most lack the enzyme catalase.

Streptokinase A bacterially produced enzyme that digests (dissolves) blood clots.

Streptolysin Toxin produced by streptococci that kills phagocytes.

Streptomycetes Gram-positive, filamentous, sporing, soil-dwelling bacteria, produces of many antibiotics.

Streptomycin An antibacterial agent that blocks protein synthesis.

Stringent Response A mechanism that inhibits the expression of all structural genes in bacteria under conditions of amino acid starvation. It involves inhibition of the synthesis of ribosomal and transfer RNAs.

Stroma The fluid-filled inner portion of a chloroplast.

Stromatolite Live or fossilized layered mats of photosynthetic prokaryotes associated with warm lagoons or hot springs.

Strongyloidiasis Parasitic disease caused by the roundworm *Stongyloides stercoralis* and a few closely related species.

Structural Gene A gene that carries information for the synthesis of a specific polypeptide.

Structural Protein A protein that contributes to the structure of cells, cell parts, and membranes.

Sty An infection at the base of an eyelash.

Subacute Disease A disease that is intermediate between an acute and a chronic disease.

Subacute Sclerosing Panencephalitis (SSPE) A complication of measles, nearly always fatal, that is due to the persistence of measles viruses in brain tissue.

Subclinical Infection See ▶ Inapparent Infection.

Subculturing The process by which cells from an existing culture are transferred to fresh medium in new containers.

Substrate (1) The substance on which an enzyme acts. (2) A surface or food source on which a cell can grow or a spore can germinate.

Substrate-Level Phosphorylation Synthesis of a nucleoside triphosphate (usually ATP) driven by the breakdown of a compound with higher phosphate transfer potential.

Suicide Inhibitor An enzyme inhibitor on which the enzyme can act catalytically but which irreversibly alters the active site of the enzyme in the process. (It is called a suicide inhibitor because the enzyme "commits suicide" by acting on it.)

Sulfate Reduction The reduction of sulfate ions to hydrogen sulfide.

Sulfonamide A synthetic, bacteriostatic agent that blocks the synthesis of folic acid. *Known also as Sulfa Drugs.*

Sulfur Cycle The cyclic movement of sulfur through an ecosystem.

Sulfur Oxidation The oxidation of various forms of sulfur to sulfate.

Sulfur Reduction The reduction of elemental sulfur to hydrogen sulfide.

Superantigens Powerful antigens, such as bacterial toxins, that activate large numbers of T cells, causing a large immune response that can cause diseases such as toxic shock.

Supercoiling For a DNA double helix, turns of the two strands around each other that either exceed or are fewer than the number of turns in the most stable helical conformation. Only a helix that is circular or else fixed at both ends can support supercoiling, See ▶ Twist.

Superhelix Density (σ) A measure of the superhelicity of a DNA molecule. It is equal to the change in linking number caused by the introduction of supercoiling divided by the linking number the DNA molecule would have in its relaxed state.

Superinfection A secondary infection from the removal of normal microbiota, allowing colonization by pathogenic, and often antibiotic-resistant, microbes.

Superoxide A highly reactive form of oxygen that kills obligate anaerobes.

Superoxide Dismutase An enzyme that converts superoxide to molecular oxygen and hydrogen peroxide.

Suppression With respect to mutations, a mutation that occurs at a different site from that of an existing mutation in a gene but restores the wild-type phenotype.

Suppressor T Cell (T_S) Possibly a type of cytotoxic or helper T cells that inhibits immune responses.

Surface Tension A phenomenon in which the surface of water behaves like a thin, invisible elastic membrane.

Surfactant A substance that reduces surface tension.

Susceptibility The vulnerability of an organism to harm by infectious agents.

Svedberg Unit (S) In ultracentrifugation, a unit used for the sedimentation coefficient; it is equal to 10^{-13} second.

Swarmer Cell Spherical, flagellated *Rhizobium* cell that invades the root hairs of leguminous plants, eventually to form nodules.

Sweat Gland Epidermal structure that empties a watery secretion through pores in the skin.

Swimmer's Itch Skin reaction to cercariae of some species of the helminth *Schistosoma.*

Symbiosis The living together of two different kinds of organisms.

Symport A membrane transport process that couples the transport of a substrate in one direction across a membrane to the transport of a different substrate in the same direction. Compare ▶ Antiport.

Symptom A disease characteristic that can be observed or felt only by the patient, such as pain or nausea.

Synchronous Growth Hypothetical pattern of growth during the log phase in which all the cells in a culture divide at the same time.

Syncytium (plural: *syncytia*) A multinucleate mass in a cell culture, for example, caused by the respiratory syncytial virus.

Syndrome A combination of signs and symptoms that occur together.

Synergism Referring to an inhibitory effect produced by two antibiotics working together that is greater than either can achieve alone.

Synthesis The step of viral replication during which new nucleic acids and viral proteins are made.

Synthetic Drug An antimicrobial agent synthesized chemically in the laboratory.

Synthetic Medium A growth medium prepared in the laboratory from materials of precise or reasonably well-defined composition.

Syphilis A sexually transmitted disease, caused by the spirochete *Treponema pallidum,* characterized by a chancre at the site of entry and often eventual neurological damage.

Systemic Blastomycosis Disease resulting from invasion by *Blastomyces dermatitides* of internal organs, especially the lungs.

Systemic Infection An infection that affects the entire body. *Known also as a Generalized Infection.*

Systemic Lupus Erythematosus A widely disseminated, systemic autoimmune disease resulting from production of antibodies against DNA and other body components.

Tapeworm Flatworm that lives in the adult stage as a parasite in the small intestine of animals.

Tartar Calcium deposition on dental plaque forming a very rough, hard crust.

Tautomers Structural isomers that differ in the location of their hydrogen and double bonds.

Taxon (plural: *taxa*) A category used in classification, such as species, genus, order, family.

Taxonomy The science of classification.

Tay-Sachs Disease A genetic disease caused by a deficiency of the lysosomal enzyme N-acetylhexosaminidase A, which is involved in sphingolipid degradation. The deficiency results in accumulation of the ganglioside sphingolipid GM_2, particularly in the brain.

T Cell See T Lymphocyte.

T-Dependent Antigen Antigen requiring helper T cell (T_H2) activity to activate B cells.

Teichoic Acid A polymer attached to peptidoglycan in Gram-positive cell walls.

Telomerase A DNA polymerase that adds a short repeating sequence to the 3′ strand at either end of a chromosomal DNA molecule, thus creating a single-strand overhand. This overhand gives room for priming the origin of a final Okazaki fragment during DNA replication so that the full length of the chromosome can be copied.

Telomeres Special DNA sequences at the ends of eukaryotic chromosomes.

Temperate Phage A bacteriophage that does not cause a virulent infection; rather, its DNA is incorporated into the host cell chromosome, as a prophage, and replicated with the chromosome.

Temperate Phages Bacterial phages that can establish a condition of lysogeny.

Template DNA used as a pattern for the synthesis of a new nucleotide polymer in replication or transcription.

Template Strand A DNA or an RNA strand that directs the synthesis of a complementary nucleic acid strand.

Temporary Parasite A parasite that feeds on and then leaves its host (such as a biting insect).

Teratogen An agent that induces defects during embryonic development.

Teratogenesis The induction of defects during embryonic development.

Terminator See Stop Codon.

Terminator Codon A codon that signals the end of the information for a particular protein. *Known also as Nonsense Codon or Stop Codon.*

Tertiary Structure (1) The folding of a protein molecule into globular shapes. (2) Large-scale folding structure in a linear polymer that is at a higher order than secondary structure. For proteins and RNA molecules, the tertiary structure is the specific three-dimensional shape into which the entire chain is folded. Compare ▶ Primary Structure, ▶ Quaternary Structure, and ▶ Secondary Structure.

Tertiary Treatment Chemical and physical treatment of sewage to produce an effluent of water pure enough to drink.

Test A shell made of calcium carbonate and common to some protists.

Testis (plural: *testes*) One of a pair of male reproductive glands that produce testosterone and sperm.

Tetanus Disease caused by *Clostridium tetani* in which muscle stiffness progresses to eventual paralysis and death. *Known also as Lockjaw.*

Tetanus Neonatorum Type of tetanus acquired through the raw stump of the umbilical cord.

Tetracycline An antibacterial agent that inhibits protein synthesis.

Tetrad Cuboidal groups of four cocci.

Thallus The body of a fungus.

Theca A tightly affixed, secreted outer layer of dinoflagellates that often contains cellulose.

Therapeutic Dosage Level Level of drug dosage that successfully eliminates a pathogenic organism if maintained over a period of time.

Thermal Death Point The temperature that kills all the bacteria in a 24-hour-old broth culture at neutral pH in 10 minutes.

Thermal Death Time The time required to kill all the bacteria in a particular culture at a specified temperature.

Thermoacidophile A member of one of the groups of the archaeobacteria that live in extremely hot, acidic environments.

Thermophile A heat-loving organism that grows best at temperatures from 50 to 60°C.

Thermophilic Anaerobic Spoilage Spoilage due to endospore germination and growth in which gas and acid are produced, making cans bulge.

Thrush Milky patches of inflammation on oral mucous membranes; a symptom of Candidiasis, caused by *Candida albicans.*

Thylakoid An internal membrane of chloroplasts that contains chlorophyll.

Thymus Gland Multilobed lymphatic organ located beneath the sternum that posses lymphocytes into T cells.

Tick Paralysis A disease characterized by fever and paralysis due to anticoagulants and toxins secreted into a tick's bite via the ectoparasite's saliva.

Tincture An alcoholic solution.

T-Independent Antigen Antigen not requiring helper T cells (T_H2) activity to activate B cells.

Tinea Barbae Barber's itch; a type of ringworm that causes lesions in the beard.

Tinea Capitis Scalp ringworm, a form of ringworm in which hyphae grow in hair follicles, often leaving circular patterns of baldness.

Tinea Corporis Body ringworm, a form of ringworm that causes ringlike lesions with a central scaly area.

Tinea Cruris Groin ringworm, a form of ringworm that occurs in skin folds in the pubic region. *Known also as Jock Itch.*

Tinea Pedis See ▶ Athlete's Foot.

Tinea Unguium A form of ringworm that causes hardening and discoloration of fingernails and toenails.

Tissue Culture Culture made from a single tissue, assuring a reasonably homogenous set of cultures in which to test the effects of a virus or to culture an organism.

Titer The quantity of a substance needed to produce a given reaction.

T-Lymphocyte Thymus-derived cell of the immune system and agent of cellular immune responses. *Known also as T Cell.*

Togavirus A small, enveloped RNA virus that multiplies in many mammalian and anthropod cells.

Tolerance A state in which antigens no longer elicit an immune response.

Tonsilitis A bacterial infection of the tonsils.

Tonsil Lymphoid tissue that contributes immune defenses in the form of B cells and T cells.

Topoisomerases Enzymes that change the supercoiling of DNA helices by either allowing the superhelical torsion to relax (thus reducing the supercoiling) or adding more twists (thus increasing the supercoiling).

Topoisomers With respect to DNA, closed circular DNA molecules that are identical except in their sense or degree of supercoiling. DNA topoisomers can be interchanged only by cutting one or both strands using topoisomerases.

TORCH Series A group of blood tests used to identify teratogenic diseases in pregnant women and newborn infants.

Total Magnification Obtained by multiplying the magnifying power of the objective lens by the magnifying power of the ocular lens.

Toxemia The presence and spread of exotoxins in the blood.

Toxic Dosage Level Amount of a drug necessary to cause host damage.

Toxic Shock syndrome (TSS) Condition caused by infection with certain toxigenic strains of *Staphylococcus aureus*; often associated with the use of super absorbent but abrasive tampons.

Toxin Any substance that is poisonous to other organisms.

Toxoid An exotoxin inactivated by chemical treatment but which retains its antigenicity and therefore can be used to immunize against the toxin.

Toxoplasmosis Disease caused by the protozoan *Toxoplasma gondii* that can cause congenital defects in newborns.

Trace Element Minerals, such as copper, iron, zinc, and cobalt ions, that are required in minute amounts for growth.

Trachea The windpipe.

Trachoma Eye disease caused by *Chlamydia trachomatis* that can result in blindness.

Transamination In the cell, the enzymatic transfer of an amino group from an amino group a keto acid. The keto acid becomes an amino acid and vice versa.

Transcription The synthesis of an RNA molecule complementary to a DNA strand; the information encoded in the base sequence of the DNA is thus "transcribed" into the RNA version of the same code. Compare ► Translation.

Transcription Factors Proteins that influence the transcription of particular genes, usually by binding to specific promoter sites.

Transduction The transfer of genetic material from one bacterium to another by a bacteriophage.

Transfer RNA (tRNA) Type of RNA that transfers amino acids from the cytoplasm to the ribosomes for placement in a protein molecule.

Transformation A change in an organism's characteristics through the transfer of naked DNA.

Transfusion Reaction Reaction that occurs when matching antigens and antibodies are present in the blood at the same time.

Transgenic (1) State of permanently changing an organism's characteristics by integrating foreign DNA (genes) into the organism. (2) Refers to an organism whose genome contains one or more DNA sequences from a different species (transgenes). Genetic engineering can be used to create transgenic animals.

Transient Microflora Microorganisms that may be present in or on an organism under certain conditions and for certain lengths of time t sites where resident microbiota.

Transition State In any chemical reaction, the high-energy or unlikely state that must be achieved by the reacting molecule(s) for the reaction to occur.

Translation The synthesis of a polypeptide under the direction of an mRNA, so that the nucleotide sequence of the mRNA is "translated" into the amino acid sequence of the protein. Compare ► Transcription.

Transmissible Spongiform Encephalopathies Prion-caused diseases resulting in brain tissue developing multiple holds such that it resembles a sponge, includes Creutzfeldt-Jakob disease, mad cow disease, kuru, scrapie and others.

Transmission The passage of light through an object.

Transmission Electron Microscope (TEM) A type of electron microscopy in which a beam of electrons passes through the object to be viewed and creates an image on a photographic plate or screen. Very thin slices of specimens are used.

Transovarian Transmission Passing of pathogen from one generation of ticks to the next as eggs leave the ovaries.

Transplantation The moving of tissue from one site to another.

Transplant Rejection Destruction of grafted tissue or of a transplanted organ by the host immune system.

Transposable Element A mobile genetic sequence that can move from one plasmid to another plasmid or chromosome.

Transposable Genetic Elements Genetic elements that are able to move from place to place within a genome. A Transposon is one type of transposable element.

Transposal of Virulence A laboratory technique in which a pathogen is passed from its normal host sequentially through many individual members of a new host species, resulting in a lessening or even total loss of its virulence in the original host.

Transposase An enzyme that is involved in the insertion of a bacterial Transposon into a target site.

Transposition The process whereby certain genetic sequences in bacteria or eukaryotes can move from one location to another.

Transposon A mobile genetic sequence that contains the genes for transposition as well as one or more other genes not related to transposition.

Traumatic Herpes Type of herpes infection in which the virus enters traumatized skin in the area of a burn or other injury.

Traveler's Diarrhea Gastrointestinal disorder generally caused by pathogenic strains of *Escherichia coli.*

Trench Fever Rickettsial disease, caused by *Rochalimaea Quintana*, resembling epidemic typhus in that is transmitted by lice and is prevalent during wars and under unsanitary conditions. *Known also as Shinbone Fever.*

Treponemes Spirochetes belonging to the genus *Treponema.*

Triacylglycerol A molecule formed from three fatty acids bonded to glycerol.

Tricarboxylic Acid Cycle See ► Citric Acid Cycle and ► Krebs Cycle.

Trichinosis A disease caused by a small nematode, *Trichinella spiralis*, that enters the digestive tract as encysted larvae in poorly cooked meat, usually pork.

Trichocyst Tentaclelike structure on ciliates for catching prey for attachment.

Trichomoniasis A parasitic urogenital disease, transmitted primarily by sexual intercourse, that causes intense itching and a copious white discharge, especially in females.

Trichuriasis Parasitic disease caused by the whipworm, *Trichuris trichiura*, that damages intestinal mucosa and causes chronic bleeding.

Trickling Filter System Procedure in which sewage is spread over a bed of rocks coated with aerobic organisms that decompose the organic matter in it.

Trophozoite Vegetative form of a protozoan such as *Plasmodium*.

Trypanosomiasis See ▶ African Sleeping Sickness.

Tube Agglutination Test Serologic test that measures antibody titers by comparing various dilutions of the patient's serum against known quantities of an antigen.

Tubercle A solidified lesion or chronic granuloma that forms in the lungs in patients with tuberculosis.

Tuberculin Hypersensitivity Cell-mediated hypersensitivity reaction that occurs in sensitized individuals when they are exposed to tuberculin.

Tuberculin Skin Test An immunological test for tuberculosis in which a purified protein derivative from the *Mycobacterium tuberculosis* is injected subcutaneously, resulting in an induration if there was previous exposure to the bacterium.

Tuberculoid Referring to the anesthetic form of Hansen's (disease leprosy) in which areas of skin lose pigment and sensation.

Tuberculosis Disease caused mainly by *Mycobacterium tuberculosis.*

Tularemia Zoonosis caused by *Francisella tularensis*, most often associated with cottontail rabbits.

Tumor An uncontrolled division of cells, often caused by viral infection.

Turbidity A cloudy appearance in a culture tube indicating the presence of organisms.

Turnover Number With respect to an enzyme-catalyzed reaction, the number of substrate molecules one enzyme molecule can process (turn over) per second when saturated with substrate. It is equivalent to the catalytic rate constant, k_{cat}.

Twist (*T*) With respect to a DNA double helix, the total number of times the two strands of the helix cross over each other, excluding writhing. It is a measure of how tightly the helix is wound. See also ▶ Linking Number and ▶ Writhe.

Tympanic Membrane Membrane separating the outer and middle ear. *Known also as the Eardrum.*

Type Strain Original reference strain of a bacterial species, descendants of a single isolation in pure culture.

Typhoidal Tularemia Septicemia that resembles typhoid fever, caused by bacteremia from tularemia lesions.

Typhoid Fever An epidemic enteric infection caused by *Salmonella typhi*, uncommon in areas with good sanitation.

Typhus Fever Rickettsial disease that occurs in a variety of forms including epidemic, endemic (murine), and scrub typhus.

Tyrocidin An antibacterial agent that disrupts cell membranes.

Ulceroglandular Referring to the form of tularemia caused by entry of *Francisella tularensis* through the skin and characterized by ulcers on the skin and enlarged regional lymph nodes.

Ultrafiltration The technique of filtering a solution under pressure through a Semipermeable membrane, which allows water and small solutes to pass through but retains macromolecules.

Ultra-High Temperature (UHT) Processing A method of sterilizing milk and dairy products by raising the temperature to 87.8°C for three seconds.

Uncoating Process in which protein coats of animal viruses that have entered ells are removed by proteolytic enzymes.

Undulant Fever See ▶ Brucellosis.

Universal Precautions A set of guidelines established by the CDC to reduce the risks of disease transmission in hospital and medical laboratory settings.

Unsaturated Fatty Acid A fatty acid that contains at least one double bond between adjacent carbon atoms.

Upper Respiratory Tract The nasal cavity, pharynx, larynx, trachea, bronchi, and larger bronchioles.

Ureaplasmas Bacteria with unusual cell walls, require sterols as a nutrient.

Ureter Tube that carries urine from the kidney to the urinary bladder.

Urethra Tube through which urine passes from the bladder to the outside during micturition (urination).

Urethritis Inflammation of the urethra.

Urethrocystitis Common term used to describe urinary tract infections involving the urethra and the bladder.

Urinalysis The laboratory analysis of urine specimens.

Urinary Bladder Storage area for urine.

Urinary System Body system that regulates the composition of body fluids and removes nitrogenous and other wastes from the body.

Urinary Tract Infection (UTI) A bacterial urogenital infection that causes Urethritis or cystitis.

Urine Water collected in the kidney tubules.

Urogenital System Body system that (1) regulates the composition of body fluids and removes certain wastes from the body and (2) enables the body to participate in sexual reproduction.

Use-Dilution Test A method of evaluating the antimicrobial properties of a chemical agent using standard preparations of certain test bacteria.

Uterine Tube A tube that conveys ova from the ovaries to the uterus. *Known also as Fallopian Tubes or Oviducts.*

Uterus The pear-shaped organ in which a fertilized ovum implants and develops.

Vaccine A substance that contains an antigen to which the immune system responds.

Vacuole A membrane-bound structure that stores materials such as food or gas in the cytoplasm or eukaryotic cells.

Vagina The female genital canal, extending from the cervix to the outside of the body.

Vaginitis Vaginal infection, often caused by opportunistic organisms that multiply when the normal vaginal Microflora are disturbed by antibiotics or other factors.

Van der Waals Radius (r) The effective radius of an atom or a molecule that defines how close other atoms or molecules can approach; it is thus the effective radius for closest molecular packing.

Variable Anything that can change in an experiment.

Varicella-Zoster Virus A herpesvirus that causes both chickenpox and shingles.

Vasodilation Dilation of the capillary and venule walls during an acute inflammation.

Vector (1) A self-replicating carrier of DNA; usually a plasmid, bacteriophage, or eukaryotic virus. (2) An organism that transmits a disease-causing organism from one host to another. (3) In genetic engineering, a DNA molecule that can be used to introduce a DNA sequence into a cell where it will be replicated and maintained. Usually a plasmid or a viral genome.

Vegetation A growth that forms on damaged heart valve surfaces in bacterial endocarditis, exposed collagen fibers elicit fibrin deposits, and transient bacteria attach to the fibrin.

Vegetative Cell A cell that is actively metabolizing nutrients.

Vehicle A nonliving carrier of an infectious agent from its reservoir to a susceptible host.

Venezuelan Equine Encephalitis Type of viral encephalitis seen in Florida, Texas, Mexico, and South America; infects horses more frequently than humans.

Verminous Intoxication An allergic reaction to toxins in the metabolic wastes of liver flukes.

Verruga Peruana One form of bartonellosis; a chronic nonfatal skin disease.

Vertical Gene Transfer Genes that pass from parents to offspring.

Vertical Transmission Direct contact transmission of disease in which pathogens are passed from parent to offspring in an egg or sperm, across the placenta, or while traversing the birth canal.

Very Low-Density Lipoprotein (VLDL) A type of lipoprotein particle that is manufactured in the liver and functions mainly to carry triacylglycerols from the liver to adipose and other tissues.

Vesicle A membrane-bound inclusion in cells.

Vibrio A comma-shaped bacterium.

Vibriosis An enteritis caused by *Vibrio parahaemolyticus*, acquired from eating contaminated fish and shellfish that have not been thoroughly cooked.

Virion A single virus particle.

Villus (plural: *villi*) A multicellular projection from the surface of a mucous membrane, functioning in absorption.

Viral Enteritis Gastrointestinal disease caused by rotaviruses, characterized by diarrhea.

Viral Hemagglutination Hemagglutination caused by binding of viruses, such as those that cause measles and influenza, to red blood cells.

Viral Meningitis Usually self-lining and nonfatal form of meningitis.

Viral Neutralization The binding of antibodies to viruses, which is used in an immunological test to determine if a patient's serum contains viruses.

Viral Pneumonia Disease caused by viruses such as respiratory syncytial virus.

Viral Specificity Refers to the specific types of cells within an organism that a virus can infect.

Viral Yield See Burst Size.

Viremia An infection in which viruses are transported in the blood but do not multiply in transit.

Viridans Group A group of streptococci that often infect the valves and lining of the heart and cause incomplete (alpha) hemolysis of red blood cells in laboratory cultures.

Virion A complete virus particle, including its envelope if it has one.

Viroid An infectious RNA particle, smaller than a virus and lacking a capsid, that causes various plant diseases.

Virulence The degree of intensity of the disease produced by a pathogen.

Virulence Factor A structural or physiological characteristic that helps a pathogen cause infection and disease.

Virulent Phage A bacteriophage that enters the lytic cycle when it infects a bacterial cell, causing eventual lysis and death of the host cell. *Known also as Lytic Phage.*

Virus A submicroscopic, parasitic, acellular microorganism composed of a nucleic acid (DNA or RNA) core inside a protein coat.

Viruses Infectious entities that contain the nucleic acid to code for their own structure but that lack the enzymatic machinery of a cell; they replicate by invading a cell and using its machinery to express the viral genome. Most viruses consist of little but nucleic acid enclosed in a protein coat; some viruses also have an outer lipid-bilayer envelope.

Visceral Larva Migrans The migration of larvae of *Toxocara* species in human tissues, where they cause damage and allergic reactions.

Vitamin A substance required for growth that the organism cannot make.

Volutin Polyphosphate granules. *Known also as Metachromatic Granule.*

Walking Pneumonia See ► Primary Atypical Pneumonia.

Wandering Macrophages Phagocytic cells that circulate in the blood or move into tissues when microbes and other foreign material are present.

Wart A growth on the skin and mucous membranes caused by infection with human papillonavirus. *Known also as Papilloma.*

Water Cycle Process by which water is recycled through precipitation, ingestion by organisms, respiration, and evaporation. *Known also as the Hydrologic Cycle.*

Water Mold A funguslike protist that produces flagellated asexual spores (zoospores) and large, motile gametes. *Known also as Oomycota.*

Wavelength The distance between successive crests or troughs of a light wave.

Western Blotting A technique for identifying proteins or protein fragments in a mixture that react with a particular antibody. The mixture is first resolved into bands by one-dimensional denaturing gel electrophoresis. The protein bands are then "blotted" onto a nitrocellulose sheet, the sheet is treated with the antibody and any bands that bind the antibody are identified. More accurately called immunoblotting.

Western Equine Encephalitis Type of viral encephalitis seen most often in the western United States; infects horses more frequently than humans.

West Nile Fever Emerging viral disease new to the U.S., transmitted by mosquitoes, causing seizures and encephalitis; lethal to crows.

Wet Mount Microscopy technique in which a drop of fluid containing the organisms (often living) is placed on a slide.

Wetting Agent A detergent solution often used with other chemical agents to penetrate fatty substances.

Whey The liquid portion (waste product) of milk resulting from bacterial enzyme addition.

Whipworm *Trichuris trichiura,* a worm that causes trichuriasis infestation of the intestine.

Whitlow A herpetic lesion on a finger that can result from exposure to oral, ocular and probably genital herpes.

Whooping Cough A highly contagious respiratory disease caused primarily by *Bordetella pertussis. Known also as Pertussis.*

Wild-Type Refers to the normal genotype found in free-living, natural members of a group of organisms.

Wort The liquid extract from mash.

Wound Botulism Rare form of botulism that occurs in deep wounds when tissue damage impairs circulation and creates anerobic conditions in which *Clostridium botulinum* can multiply.

Writhe (*W*) With respect to a supercoiled DNA helix, the number of times the helix as a whole crosses over itself – that is, the number of superhelical turns that are present. See also ► Linking Number and ► Twist.

Xenobiotic An organic compound that is not produced by the organism in which it is found.

Xenograph A graft between individuals of different species.

X-Ray Diffraction A technique that is used to determine the three-dimensional structure of molecules, including macromolecules. A crystal or fiber of the substance is illuminated with a beam of x-rays, and the repeating elements of the structure scatter the x-rays to form a diffraction pattern that gives information on the molecule's structure. See also ► Diffraction Pattern.

Yeast Artificial Chromosomes (YACs) Artificial chromosomes used for cloning and maintaining large fragments of genomic DNA for investigational purposes, A YAC is constructed by recombinant DNA techniques from a year Centromere, two telomeres (chromosome ends), selectable markers, and cloned DNA in the megabase range.

Yeast Extract Substance from yeast containing vitamins, coenzymes, and nucleosides; used to enrich media.

Yellow Fever Viral systemic disease found in tropical areas, carried by the mosquito *Aedes aegypti.*

Yersiniosis Severe enteritis caused by *Yersinia enterocolitica.*

Z-DNA A DNA duplex with a specific left-hand helical structure. In vitro, it tends to be the most stable form from DNA duplexes that have alternating purines and pyrimidines, especially under conditions of cytosine methylation or negative supercoiling.

Ziehl-Neelsen Acid-Fast Stain A differential stain for organisms that are not decolorized by acid in alcohol, such as the bacteria that cause Hansen's disease (leprosy) and tuberculosis.

Zone of Inhibition A clear area that appears on agar in the disk diffusion method, indicating where the agent has inhibited growth of the organism.

Zoonosis (plural: *zoomoses*) A disease that can be transmitted from animals to humans.

Zygomycosis Disease in which certain fungi of the genera *Mucor* and *Rhizopus* invade lungs, the central nervous system, and tissues of the eye orbit.

Zygomycota See ▶ Bread Mold.

Zygospore In bread molds, a thick-walled, resistant, spore-producing structure enclosing a zygote.

Zygote A cell formed by the union of gametes (egg and sperm).

Appendix D2

Bacteria

The classification below is based on that in the 3rd edition of Microbiology (Bernard D David, Harper & Row, New York, pp. 27–28). The extent of classification is to the genus and, in selected cases families, but the specie are listed in the text.

Main Groups of Bacteria*

Gram-Positive Eubacteria

The Eubacteria is a procaryotic microorganism with a set of characters that unite its extraordinarily diverse taxa. Unlike the Archaea, the Eubacteria have been known and studied for more than 150 years. This is because all known bacterial pathogens are Eubacteria (can use of the term bacteria as a descriptive term that is a synonym of prokaryote). Also, some of them like *Lactobacillus* are otherwise economically important. Perhaps more importantly, many of them inhabit environments that are easily studied and sampled. The Eubacteria differ from the Archaea in the form and structure of their ribosomes, the type and linkage of their lipids, the structure of their cell covering, and the type of RNA polymerase (Margulis and Schwartz 1998). Traditionally, the Eubacteria have been separated into the Gram positive and Gram negative groups, based upon a standard stain technique. As it turns out, the way a cell stains is related to the type and structure of the cell wall. Gram positive cells have a single membrane with a murien or peptidoglycan wall to the outside of the single membrane. Gram negative cells have an inner membrane and an outer membrane with a murein layer sandwiched between them. The system of Margulis and Schwartz (1998) is based on the fundamental separation of gram positive and gram negative cells (called Firmicutes and Gracilicutes, respectively). Phylogenies based on small subunit rRNA, however, show that the eubacteria are marked by 10 or 11 deep clades that we interpret as kingdoms. This could just be the tip of the iceberg with respect to their true diversity. Garrity et al. (2001) separate the Eubacteria (a group that they call "Bacteria") into 23 groups. Also, the problems of lateral gene transfer further blur the distinctions of the groups. We present a tentative system for the Eubacteria with 9 kingdoms (see below). This system is based largely on Margulis and Schwartz (1998), with modifications from Garrity et al. (2001, 2003, and 2005), Tudge (2000), and Black (2002).

Cell Shape: Cocci
Motility: Nearly all permanently immotile

Other distinguishing characteristics	Genera	Families
Cells in cubical packets	Sarcina	Micrococcaceae
Cells irregularly arranged	Micrococus	
	Staphylooccus	
Cells in chains	Streptoccoccus	Streptococcaceae
Lactic fermentation of sugars	Leuconostoc	

Cell Shape: Straight rods
Motility: Nearly all permanently immotile

Other distinguishing characteristics	Genera	Families
Lactic fermentation of sugars	Lactobacillus	Lactobacillaceae
Propionic fermentation of sugars	Propionibacterium	Propionibacteriaceae
Oxidative, weakly fermentative	Corynebacterium	
	Listeria	
	Erysipelothrix	

Cell Shape: Straight rods
Motility: Motile with pertrichous flagella, and related immotile forms

Other distinguishing characteristics	Genera	Families
Aerobic	Bacillus	
Endorspores produced		Bacillaceae
Anaerobic	Clostridum	

Gram-Negative Bacteria, Excluding Photosynthetic Forms

Celzl shape: Cocci
Motility: Permanently immotile

Other distinguishing characteristics	Genera	Families
Aerobic	Neisseria	Neisseriaceae
Anerobic	Veillonella	
	Brucella	Brucellacease
	Bordetella	
	Pasteurella	
	Hamophilus	

Cell Shape: Straight rods
Motility: Motile with peritrichous flagella, and related immotile forms

Other distinguishing characteristics		Genera	Families
Aerobic		Bacillus	
Facultative – anerobic	Mixed acid Fermentation of sugars	Escherichia	Enterobacteriaceae
		Erwinia	
		Shigella	
		Salmonella	
Aerobic		Bacillus	
		Proteus	
		Yersinia	

Other distinguishing characteristics		Genera	Families
Butylene glycol fermentation		Enterobacter	Azotobacteraceae
		Serrati	
Aerobic	Free-living nitrogen fixers	Azotobacter	
	Symbiotic nitrogen fixers	Rhizobium	Rhizobiaceae

Cell Shape: Straight rods
Motility: Motile with polar flagella

Other distinguishing characteristics		Genera	Families
Aerobic		Bacilus	
Aerobic	Oxide inorganic compounds	Nitrosomonas	Nitrobacteraceae
		Nitrobacter	
			Thiobacillus
Facultative aneobic	Oxidize organic compounds	Pseudomonas	Pseudomonadaceae
		Acetobacter	
		Photobacterium	
Aerobic		Bacilus	
		Zymomonas	
		Aeromonas	

Cell Shape: Curved rods
Motility: Motile with polar flagella

Other distinguishing characteristics		Genera	Families
Aerobic		Bacilus	
Comma-shaped	Aerobic	Vibro	Spirllaceae
Spiral	Anaerobic	Desulfovibrio	
		Spirillum	

Other Major Groups

Other distinguishing characteristics	Genera	Families
Acid-fast rods	Mycobacterium	Actinomycetales
Ray-forming rods (actinomycetes)	Actinomyces	
	Nocardia	
	Streptomyces	
Spiral organisms, motile	Treponema	Spirochetales
	Borrelia	
	Leptospira	
	Spirocheta	
Small, pleomorphic; lack rigid wall	Mycoplasma	Mycoplasmataceae
Small intracellular parasites	Rickettsia	Rickettsiaceae
	Coxiella	
Small intracellular parasites, readily filterable	Chlamydia	Chlamydiaceae
Intracellular parasites; borderline with protoza	Bartonella	Bartonellaceae

*Some of these traditional names have been officially replaced. (Modified from Stanier RY et al: The Microbial World. Englewood Clifffs, NJ, Prentice Hall, 1963).

Bacteria are classified as follows:

Kingdom: Monera (Prokaryotae)
Division/Phylum: Gracillcutes
Subphylum:
Class: Scotobacteria
Order: Spirochaetales
Family: Spirochaetaceae
Genus: Treonema
Specific Epithet: pallidum

Subspecies(strain):
Sources:

1. Bergey's Manual of Systematic Bacteriology, 2nd edn., April 2001, Springer-Verlag, New York.
2. The web page "www/bacterio.cict.fr/" is a source for the complete identification of all bacteria and viruses using Bergey Classification of Bacteria.

Main Groups of Bacteria

The classification below is based on that in the 3rd edition of Microbiology (Bernard D David, Harper & Row, New York, pp. 27–28). The extent of classification is to the genus and, in selected cases families, but the specie are listed in the text.
Main Groups of Bacteria*

Gram-Positive Eubacteria

Cell Shape: Cocci
Motility: Nearly all permanently immotile

Other distinguishing characteristics	Genera	Families
Cells in cubical packets	Sarcina	Micrococcaceae
Cells irregularly arranged	Micrococus	
	Staphylooccus	
Cells in chains	Streptoccoccus	Streptococcaceae
Lactic fermentation of sugars	Leuconostoc	

Cell Shape: Straight rods
Motility: Nearly all permanently immotile

Other distinguishing characteristics	Genera	Families
Lactic fermentation of sugars	Lactobacillus	Lactobacillaceae
Propionic fermentation of sugars	Propionibacterium	Propionibacteriaceae
Oxidative, weakly fermentative	Corynebacterium	
	Listeria	
	Erysipelothrix	

Cell Shape: Straight rods
Motility: Motile with pertrichous flagella, and related immotile forms

Other distinguishing Characteristics	Genera	Families
Aerobic	Bacillus	
Endorspores produced		Bacillaceae
Anaerobic	Clostridum	

Other distinguishing characteristics		Genera	Families
Aerobic	Free-living nitrogen fixers	Azotobacter	
	Symbiotic nitrogen fixers	Rhizobium	Rhizobiaceae

Gram-Negative Bacteria, Excluding Photosynthetic Forms

Cell shape: Cocci
Motility: Permanently immotile

Other distinguishing characteristics	Genera	Families
Aerobic	Neisseria	Neisseriaceae
Anerobic	Veillonella	
	Brucella	Brucellacease
	Bordetella	
	Pasteurella	
	Hamophilus	

Cell Shape: Straight rods
Motility: Motile with peritrichous flagella, and related immotile forms

Other distinguishing characteristics	Genera	Families
Aerobic	Bacillus	
Facultative-anerobic	Escherichia	Enterobacteriaceae
Mixed acid Fermentation of sugars	Erwinia	
	Shigella	
	Salmonella	
Aerobic	Bacillus	
Butylene glycol fermentation	Enterobacter	Azotobacteraceae
	Serrati	

Cell Shape: Straight rods
Motility: Motile with polar flagella

Other distinguishing characteristics	Genera	Families
Aerobic	Bacilus	
Aerobic	Nitrosomonas	Nitrobacteraceae
	Nitrobacter	
Oxide inorganic compounds		Thiobacillus
	Pseudomonas	Pseudomonadaceae
Oxidize organic compounds	Acetobacter	
Facultative anerobic	Photobacterium	
Aerobic	Bacilus	
	Zymomonas	
	Aeromonas	

Cell Shape: Curved rods
Motility: Motile with polar flagella

Other distinguishing characteristics		Genera	Families
	Aerobic	Bacilus	
Comma-shaped	Aerobic	Vibro	Spirllaceae
Spiral	Anaerobic	Desulfovibrio	
		Spirillum	

Other Major Groups

Other distinguishing characteristics	Genera	Families
Acid-fast rods	Mycobacterium	Actinomycetales
Ray-forming rods (actinomycetes)	Actinomyces	
	Nocardia	
	Streptomyces	
Spiral organisms, motile	Treponema	Spirochetales
	Borrelia	
	Leptospira	
	Spirocheta	
Small, pleomorphic; lack rigid wall	Mycoplasma	Mycoplasmataceae
Small intracellular parasites	Rickettsia	Rickettsiaceae
	Coxiella	
Small intracellular parasites, readily filterable	Chlamydia	Chlamydiaceae
Intracellular parasites; borderline with protoza	Bartonella	Bartonellaceae

*Some of these traditional names have been officially replaced. (Modified from Stanier RY et al: The Microbial World. Englewood Clifffs, NJ, Prentice Hall, 1963).

Bacteria are classified as follows:

Kingdom: Monera (Prokaryotae)
Division/Phylum: Gracillcutes
Subphylum:
Class: Scotobacteria
Order: Spirochaetales
Family: Spirochaetaceae
Genus: Treonema
Specific Epithet: pallidum

Subspecies(strain):
Sources:

1. Bergey's Manual of Systematic Bacteriology, 2nd edn., April 2001, Springer-Verlag, New York.
2. The web page "www/bacterio.cict.fr/" is a source for the complete identification of all bacteria and viruses using Bergey Classification of Bacteria.

Classification of Bacteria

The following information came from: Garrity, George M, Winters, Matthew and Searles, Denise B, Taxonoic Outline of the Procaryotic Genera, Bergey's manual of systematic bacteriology, 2nd Edition, April 2001, Sprinter-Verlag, New York.
Domain "Archaea"

Phylum A1. *Crenarchaeota phy. nov.*
Class I. *Thermoprotei class. mov.*
Order I. *Thermoproteales*
Family I. Thermoproteaceae
Genus I. Thermoproteus
Genus II. Caldivirga
Genus III. Pyrobaculum
Genus IV. Thermocladium
Family II. Thermofilaceae
Genus I. Thermofilum
Order II. *Desulfurococcales ord. nov.*
Family I. Desulfurococcaceae
Genus I. Desulfurococcus
Genus II. Acidolobus
Genus III. Aeropyrum
Genus IV. Ignicoccus
Genus V. Staphylothermus
Genus VI. Stetteria
Genus VII. Sulfophobococcus
Genus VIII. Thermodiscus gen. nov.
Genus IX. Thermosphaera
Family II. Pyrodictiaceae
Genus I. Pyrodictium
Genus II. Hyperthermus
Genus III. Pyrolobus
Order III. *Sulfolobales*
Family I. Sulfolobaceae
Genus I. Sulfolobus
Genus II. Acidianus
Genus III. Metallosphaera
Genus IV. Stygiolobus
Genus V. Sulfurisphaera
Genus VI. Sulfurococcus
Phylum AII. *Euryarchaeota phy. nov.*
Class I. *Methanobacteria class. nov.*
Order I. *Methanobacteriales*
Family I. Methanobacteriaceae
Genus I. Methanobacterium
Genus II. Methanobrevibacter
Genus III. Methanosphaera
Genus IV. Methanothermobacter

Family II. Methanothermaceae
Genus I. Methanothermus
Class II. *Methanococci class. nov.*
Order I. *Methanococcales*
Family I. Methanococcaceae
Genus I. Methanococcus
Genus II. Methanothermococcus gen. nov.
Family II. Methanocaldococcaceae fam. nov.
Genus I. Methanocaldococus gen. nov.
Genus II. Methanotorris gen. nov.
Order II. *Methanomicrobiales*
Family I. Methanomicrobiaceae
Genus I. Methanomicrobium
Genus II. Methanoculleus
Genus III. Methanofollis
Genus IV. Methanogenium
Genus V. Methanolacinia
Genus VI. Methanoplanus
Family II. Methanocorpusculaceae
Genus I. Methanocorpusculum
Family III. Methanospirillaceae fam. nov.
Genus I. Methanospirillum
Genera incertae sedis
Genus I. Methanocalculus
Order III. *Methanosarcinales ord. nov.*
Family I. Methanosarcinaceae
Genus I. Methanosarcina
Genus II. Methanococcoides
Genus III. Methanohalobium
Genus IV. Methanohalophilus
Genus V. Methanolobus
Genus VI. Methanomicrococcus
Genus VII. Methanosalsum gen. nov.
Family II. Methanosaetaceae fam. nov.
Genus I. Methanosaeta
Class III. *Halobacteria class. nov.*
Order I. *Halobacteriales*
Family I. Halobacteriaceae
Genus I. Halobacterium
Genus II. Haloarcula
Genus III. Halobaculum
Genus IV. Halococcus
Genus V. Haloferax
Genus VI. Halogeometricum
Genus VII. Halorhabdus
Genus VIII. Halorubrum
Genus IX. Haloterrigena
Genus X. Natrialba
Genus XI. Natrinema
Genus XII. Natronobacterium
Genus XIII. Natronococcus
Genus XIV. Natronomonas
Genus XV. Natronorubrum
Class IV. *Thermoplasmata class. nov.*
Order I. *Thermoplasmatales ord. nov.*
Family I. Thermoplasmataceae fam. nov.
Genus I. Thermoplasma
Family II. Picrophilaceae
Genus I. Picrophilus
Family III. Ferroplasmatacaea
Genus I. Ferroplasma
Class V. *Thermococci class. nov.*
Order I. *Thermococcales*
Family I. Thermococcaceae
Genus I. Thermococcus
Genus II. Palaeococcus
Genus III. Pyrococcus
Class VI. *Archaeoglobi class. nov.*
Order I. *Archaeoglobales ord. nov.*
Family I. Archaeoglobaceae fam. nov.
Genus I. Archaeoglobus
Genus II. Ferroglobus
Class VII. *Methanopyri class. nov.*
Order I. *Methanopyrales ord. nov.*
Family I. Methanopyraceae fam. nov.
Genus I. Methanopyrus

Domain "Bacteria"

Phylum BI. *Aquificae phy. nov.*
Class I. *Aquificae class. nov.*
Order I. *Aquificales ord. nov.*
Family I. Aquificaceae fam. nov.
Genus I. Aquifex
Genus II. Calderobacterium
Genus III. Hydrogenobacter
Genus IV. Thermocrinis
Genera incertae sedis
Genus I. Desulfurobacterium
Phylum BII. *Thermotogae phy. nov.*
Class I. *Thermotogae class. nov.*
Order I. *Thermotogales ord. nov.*
Family I. Thermotogaceae fam. nov.
Genus I. Thermotoga
Genus II. Fervidobacterium
Genus III. Geotoga
Genus IV. Petrotoga
Genus V. Thermosipho
Phylum BIII. *Thermodesulfobacteria phy. nov.*
Class I. *Thermodesulfobacteria class. nov.*
Order I. *Thermodesulfobacteriales ord. nov.*
Family I. Thermodesulfobacteriaceae fam. nov.
Genus I. Thermodesulfobacterium

Phylum BIV. *"Deinococcus-Thermus"*
Class I. *Deinococci class. nov.*
Order I. *Deionococcales*
Family I. Deinococcaceae
Genus I. Deinococcus
Order II. *Thermales ord. nov.*
Family I. Thermaceae fam. nov.
Genus I. Thermus
Genus II. Meiothermus
Phylum BV. *Chrysiogenetes phy. nov.*
Class I. *Chrysiogenetes class. nov.*
Order I. *Chrysiogenales ord. nov.*
Family I. Chrysiogenaceae fam. nov.
Genus I. Chrysiogenes
Phylum BVI. *Chloroflexi phy. nov.*
Class I. *"Chloroflexi"*
Order I. *"Chloroflexales"*
Family I. "Chloroflexaceae"
Genus I. Chloroflexus
Genus II. Chloronema
Genus III. Heliothrix
Family II. Oscillochloridaceae
Genus I. Oscillochloris (moved)
Order II. *"Herpetosiphonales"*
Family I. "Herpetosiphonaceae"
Genus I. Herpetosiphon
Phylum BVII. *Thermomicrobia phy. nov.*
Class I. *Thermomicrobia class. nov.*
Order I. *Thermomicrobiales ord. nov.*
Family I. Thermomicrobiaceae fam. nov.
Genus I. Thermomicrobium
Phylum BVIII. *Nitrospira phy. nov.*
Class I. *"Nitrospira"*
Order I. *"Nitrospirales"*
Family I. "Nitrospiraceae"
Genus I. Nitrospira
Genus II. Leptospirillum
Genus III. Magnetobacterium
Genus IV. Thermodesulfovibrio
Phylum BIX. *Deferribacteres phy. nov.*
Class I. *Deferribacteres class. nov.*
Order I. *Deferribacterales ord. nov.*
Family I. Deferribacteraceae fam. nov.
Genus I. Deferribacter
Genus II. Denitrovibrio
Genus III. Flexistipes
Genus IV. Geovibrio
Genera incertae sedis
Genus I. Synergistes
Phylum BX. *Cyanobacteria*
Class I. *"Cyanobacteria"*

Subsection I.
Family I.
Form genus I. Chamaesiphon
Form genus II. Chroococcus
Form genus III. Cyanobacterium
Form genus IV. Cyanobium
Form genus V. Cyanothece
Form genus VI. Dactylococcopsis
Form genus VII. Gloeobacter
Form genus VIII. Gloeocapsa
Form genus IX. Gloeothece
Form genus X. Microcystis
Form genus XI. Prochlorococcus
Form genus XII. Prochloron
Form genus XIII. Synechococcus
Form genus XIV. Synechocystis
Subsection II.
Family I.
Form genus I. Cyanocystis
Form genus II. Dermocarpella
Form genus III. Stanieria
Form genus IV. Xenococcus
Family II.
Form genus I. Chroococcidiopsis
Form genus II. Myxosarcina
Form genus III. Pleurocapsa
Subsection III.
Family I.
Form genus I. Arthrospira
Form genus II. Borzia
Form genus III. Crinalium
Form genus IV. Geitlerinema
Genus V. Halospirulina
Form genus VI. Leptolyngbya
Form genus VII. Limnothrix
Form genus VIII. Lyngbya
Form genus IX. Microcoleus
Form genus X. Oscillatoria
Form genus XI. Planktothrix
Form genus XII. Prochlorothrix
Form genus XIII. Psuedanabaena
Form genus XIV. Spirulina
Form genus XV. Starria
Form genus XVI. Symploca
Genus XVII. Trichodesmium
Form genus XVIII. Tychonema
Subsection IV.
Family I.
Form genus I. Anabaena
Form genus II. Anabaenopsis
Form genus III. Aphanizomenon

Form genus IV. Cyanospira
Form genus V. Cylindrospermopsis
Form genus VI. Cylindrospermum
Form genus VII. Nodularia
Form genus VIII. Nostoc
Form genus IX. Scytonema
Family II.
Form genus I. Calothrix
Form genus II. Rivularia
Form genus III. Tolypothrix
Subsection V.
Family I.
Form genus I. Chlorogloeopsis
Form genus II. Fischerella
Form genus III. Geitleria
Form genus IV. Iyengariella
Form genus V. Nostochopsis
Form genus VI. Stigonema
Phylum BXI. *Chlorobi phy. nov.*
Class I. *"Chlorobia"*
Order I. *Chlorobiales*
Family I. Chlorobiaceae
Genus I. Chlorobium
Genus II. Ancalochloris
Genus III. Chloroherpeton
Genus IV. Pelodictyon
Genus V. Prosthecochloris
Phylum BXII. *Proteobacteria phy. nov.*
Class I. *"Alphaproteobacteria"*
Order I. *Rhodospirillales*
Family I. Rhodospirillaceae
Genus I. Rhodospirillum
Genus II. Azospirillum
Genus III. Magnetospirillum
Genus IV. Phaeospirillum
Genus V. Rhodocista
Genus VI. Rhodospira
Genus VII. Rhodothalassium
Genus VIII. Rhodovibrio
Genus IX. Roseospira
Genus X. Skermanella
Family II. Acetobacteraceae
Genus I. Acetobacter
Genus II. Acidiphilium
Genus III. Acidisphaera
Genus IV. Acidocella
Genus V. Acidomonas
Genus VI. Asaia
Genus VII. Craurococcus
Genus VIII. Gluconacetobacter
Genus IX. Gluconobacter
Genus X. Paracraurococcus
Genus XI. Rhodopila
Genus XII. Roseococcus
Genus XIII. Stella
Genus XIV. Zavarzinia
Order II. *Rickettsiales*
Family I. Rickettsiaceae
Genus I. Rickettsia
Genus II. Orientia
Genus III. Wolbachia
Family II. Ehrlichiaceae
Genus I. Ehrlichia
Genus II. Aegyptianella
Genus III. Anaplasma
Genus IV. Cowdria
Genus V. Neorickettsia
Genus VI. Xenohaliotis
Family III. "Holosporaceae"
Genus I. Holospora
Genus II. Caedibacter
Genus III. Lyticum
Genus IV. Polynucleobacter
Genus V. Pseudocaedibacter
Genus VI. Symbiotes
Genus VII. Tectibacter
Genus VIII. Odyssella
Order III. *"Rhodobacterales"*
Family I. "Rhodobacteraceae"
Genus I. Rhodobacter
Genus II. Ahrensia
Genus III. Amaricoccus
Genus IV. Antarctobacter
Genus V. Gemmobacter
Genus VI. Hirschia
Genus VII. Hyphomonas
Genus VIII. Maricaulis
Genus IX. Methylarcula
Genus X. Octadecabacter
Genus XI. Paracoccus
Genus XII. Rhodovulum
Genus XIII. Roseibium
Genus XIV. Roseinatronobacter
Genus XV. Roseivivax
Genus XVI. Roseobacter
Genus XVII. Roseovarius
Genus XVIII. Rubrimonas
Genus XIX. Ruegaria
Genus XX. Sagittula
Genus XXI. Staleya
Genus XXII. Stappia
Genus XXIII. Sulfitobacter

Order IV. *"Sphingomonadales"*
Family I. Sphingomonadaceae
Genus I. Sphingomonas
Genus II. Blastomonas
Genus III. Erythrobacter
Genus IV. Erythromicrobium
Genus V. Erythromonas
Genus VI. Porphyrobacter
Genus VII. Rhizomonas
Genus VIII. Sandaracinobacter
Genus IX. Zymomonas
Order V. *Caulobacterales*
Family I. Caulobacteraceae
Genus I. Caulobacter
Genus II. Asticcacaulis
Genus III. Brevundimonas
Genus IV. Phenylobacterium
Order VI. *"Rhizobiales"*
Family I. Rhizobiaceae
Genus I. Rhizobium
Genus II. Agrobacterium
Genus III. Carbophilus
Genus IV. Chelatobacter
Genus V. Ensifer
Genus VI. Sinorhizobium
Family II. Bartonellaceae
Genus I. Bartonella
Family III. Brucellaceae
Genus I. Brucella
Genus II. Mycoplana
Genus III. Ochrobactrum
Family IV. "Phyllobacteriaceae
Genus I. Phyllobacterium
Genus II. Allorhizobium
Genus III. Aminobacter
Genus IV. Aquamicrobium
Genus V. Defluvibacter
Genus VI. Mesorhizobium
Genus VII. Pseudaminobacter
Family V. "Methylocystaceae"
Genus I. Methylocystis
Genus II. Methylopila
Genus III. Methylosinus
Family VI. "Beijerinckiaceae"
Genus I. Beijerinckia
Genus II. Chelatococcus
Genus III. Derxia
Genus IV. Methylocella
Family VII. "Bradyrhizobiaceae"
Genus I. Bradyrhizobium
Genus II. Afipia
Genus III. Agromonas
Genus IV. Blastobacter
Genus V. Bosea
Genus VI. Nitrobacter
Genus VII. Oligotropha
Genus VIII. Rhodopseudomonas
Family VIII. Hyphomicrobiaceae
Genus I. Hyphomicrobium
Genus II. Ancalomicrobium
Genus III. Ancylobacter
Genus IV. Angulomicrobium
Genus V. Aquabacter
Genus VI. Azorhizobium
Genus VII. Blastochloris
Genus VIII. Devosia
Genus IX. Dichotomicrobium
Genus X. Filomicrobium
Genus XI. Gemmiger
Genus XII. Labrys
Genus XIII. Methylorhabdus
Genus XIV. Pedomicrobium
Genus XV. Prosthecomicrobium
Genus XVI. Rhodomicrobium
Genus XVII. Rhodoplanes
Genus XVIII. Seliberia
Genus XIX. Starkeya
Genus XX. Xanthobacter
Family IX. "Methylobacteriaceae"
Genus I. Methylobacterium
Genus II. Protomonas
Genus III. Roseomonas
Family X. "Rhodobiaceae"
Genus I. Rhodobium
Class II. *"Betaproteobacteria"*
Order I. *"Burkholderiales"*
Family I. "Burkholderiaceae"
Genus I. Burkholderia
Genus II. Cupriavidus
Genus III. Lautropia
Genus IV. Pandoraea
Genus V. Thermothrix
Family II. "Ralstoniaceae"
Genus I. Ralstonia
Family III. "Oxalobacteraceae"
Genus I. Oxalobacter
Genus II. Duganella
Genus III. Herbaspirillum
Genus IV. Janthinobacterium
Genus V. Massilia
Genus VI. Telluria
Family IV. Alcaligenaceae

Genus I. Alcaligenes
Genus II. Achromobacter
Genus III. Bordetella
Genus IV. Pelistega
Genus V. Sutterella
Genus VI. Taylorella
Family V. Comamonadaceae
Genus I. Comamonas
Genus II. Acidovorax
Genus III. Aquabacterium
Genus IV. Brachymonas
Genus V. Delftia
Genus VI. Hydrogenophaga
Genus VII. Ideonella
Genus VIII. Leptothrix
Genus IX. Polaromonas
Genus X. Rhodoferax
Genus XI. Roseateles
Genus XII. Rubrivivax
Genus XIII. Sphaerotilus
Genus XIV. Tepidimonas
Genus XV. Thiomonas
Genus XVI. Variovorax
Order II. *"Hydrogenophilales"*
Family I. "Hydrogenophilaceae"
Genus I. Hydrogenophilus
Genus II. Thiobacillus
Order III. *"Methylophilales"*
Family I. "Methylophilaceae"
Genus I. Methylophilus
Genus II. Methylobacillus
Genus III. Methylovorus
Order IV. *"Neisseriales"*
Family I. Neisseriaceae
Genus I. Neisseria
Genus II. Alysiella
Genus III. Aquaspirillum
Genus IV. Catenococcus
Genus V. Chromobacterium
Genus VI. Eikenella
Genus VII. Formivibrio
Genus VIII. Iodobacter
Genus IX. Kingella
Genus X. Microvirgula
Genus XI. Prolinoborus
Genus XII. Simonsiella
Genus XIII. Vitreoscilla
Genus XIV. Vogesella
Order V. *"Nitrosomonadales"*
Family I. "Nitrosomonadaceae"
Genus I. Nitrosomonas
Genus II. Nitrosospira
Family II. Spirillaceae
Genus I. Spirillum
Family III. Gallionellaceae
Genus I. Gallionella
Order VI. *"Rhodocyclales"*
Family I. "Rhodocyclaceae"
Genus I. Rhodocyclus
Genus II. Azoarcus
Genus III. Azonexus
Genus IV. Azospira
Genus V. Azovibrio
Genus VI. Ferribacterium
Genus VII. Propionibacter
Genus VIII. Propionivibrio
Genus IX. Thauera
Genus X. Zoogloea
Class III. *"Gammaproteobacteria"*
Order I. *"Chromatiales"*
Family I. Chromatiaceae
Genus I. Chromatium
Genus II. Allochromatium
Genus III. Amoebobacter
Genus IV. Halochromatium
Genus V. Halothiobacillus
Genus VI. Isochromatium
Genus VII. Lamprobacter
Genus VIII. Lamprocystis
Genus IX. Marichromatium
Genus X. Nitrosococcus
Genus XI. Pfennigia
Genus XII. Rhabdochromatium
Genus XIII. Thermochromatium
Genus XIV. Thioalkalicoccus
Genus XV. Thiocapsa
Genus XVI. Thiococcus
Genus XVII. Thiocystis
Genus XVIII. Thiodictyon
Genus XIX.
Genus XX. Thiohalocapsa
Genus XXI. Thiolamprovum
Genus XXII. Thiopedia
Genus XXIII. Thiorhodococcus
Genus XXIV. Thiorhodovibrio
Genus XXV. Thiospirillum
Family II. Ectothiorhodospiraceae
Genus I. Ectothiorhodospira
Genus II. Arhodomonas
Genus III. Halorhodospira
Genus IV. Nitrococcus
Genus V. Thiorhodospira

Order II. *Acidithiobacillales*
- *Family I. Acidithiobacillaceae*
  - *Genus I. Acidithiobacillus*
- *Family II. Thermithiobacillaceae*
  - *Genus I. Thermithiobacillus*

Order III. *"Xanthomonadales"*
- *Family I. "Xanthomonadaceae"*
  - *Genus I. Xanthomonas*
  - *Genus II. Frateuria*
  - *Genus III. Luteimonas*
  - *Genus IV. Lysobacter*
  - *Genus V. Nevskia*
  - *Genus VI. Pseudoxanthomonas*
  - *Genus VII. Rhodanobacter*
  - *Genus VIII. Stenotrophomonas*
  - *Genus IX. Xylella*

Order IV. *"Cardiobacteriales"*
- *Family I. Cardiobacteriaceae*
  - *Genus I. Cardiobacterium*
  - *Genus II. Dichelobacter*
  - *Genus III. Suttonella*

Order V. *"Thiotrichales"*
- *Family I. "Thiotrichaceae"*
  - *Genus I. Thiothrix*
  - *Genus II. Achromatium*
  - *Genus III. Beggiatoa*
  - *Genus IV. Leucothrix*
  - *Genus V. Macromonas*
  - *Genus VI. Thiobacterium*
  - *Genus VII. Thiomargarita*
  - *Genus VIII. Thioploca*
  - *Genus IX. Thiospira*
- *Family II. "Piscirickettsiaceae"*
  - *Genus I. Piscirickettsia*
  - *Genus II. Cycloclasticus*
  - *Genus III. Hydrogenovibrio*
  - *Genus IV. Methylophaga*
  - *Genus V. Thiomicrospira*
- *Family III. "Francisellaceae"*
  - *Genus I. Francisella*

Order VI. *"Legionellales"*
- *Family I. Legionellaceae*
  - *Genus I. Leionella*
- *Family II. "Coxiellaceae"*
  - *Genus I. Coxiella*
  - *Genus II. Rickettsiella*

Order VII. *"Methylococcales"*
- *Family I. Methylocaccaceae*
  - *Genus I. Methylococcus*
  - *Genus II. Methylobacter*
  - *Genus III. Methylocaldum*
  - *Genus IV. Methylomicrobium*
  - *Genus V. Methylomonas*
  - *Genus VI. Methylosphaera*

Order VIII. *"Oceanspirillales"*
- *Family I. "Oceanospirillaceae"*
  - *Genus I. Oceanospirillum*
  - *Genus II. Balneatrix*
  - *Genus III. Fundibacter*
  - *Genus IV. Marinomonas*
  - *Genus V. Marinospirillum*
  - *Genus VI. Neptunomonas*
- *Family II. Halomonadaceae*
  - *Genus I. Halomonas*
  - *Genus II. Alcanivorax*
  - *Genus III. Carnimonas*
  - *Genus IV. Chromohalobacter*
  - *Genus V. Deleya*
  - *Genus VI. Zymobacter*

Order IX. *Pseudomonadales*
- *Family I. Pseudomonadaceae*
  - *Genus I. Pseudomonas*
  - *Genus II. Azomonas*
  - *Genus III. Azotobacter*
  - *Genus IV. Cellvibrio*
  - *Genus V. Chryseomonas*
  - *Genus VI. Flaviomonas*
  - *Genus VII. Lampropedia*
  - *Genus VIII. Mesophilobacter*
  - *Genus IX. Morococcus*
  - *Genus X. Oligella*
  - *'Genus XI. Rhizobacter*
  - *Genus XII. Rugamonas*
  - *Genus XIII. Serpens*
  - *Genus XIV. Thermoleophilum*
  - *Genus XV. Xylophilus*
- *Family II. Moraxellaceae*
  - *Genus I. Moraxella*
  - *Genus II. Acinetobacter*
  - *Genus III. Psychrobacter*

Order X. *"Alteromonadales"*
- *Family I. "Alteromonadaceae"*
  - *Genus I. Alteromonas*
  - *Genus II. Allishewanella*
  - *Genus III. Colwellia*
  - *Genus IV. Ferrimonas*
  - *Genus V. Glaciecola*
  - *Genus VI. Idiomarina*
  - *Genus VII. Marinobacter*
  - *Genus VIII. Marinobacterium*
  - *Genus IX. Microbulbifer*
  - *Genus X. Moritella*

Genus XI. Pseudoalteromonas
Genus XII. Shewanella
Order XI. "*Vibrionales*"
Family I. Vibrionaceae
Genus I. Vibrio
Genus II. Allomonas
Genus III. Enhydrobacter
Genus IV. Listonella
Genus V. Photobacterium
Genus VI. Salinivibrio
Order XII. "*Aeromonadales*"
Family I. Aeromonadaceae
Genus I. Aeromonas
Genus II. Oceanomonas
Genus III. Tolumonas
Family II. Succinivibrionaeae
Genus I. Succinivibrio
Genus II. Anaerobiospirillum
Genus III. Ruminobacter
Genus IV. Succinimonas
Order XIII. "*Enterobacteriales*"
Family I. Enterobacteriaceae
Genus I. Enterobacter
Genus II. Alterococcus
Genus III. Arsenophonus
Genus IV. Brenneria
Genus V. Buchnera
Genus VI. Budvicia
Genus VII. Buttiauxella
Genus VIII. Calymmatobacterium
Genus IX. Cedecea
Genus X. Citrobacter
Genus XI. Edwardsiella
Genus XII. Erwinia
Genus XIII. Escherichia
Genus XIV. Ewingella
Genus XV. Hafnia
Genus XVI. Klebsiella
Genus XVII. Kluyvera
Genus XVIII. Leclercia
Genus XIX. Leminorela
Genus XX. Moellerella
Genus XXI. Morganella
Genus XXII. Obesumbacterium
Genus XXIII. Pantoea
Genus XXIV. Pectobacterium
Genus XXV. Photorhabdus
Genus XXVI. Plesiomonas
Genus XXVII. Pragia
Genus XXVIII. Proteus
Genus XXIX. Providencia
Genus XXX. Rahnella
Genus XXXI. Saccharobacter
Genus XXXII. Salmonella
Genus XXXIII. Serratia
Genus XXXIV. Shigella
Genus XXXV. Sodalis
Genus XXXVI. Tatumella
Genus XXXVII. Trabulsiella
Genus XXXVIII. Wigglesworthia
Genus XXXIX. Xenorhabdus
Genus XL. Yersinia
Genus XLI. Yokenella
Order XIV. "*Pasteurellales*"
Family I Pasteurellaceae
Genus I. Pasteurella
Genus II. Actinobacillus
Genus III. Hamemorphilus
Genus IV. Lonepinella
Genus V. Mannheimia
Genus VI. Phocoenobacter
Class IV. "*Deltaproteobacteria*"
Order I. "*Desulfurellales*"
Family I. "Desulfurellaceae"
Genus I. Desulfurella
Genus II. Hippea
Order II. "*Desulfovibrionales*"
Family I. "Desulfovibrionacea"
Genus I. Desulfovibrio
Genus II. Bilophila
Genus III. Lawsonia
Family II. "Desulfomicrobiacea"
Genus I. Desulfomicrobium
Family III. "Desullfohalobiaceae"
Genus I. Desulfohalobium
Genus II. Desulfomonas
Genus III. Desulfonatronovibrio
Family IV. "Desulfonatronumaceae"
Genus I. Desullfonatronum
Order III. "*Desulfobacterales*
Family I. "Desulfobacteraceae"
Genus I. Desulfobacter
Genus II. Desulfobacerium
Genus III. Desulfobacula
Genus IV. Desulfocella
Genus V. Desulfococcus
Genus VI. Desulfofaba
Genus VII. Desulfofrigus
Genus VIII. Desulfonema
Genus IX. Desulfosarcina
Genus X. Desulfospira
Genus XI. Desullfotalea

Genus XII. Desulfotignum
Family II. "Desulfobulbaceae"
Genus I. Desulfobulbus
Genus II. Desulfocapsa
Genus III. Desulfofustus
Genus IV. Desulforhopalus
Family III. "Nitrospinaceae"
Genus I. Nitrospina
Genus II. Desulfobacca
Genus III. Desulfomonile
Order IV. *"Desulfuromonadales"*
Family I. "Desulfuromonadaceae"
Genus I. Desulfuromonas
Genus II. Desulfuromusa
Family II. "Geobacteraceae"
Genus I. Geobacter
Family III. "Pelobacteraceae"
Genus I. Pelobacter
Genus II. Malonomonas
Genus III. Trichlorobacter
Order V. *"Syntrohobacterales"*
Family I. "Syntrophobacteraceae"
Genus I. Syntrophobacter
Genus II. Desulfacinum
Genus III. Desulforhabdus
Genus IV. Desulfovirga
Genus V. Thermodesulforhabdus
Family II. "Syntrophaceae"
Genus I. Syntroiphus
Genus II. Smithella
Order VI. *"Bdelloovibrionales"*
Family I. "Bdellovibrionaceae"
Genus I. Bdellovibrio
Genus II. Bacteriovorax
Genus III. Micavibrio
Genus IV. Vampirovibrio
Order VII. *Myxococcales*
Family I. Myxococcaceae
Genus Myxococccus
Genus II. Angiococcus
Family II. Archangiaceae
Genus Archangium
Family III. Cystobacteraceae
Genus I. Cystobacter
Genus II. Melittangium
Genus III. Stigmatella
Family IV. Polyangiaceae
Genus I. Polyangium
Genus II. Chondromyces
Genus III. Nannocystis
Class V. *"Epsilonproteobacteria"*
Order I. *"Campylobacterales"*
Family I. Campylobacteraceae
Genus I. Campylobacter
Genus II. Arobacter
Genus III. Sulfurospirillum
Genus IV. Thiovulum
Family II. "Helicobavteraceae"
Genus I. Helicobacter
Genus II. Wolinella
Phylum BXIII. *Firmicutes*
Class I. *"Clostridia"*
Order I. *Clostridiales*
Family I. Clostridiaceae
Genus I. Clostridium
Genus II. Acetivibrio
Genus III. Acidaminobacter
Genus IV. Anaerobacter
Genus V. Caloramator
Genus VI. Coprobacillus
Genus VII. Natronincola
Genus VIII. Oxobacter
Genus IX. Sarcina
Genus X. Sporobacter
Genus XI. Thermobrachium
Genus XII. Thermohalobacter
Genus XIII. Tindallia
Family II. "Lachnospiraceae"
Genus I. Lachnospira
Genus II. Acetitomaculum
Genus III. Anaerofilum
'Genus IV. Butyrivibrio
Genus V. Catenibacterium
Genus VI. Catonella
Genus VII. Coprococcus
Genus VIII. Johnsonella
Genus IX. Pseudobutyrivibrio
Genus X. Roseburia
Genus XI. Ruminococcus
Genus XII. Sporobacterium
Family III. "Peptostreptococcaceae"
Genus I. Peptostreptococcus
Genus II. Filifactor
Genus III. Finegoldia
Genus IV. Fusibacter
Genus V. Helcococcus
Genus VI. Micromonas
Genus VII. Tissierella
Family IV. "Eubacteriaceae"
Genus I. Eubacterium
Genus II. Acetobacterium
Genus III. Anaerovorax

Genus IV. Mogibacterium
Genus V. Pseudoramibacter
Family V. Peptococcaceae
Genus I. Peptococcus
Genus II. Anaeroarcus
Genus III. Anaerosinus
Genus IV. Anaerovibrio
Genus V. Carboxydothermus
Genus VI. Centipeda
Genus VII. Dehalobacter
Genus VIII. Dendrosporobacter
Genus IX. Desulfitobacterium
Genus X. Desulfonispora
Genus XI. Desulfosporosinus
Genus XII. Desulfotomaculum
Genus XIII. Mitsuokella
Genus XIV. Propionispira
Genus XV. Succinispira
Genus XVI. Syntrophobotulus
Genus XVII. Thermoterrabacterium
Family VI. "Heliobacteriaceae"
Genus I. Heliobacterium
Genus II. Heliobacillus
Genus III. Heliophilum
Genus IV. Heliorestis
Family VII. "Acidaminococcaceae"
Genus I. Acidaminococcus
Genus II. Acetonema
Genus III. Anaeromusa
Genus IV. Dialister
Genus V. Megasphaera
Genus VI. Papillibacter
Genus VII. Pectinatus
Genus VIII. Phascolarctobacterium
Genus IX. Quinella
Genus X. Schwartzia
Genus XI. Selenomonas
Genus XII. Sporomusa
Genus XIII. Succiniclasticum
Genus XIV. Veillonella
Genus XV. Zymophilus
Family VIII. Syntrophomonadaceae
Genus I. Syntrophomonas
Genus II. Acetogenium
Genus III. Aminobacterium
Genus IV. Aminomonas
Genus V. Anaerobaculum
Genus VI. Anaerobranca
Genus VII. Caldicellulosiruptor
Genus VIII. Dethiosulfovibrio
Genus IX. Pelospora
Genus X. Syntrophospora
Genus XI. Syntrophothermus
Genus XII. Thermaerobacter
Genus XIII. Thermanaerovibrio
Genus XIV. Thermohydrogenium
Genus XV. Thermosyntropha
Order II. *"Thermoanaerobacteriales"*
Family I. "Thermoanaerobacteriacea"
Genus I. Thermoanerobacterium
Genus II. Ammonifex
Genus III. Carboxydobrachium
Genus IV. Coprothermobacter
Genus V. Moorella
Genus VI. Sporotomaculum
Genus VII. Thermacetogenium
Genus VIII. Thermoanaerobacter
Genus IX. Theremoanaerobium
Order III. *Haloanaerobiales*
Family I. Haloanaerobiaceae
Genus I. Haloanaerobium
Genus II. Halocella
Genus III. Halothermothrix
Genus IV. Natroniella
Family II. Halobacteroidaceae
Genus I. Halobacteroides
Genus II. Acetohalobium
Genus III. Haloanaerobacter
Genus IV. Orenia
Genus V. Sporohalobacter
Class II. *Mollicutes*
Order I. *Mycoplasmatales*
Family I. Mycoplasmataceae
Genus I. Mycoplasma
Genus II. Eperythrozoon
Genus III. Haemobartonella
Genus IV. Ureaplasma
Order II. *Entomoplasmatales*
Family I. Entomoplasmataceae
Genus I. Entomoplasma
Genus II. Mesoplasma
Family II. Spiroplasmataceae
Genus I. Spiroplasma
Order III. *Acholeplasmatales*
Family I. Acholeplasmataceae
Genus I. Acholeplasma
Order IV. *Anaeroplasmatales*
Family I. Anaeroplasmataceae
Genus I. Anaeroplasma
Genus II. Asteroleplasma
Order V. *Incertae sedis*
Family I. "Erysipelotrichaceae"

Genus I. Erysipelothrix
Genus II. Bulleidia
Genus III. Holdermania
Genus IV. Solobacterium
Class III. "*Bacilli*"
Order I. *Bacillales*
Family I. Bacillaceae
Genus I. Bacillus
Genus II. Amphibacillus
Genus III. Anoxybacillus
Genus IV. Exiguobacterium
Genus V. Gracilibacillus
Genus VI. Halobacillus
Genus VII. Saccharococus
Genus VIII. Salibacillus
Genus IX. Virgibacillus
Family II. Planococcaceae
Genus I. Planococcus
Genus II. Filibacter
Genus III. Kurthia
Genus IV. Sporosarcina
Family III. Caryophanaceae
Genus I. Caryophanon
Family IV. "Listeriaceae"
Genus I. Listeria
Genus II. Brochothrix
Family V. "Staphylococcaceae"
Genus I. Staphylococcus
Genus II. Gemella
Genus III. Macrococcus
Genus IV. Salinicoccus
Family VI. "Sporolactobacillaceae"
Genus I. Sporolactobacillus
Genus II. Marinococcus
Family VII. "Paenibacillaceae"
Genus I. Paenibacillus
Genus II. Ammoniphilus
Genus III. Aneurinibacillus
Genus IV. Brevibacillus
Genus V. Oxalophagus
Genus VI. Thermicanus
Genus VII. Thermobacillus
Family VIII. "Alicyclobacillaceae"
Genus I. Alicyclobacillus
Genus II. Pasteuria
Genus III. Sulfobacillus
Family IX. "Thermoactinomycetaceae"
Genus I. Thermoactinomyces
Order II. "*Lactobacillales*"
Family I. Lactobacillaceae
Genus I. Lactobacillus
Genus II. Paralactobacillus
Genus III. Pediococcus
Family II. "Aerococcaceae"
Genus I. Aerococcus
Genus II. Abiotrophia
Genus III. Dolosicoccus
Genus IV. Eremococcus
Genus V. Facklamia
Genus VI. Globicatella
Genus VII. Ignavigranum
Family III. "Carnobacteriaceae"
Genus I. Carnobacterium
Genus II. Agitococcus
Genus III. Alloiococcus
Genus IV. Desemzia
Genus V. Dolosigranulum
Genus VI. Granulicatella
Genus VII. Lactosphaera
Genus VIII. Trichococcus
Family IV. "Enterococcaceae"
Genus I. Enterococcus
Genus II. Atopobacter
Genus III. Melissococcus
Genus IV. Tetragenococcus
Genus V. Vagococcus
Family V. "Leuconostocaceae"
Genus I. Leuconostoc
Genus II. Oenococcus
Genus III. Weissella
Family VI. Streptococcaceae
Genus I. Streptococcus
Genus II. Lactococcus
Family VII. Incertae sedis
Genus I. Acetoanaerobium
Genus II. Oscillospira
Genus III. Syntrophococcus
Phylum BXIV. Actinobacteria phy. nov.
Class I. Actinobacteria
Subclass I. *Acidimicrobidae*
Order I. *Acidimicrobiales*
Suborder I. "*Acidimicrobineae*"
Family I. Acidimicrobiaceae
Genus I. Acidimicrobium
Subclass II. *Rubrobacteridae*
Order I. *Rubrobacterales*
Suborder II. "*Rubrobacterineae*"
Family I. Rubrobacteraceae
Genus I. Rubrobacter
Subclass III. *Coriobacteridae*
Order I. *Coriobacteriales*
Suborder III. "*Coriobacterineae*"

Family I. Coriobacteriaceae
Genus I. Coriobacterium
Genus II. Atopobium
Genus III. Collinsella
Genus IV. Cryptobacterium
Genus V. Denitrobacterium
Genus VI. Eggerthella
Genus VII. Slackia
Subclass IV. *Sphaerobacteridae*
Order I. *Sphaerobacterales*
Suborder IV. *"Sphaerobacterineae"*
Family I. Sphaerobacteraceae
Genus I. Sphaerobacter
Subclass V. *Actinobacteridae*
Order I. *Actinomycetales*
Suborder V. *Actinomycineae*
Family I. Actinomycetaceae
Genus I. Actinomyces
Genus II. Actinobaculum
Genus III. Arcanobacterium
Genus IV. Mobiluncus
Suborder VI. *Micrococcineae*
Family I. Micrococcanceae
Genus I. Micrococcus
Genus II. Arthrobacter
Genus III. Kocuria
Genus IV. Nesterenkonia
Genus V. Renibacterium (moved)
Genus VI. Rothia
Genus VII. Stomatococcus
Family II. Bogoriellaceae
Genus I. Bogoriella (moved)
Family III. Rarobacteraceae
Genus I. Rarobacter (moved)
Family IV. Sanguibacteraceae
Genus I. Sanguibacter
Family V. Brevibacteriaceae
Genus I. Brevibacterium
Family VI. Cellulomonadaceae
Genus I. Cellulomonas
Genus II. Oerskovia
Family VII. Dermabacteraceae
Genus I. Dermabacter
Genus II. Brachybacterium
Family VIII. Dermatophilaceae
Genus I. Dermatophilus
Family IX. Dermacoccaceae
Genus I. Dermacoccus (moved)
Genus II. Demetria (moved)
Genus III, Kytococcus (moved)
Family X. Intrasporangiaceae
Genus I. Intrasporangiun
Genus II. Janibacter
Genus III. Ornithinicoccus
Genus IV. Ornithinimicrobium
Genus V. Nostocoidia
Genus VI. Terrabacter
Genus VII. Terracoccus
Genus VIII. Tetrasphaera
Family XI. Jonesiaceae
Genus I. Jonesia
Family XII. Microbacteriaceae
Genus I. Microbacterium
Genus II. Agrococcus
Genus III. Agromyces
Genus IV. Aureobacterium
Genus V. Clavibacter
Genus VI. Cryobacterium
Genus VII. Curtobacterium
Genus VIII. Frigoribavterium
Genus IX. Leifsonia
Genus X. Leucobacter (moved)
Genus XI. Rathayibacter
Genus XII. Subtercola
Family XIII. "Beutenbergiaceae"
Genus I. Beutenbergia
Family XIV. Promicromonosporaceae
Genus I. Promicromonospora
Suborder VII. *Corynebacterineae*
Family I. Corynebacteriaceae
Genus I. Corynebacterium
Family II. Dietziaceae
Genus I. Dietzia
Family III. Gordoniaceae
Genus I. Gordonia
Genus II. Skermania
Family IV. Mycobacteriaceae
Genus I. Mycobacterium
Family V. Nocardiaceae
Genus I. Nocardia
Genus II. Rhodococcus
Family VI. Tsukamurellaceae
Genus I. Tsukamurella
Family VII. "Williamsiaceae"
Genus I. Williamsia
Suborder VIII. *MKicromonosporineae*
Family I. Micromonosporaceae
Genus I. Micromonospora
Genus II. Acinoplanes
Genus III. Catallatospora

Genus IV. Catenuloplanes
Genus V. Couchinoplanes
Genus VI. Dactylosporangium
Genus VII. Pilimelia
Genus VIII. Spirilliplanes
Genus IX. Verrucosispora
Suborder IX. *Propionibacterineae*
Family I. Propionibacteriaceae
Genus I. Propionibacterium
Genus II. Luteococcus
Genus III. Microlunatus
Genus IV. Propioniferax
Genus V. Tessaracoccus
Family II. Nocardioidaceae
Genus I. Nocardioides
Genus II. Aeromicrobium
Genus III. Friedmanniella
Genus IV. Hongia
Genus V. Kribella
Genus VI. Micropruina
Genus VII. Marmoricola
Suborder X. *Pseudonocardineae*
Family I. Pseudonocardiaceae
Genus I. Pseudonocardia
Genus II. Actinoalloteichus
Genus III. Actinopolyspora
Genus IV. Amycolatopsis
Genus V. Kibdelosporangium
Genus VI. Kutzneria
Genus VII. Prauserella
Genus VIII. Saccharomonospora
Genus IX. Saccharopolyspora
Genus X. Streptoalloteichus
Genus XI. Thermobispora
Genus XII. Thermocrispum
Family II. Actinosynnemataceae
Genus I. Actinosynnema
Genus II. Actinokineospora
Genus III. Lentzea
Genus IV. Saccharothrix
Suborder XI. *Streptomycineae*
Family I. Streptomycetaceae
Genus I. Strepstomyces
Genus II. Kitasatospora
Genus III. Streptoverticillium
Suborder XII. *Streptosporangineae*
Family I. Streptosporangiaceae
Genus I. Streptosporangium
Genus II. Acrocarpospora
Genus III. Herbidospora
Genus IV. Microbispora
Genus V. Microtetraspora
Genus VI. Nonomuraea
Genus VII. Planobispora
Genus VIII. Planomonospora
Genus IX. Planopolyspora
Genus X. Planotetraspora
Family II. Nocardiopsaceae
Genus I. Nocardiopsis
Genus II. Thermobifida
Family III. Thermomonosporaceae
Genus I. Thermomonospora
Genus II. Actinomadura
Genus III. Spirillospora
Suborder XIII. *Frankineae*
Family I. Frankiaceae
Genus I. Frankia
Family II. Geodermatophilaceae
Genus I. Geodermatophilus
Genus II. Blastococcus
Genus III. Modestobacter
Family III. Microsphaeraceae
Genus I. Microsphaera
Family IV. Sporichthyaceae
Genus I. Sporichthya
Family V. Acidothermaceae
Genus I. Acidothermus
Family VI. "Kineosporiaceae"
Genus I. Kineosporia
Genus II. Cryptosporangium
Genus III. Kineococcus
Suborder XIV. *Glycomycineae*
Family I. Glycomycetaceae
Genus I. Glycomyes
Order II. *Bifidobacteriales*
Family I. Bifidobacteriaceae
Genus I. Bifidobacterium
Genus II. Falcivibrio
Genus III. Gardnerella
Family II. Unknown Affiliation
Genus I. Actinobispora
Genus II. Actinocorallia
Genus III. Excellospora
Genus IV. Pelczaria
Genus V. Turicella
Phylum BXV. *Planctomycetes phy. nov.*
Class I. *"Planctomycetacia"*
Order I. *Planctomycetales*
Family I. Planctomycetaceae
Genus I. Planctomyces

Genus II. Gemmata
Genus III. Isophaera
Genus IV. Pirellula
Phylum BXVI. *Chlamydiae phy. nov.*
Class I. *"Chlamydiae"*
Order I. *Chlamydiales*
Family I. Chlamydiaceae
Genus I. Chlamydia
Genus II. Chlamydophila
Family II. Parachlamydiaceae
Genus I. Parachlamydia
Family III. Simkaniaceae
Genus I. Simkania
Family IV. Waddliaceae
Genus I. Waddlia
Phylum BXVII. *Spirochaetes phy. nov.*
Class I. *"Spirochaetes"*
Order I. *Spirochaetales*
Family I. Spirochaetaceae
Genus I. Spirochaeta
Genus II. Borrelia
Genus III. Brevinema
Genus IV. Clevelandina
Genus V. Cristispira
Genus VI. Diplocalyx
Genus VII. Hollandina
Genus VIII. Pillotina
Genus IX. Treponema
Family II. "Serpulinaceae"
Genus I. Serpulina
Genus II. Brachyspira
Family III. Leptospiraceae
Genus I. Leptonema
Genus II. Leptospira
Phylum BXVIII. *Fibrobacteres phy. nov.*
Class I. *"Fibrobacteres"*
Order I. *"Fibrobacterales"*
Family I. "Fibrobacteraceae"
Genus I. Fibrobacter
Phylum BXIX. *Acidobacteria phy. nov.*
Class I. *"Acidobacteria"*
Order I. *"Acidobacteriales"*
Family I. "Acidobacteriaceae"
Genus I. Acidobacterium
Genus II. Geothrix
Genus III. Holophaga
Phylum BXX. *Bacteroidetes phy. nov.*
Class I. *"Bacteroidetes"*
Order I. *"Bacteroidales"*
Family I. Bacteroidaceae
Genus I. Bacteroides
Genus II. Acetofilamentum
Genus III. Acetomicrobium
Genus IV. Acetothermus
Genus V. Anaerorhabdus
Genus VI. Megamonas
Family II. "Rikenellaceae"
Genus I. Rikenella
Genus II. Marinilabilia
Family III. "Porphyromonadaceae"
Genus I. Porphyromonas
Genus II. Dysgonomonas
Family IV. "Prevotellaceae"
Genus I. Prevotella
Class II. *"Flavobacteria"*
Order I. *"Flavobacteriales"*
Family I. Flavobacteriaceae
Genus I. Flavobacterium
Genus II. Bergeyella
Genus III. Capnocytophaga
Genus IV. Cellulophaga
Genus V. Chryseobacterium
Genus VI. Coenonia
Genus VII. Empedobacter
Genus VIII. Gelidibacter
Genus IX. Ornithobacterium
Genus X. Polaribacter
Genus XI. Psychroflexus
Genus XII. Psychroserpens
Genus XIII. Riemerella
Genus XIV. Saligentibacter
Genus XV. Weeksella
Family II. "Myroidaceae"
Genus I. Myroides
Genus II. Psychromonas
Family III. "Blattabacteriaceae"
Genus I. Blattabacterium
Class III. *"Sphingobacteria"*
Order I. *"Sphingobacteriales"*
Family I. Sphingobacteriaceae
Genus I. Sphingobacterium
Genus II. Pedobacter
Family II. "Saprospiraceae"
Genus I. Saprospira
Genus II. Haliscomenobacter
Genus III. Lewinella
Family III. "Flexibacteraceae"
Genus I. Flexibacter
Genus II. Cyclobacterium
Genus III. Cytophaga
Genus IV. Dyadobacter
Genus V. Flectobacillus

Genus VI. *Hymenobacter*
Genus VII. *Meniscus*
Genus VIII. *Microscilla*
Genus IX. *Runella*
Genus X. *Spirosoma*
Genus XI. *Sporocytophaga*
Family IV. *"Flammeovirgaceae"*
Genus I. *Flammeovirga*
Genus II. *Flexithrix*
Genus III. *Persicobacter*
Genus IV. *Thermonema*
Family V. *Crenotrichaceae*
Genus I. *Crenothrix*
Genus II. *Chitinophaga*
Genus III. *Rhodothermus*
Genus IV. *Toxothrix*
Phylum BXXI. *Fusobacteria phy. nov.*
Class I. *"Fusobacteria"*
Order I. *"Fusobacteriales"*
Family I. *"Fusobacteriaceae"*
Genus I. *Fusobacterium*
Genus II. *Ilyobacter*
Genus III. *Leptrotrichia*
Genus IV. *Propionigenium*
Genus V. *Sebaldella*
Genus VI. *Streptobacillus*
Family II. *Incertae sedis*
Genus I. *Cetobacterium*
Phylum BXXII. *Verrucomicrobia phy. nov.*
Class I. *Verrucomicrobiae*
Order I. *Verrucomicrobiales*
Family I. *Verrucomicrobiaceae*
Genus I. *Verrucomicrobium*
Genus II. *Prosthecobacter*
Family II. *"Xipinematobacteriaceae"*
Genus I. *Xiphinematobacter*
Phylum BXXIII. *Dictyoglomus phy. nov.*
Class I. *"Dictyoglomi"*
Order I. *"Dictyoglomales"*
Family I. *"Dictyoglomaceae"*
Genus I. *Dictyoglomus*

Classification of Viruses

The classification of viruses has undergone great change, as has bacterial taxonomy. Most viruses have not even been classified due to a lack of data concerning their reproduction and molecular biology. Estimates suggest that more than 30,000 viruses are being studies in laboratories and reference centers worldwide.

The classification and viral information presented here follows the outline given in Chap. 10 (Tables 10.1 and 10.2) of Bergey's Manual® of Systematic Bacteriology, (G Garrity, M Winters, D Searles 2nd edn., April 2001, Springer-Verlag, New York). Information can also be found in Human Virology: A test for Students of Medicine, Dentistry, and Microbiology (L. Collier, J. Oxford, 1993 Oxford University Press), and in Virology (J. Levy, H Fraenkel-Conrat, R. Owens, 2nd edn., 1994, Prentice Hall).

The 21 families of viruses listed here are primarily those that infect vertebrates. Thus, these families represent only a small part of the 71 families and more than 3000 viruses recognized in *Virus Taxonomy: Seventh Report of the International Committee on Taxonomy of Viruses*, Van Regenmortel HV, Bishop DHL, Van Regenmortel MH, and Fauquet CM eds., Academic Press, San Dieago, CA 2000 (ISBN 0123702003).

Family: Picornaviridae

Genera:

Enterovirus (gastrointestinal viruses, poliovirus, coxsackie viruses A and B, echoviruses)
Hapatovirus (hepatitis A virus)
Cardiovirus (encephalomyocarditis virus of mice and other rodents)
Rhinovirus (upper respiratory tract viruses, common cold viruses)
Aphthovirus (foot-and-mouth disease virus)

Naked, polyhedral, positive-sense, ssRNA. Synthesis and maturation take place in the host ell cytoplasm. Viruses are released via cell lysis.

Family: Caliciviridae

Genus:

Calicivirus (Norwalk viruses and similar viruses causing gastroenteritis, hepatitis E virus)

Naked, polyhedral, positive-sense ssRNA. Synthesis and maturation takes place in the host cell cytoplasm. Viruses are released via cell lysis.

Family: Togaviridae

Genera:

Alphavirus (eastern, western, and Venezuelan equine encephalitis viruses, Semliki forest virus)
Rubivirus (rubella virus)
Arterivirus (equine arteritis virus, simian hemorrhagic fever virus)

Enveloped, polyhedral, positive-sense ssRNA. Synthesis occurs in the host cell cytoplasm; maturation involves budding of nucleocapsids through the host cell plasma membrane. Viruses are released via cell lysis (*Arterivirus*). Many replicate in arthropods and vertebrates.

Family: Flaviviridae
Genera:

Flavivirus (yellow-fever virus, dengue fever virus, St. Louis and Japanese encephalitis viruses, tickborne encephalitis virus)
Pestivirus (bovine diarrhea virus, hop cholera virus) Hapatitis C virus.

Enveloped, polyhedral, positive-sense ssRNA. Synthesis occurs in the host cell cytoplasm; maturation involves budding through host cell endoplasmic reticulum and Golgi apparatus membranes. Most replicate in arthropods.

Family: Coronavirdae
Genus:

Coronavirus (common cold viruses, avian infectious bronchitis virus, feline infectious peritonitis virus, mouse hepatitis virus)

Enveloped, helical, positive-sense, ssRNA. Synthesis occurs in the host cell cytoplasm; maturation involves budding through membranes and the endoplasmic reticulum, and Golgi apparatus. Viruses are released via cell lysis.

Family: Rhabdoviridae
Genera:

Vesiculovirus (vesicular stomatitis-like virus)
Lyssavirus (rabies and rabieslike viruses)
[Unnamed] (proposed, for bovine ephemeral feverlike viruses)

Enveloped, helical, negative-sense, ssRNA. Synthesis occurs in the host cell nucleus; maturation occurs via budding from the host cell plasma membrane. Many replicate in arthropods.

Family: Filoviridae
Genus:

Filovirus (Marburg and Ebola viruses)

Enveloped; long, filamentous forms, sometimes with branching, and sometimes U-shaped, 6-shaped, or circular; negative-sense, ssRNA. Synthesis occurs in the host cell cytoplasm; maturation involves budding from the host cell plasma membrane. Viruses are released via cell lysis. These viruses are "Biosafety Level 4" pathogens – they must be handled in the laboratory under maximum containment conditions.

Family: Paramyxoviridae
Genera:

Paramyxovirus (parainfluenza viruses 1–4, mumps virus, Newcastle disease virus)
Morbillivirus (measles and measleslike viruses, canine distemper virus)
Pneumovirus (respiratory syncytial virus)

Enveloped, helical, negative-sense, ssRNA. Synthesis occurs in the host cell cytoplasm; maturation involves budding through the host cell plasma membrane. viruses are released via cell lysis. Morbilliviruses can cause persistent infections.

Family: Orthomyxoviridae
Genera:

Influenzavirus A and B (influenza viruses A and B)
Influenzavirus C (influenza C virus)

Enveloped, helical, negative-sense, ssRNA (eight segments). Synthesis occurs in the host cell nucleus; maturation takes place in the host cell cytoplasm. Viruses are released through budding from the host cell's plasma membrane. These viruses can reassort genes curing mixed infections.

Family Bunyaviridae
Genera:

Bunyavirus (Bunyamwera supergroup)
Phlebovirus (sandfly fever viruses)
Nairovirus (Nairobi sheep diseaselike viruses)
Uukuvirus (Uukuniemi-like viruses)
Hantavirus (hemorrhagic fever viruses, Korean hemorrhagic fever, Sin Nombre hantavirus)

Enveloped, spherical, negative-sense, ssRNA (three segments; *Phlebovirus* ambisense ssRNA). Synthesis occurs in the host cell cytoplasm; maturation occurs within the Golgi apparatus. Viruses are released via cell lysis. Closely related viruses can reassort genes during mixed infections.

Family: Arenaviridae
Genus:

Arenavirus (Lassa fever virus, lymphocytic choriomeningitis virus, Machupo virus, Junin virus)

Enveloped, helical, ambisense, ssRNA (two segments). Synthesis occurs in the host cell cytoplasm; maturation involves budding from the host cell plasma membrane. Virions contain ribosomes. The human pathogens Lassa, Machuppo, and Junin viruses are "Biosafety Level 4" pathogens – they must be handled in the laboratory under maximum containment conditions.

Family: Reoviridae
Genera:

Orthoreovirus (reoviruses 1, 2, and 3)
Orbivirus (Orungo virus)
Rotavirus (human rotaviruses)
Cypovirus (cytoplasmic polyhidrosis viruses)
Coltivirus (Colorado tick fever virus)
Plant Reovirus 1/3 (plant reoviruses subgroups 1, 2, and 3)

Each genus differs in morphology and physiochemical details. In general, virions are naked, polyhedral, dsRNA (10–12 segments). Synthesis and maturation takje place in the host cell cytoplasm. Viruses are released via cell lysis. Virions contain ribosomes.

Family: Birnaviridae
Genera:

Birnavirus (infectious pancreatic necrosis virus of fish and infections bursal disease virus of fowl)

Naked, polyhedral, dsRNA (two segments). Synthesis and maturation take place in the host cell cytoplasm. Viruses are released via cell lysis.

Family: Retroviridae
Genera:

MLV-related virus (spleen necrosis virus, mouse and feline leukemia viruses)
Mammalian type-B (mouse mammary tumor virus)
Type D (squirrel monkey retrovirus)
ALV-related virus (avian leukemia virus, rous sarcoma virus)
HTL-BLV group (human T cell leukemia virus HTLV-1, HTLV-II, bovine leukemia virus)
Spumavirus (the foamy viruses)
Lentivirus (human, feline, simian, and bovine immunodeficiency viruses)

Enveloped, spherical, negative-sense, ssRNA (two identical strands). Synthesis occurs in the host cell cytoplasm; maturation involves budding through the host cell plasma membrane. These viruses contain the enzyme reverse transcriptase. The retroviruses (except the *Spumavirus* and *Lentivirus* genera) represent the RNA tumor viruses, causing leukemias, carcinomas, and sarcomas.

Family: Hepadnaviridae
Genera:

Orthohepadnavirus (hepatitis B virus)
Avihepadnavirus (duck hepatitis virus)

Enveloped, polyhedral, partially dsDNA. Synthesis and maturation take place in the host cell nucleus. Surface antigen production occurs in the cytoplasm. Persistence is common and is associated with chronic disease and neoplasia.

Family: Parvoviridae
Genera:

Parvovirus (feline leucopenia virus, canine parvovirus)
Dependovirus (adeno-associated viruses)
Densovirus (insect parvoviruses)
Erythrovirus (human erythrovirus B19)

Naked, polyhedral, negative-sense, ssDNA (*Parvovirus*) or positive-sense and negative-sense, ssDNA (other genera). Synthesis and maturation occur in rapidly dividing host cells, specifically in the host cell nucleus. Viruses are released via cell lysis.

Family: Papovaviridae
Genera:

Papillomarirus (wart viruses, genital condylomas, DNA tumor viruses)
Polyomavirus (human polyoma-like viruses, SV-40)

Naked, polyhedral, dsDNA. Synthesis and maturation take place in the host cell nucleus. Viruses are released via cell lysis.

Family: Adenoviridae
Genera:

Mastadenovirus (human adenoviruses A–F, infectious canine hepatitis virus)
Aviadenovirus (avian adenoviruses)

Naked, polyhedral, dsDNA. Synthesis and maturation take place in the host cell nucleus. Viruses are released via cell lysis.

Family: Herpesviridae
Subfamily: Alphaherpesvirinae
Genera:

Simplexvirus (herpes simplex viruses 1 and 2)
Varicellovirus (varicella-zoster virus)

Subfamily: Betaherpesvirinae
Genera:

Cytomegalovirus (human cytomegalovirus)
Muromegalovirus (murine cytomegalovirus)

Subfamily: Gammaherpesvirinae
Genera:

Lymphocryptovirus (Epstein-Barr viruses)
Rhadinovirus (saimiri-ateles-like virus)

Enveloped, polyhedral, dsDNA. Synthesis and maturation occur in the host cell nucleus, with budding through the nuclear envelope. Although most herpesviruses cause persistent infections, virions can be released by rupture of the host cell plasma membrane.

Family: Poxviridae
Subfamily: Chordopoxvirinae
Genera:

Orthopoxvirus (vaccinia and variola viruses, cowpox virus)
Parapoxvirus (orf virus, pseudocowpox virus)
Avipoxvirus (fowlpox virus)
Capripoxvirus (sheep pox virus)
Leporipoxvirus (myxoma virus)
Suipoxvirus (swinepox virus)
Yatapoxvirus (yabapox virus and tanapox virus)
Molluscipoxvirus (mollescum contagiosum virus)

Subfamily: Entomopoxvirinae
Genus:

Entomopoxvirus A/B/C (poxviruses of insects)

External envelope, large, brick-shaped (or ovoid), dsDNA. Synthesis and maturation take place in the portion of the host cell cytoplasm called viroplasm ("viral factories.") Viruses are released via cell lysis.

Family: Irdoviridae
Genera:

Iridovirus (small iridescent insect viruses)
Chloriridovirus (large iridescent insect viruses)
Ranavirus (frog viruses)
Lymphocystivirus (lymphocystis viruses of fish)

Enveloped (missing on some insect viruses), polyhedral, dsDNA. Synthesis occurs in both the host cell nucleus and cytoplasm. Most virions remain cell-associated.

Word Roots Commonly Encountered in Microbiology

a-, an- not, without absence abiotic, not living; anaerobic, in the absence of air
acantho- thorn or spinelike *Acanthamoeba*, an amoeba with spinelike projections
actino- having rays *Actinomyces*, a bacterium forming colonies that look like sunbursts
aero- air aerobic, in the presence of air
agglutino- clumping or sticking together hemagglutinatin, clumping of blood cells
albo- white *Candida albicans*, a white fungus
amphi- around, doubly, both Amphitrichous describes flagella found at both ends of a bacterial cell
ant-, anti- against, versus Antibacterial compounds kill bacteria
archaeo- ancient Archaeobacteria are thought to resemble ancient forms of life
artho- joint arthritis, inflammation of joints
asco- sac, bag Ascospores are held in a saclike container, the ascus
-ase denotes enzyme lipase, an enzyme attacking lipids
aureo- golden *Staphylococcus aureus* has gold-colored colonies
auto- self Autotrophs, self-feeding organisms
bacillo- rod bacillus, rod-shaped bacterium
basid- base, foundation basidium, fungal cell bearing spores at its end
bio- life biology, the study of living things
blast- bud bladstospore, spore formed by budding
bovi- cow *Mycobacterium bovis*, bacterium causing tuberculosis in cattle
brevi- short *Lactobacillus brevis*, a bacterium with short rod-shaped cells
butyr- butter Butyric acid gives rancid butters its unpleasant odor
campylo- curved *Campylobaccter*, a curved bacterium
carcino- cancer A carcinogen causes cancer.
caryo-, karyo- center, kernel Prokaryotic cells lack a true, discrete nucleus
caseo- cheese caseous, cheeselike lesions
caul- stalk, stem *Caulobacter*, a stalked bacterium
ceph-, cephalo- of the head or brain encephalitis, inflammation of the brain
chlamydo- cloaked hidden *Chlamydia* are difficult bacteria to detect
chloro- green chlorophyll, a green pigment

chromo- colored Metachromatic granules stain various colors within a cell
chryso- golden *Streptomyces chryseus*, a bacterium forming golden colonies
-cide to kill Fungicide kills fungi
co-, con- with, together congenital, existing from birth
cocc- berry *Streptococcus*, spherical bacteria in chains
coeno- shared in common coenocytic, many nuclei not separated by septa
col-, colo- colon coliform bacteria, found in the colon large intestine)
conidio- dust conidiam tiny dustlike spores produced by fungi
coryne- club *Corynebacterium diphtheriae*, club-shaped bacterium
-cul little, tiny molecule, a tiny mass
cut-, -cut skin cutaneous, of the skin
cyan- blue cyanobacteria, formerly called the blue-green algae
cyst, -cyst bladder cystitis, inflammation of the urinary bladder
cyt, -cyte cell leukocyte, white blood cell
de- lack of removal decolorize, to remove color
dermato- skin dermatitis, inflammation of the skin
di-, diplo- two, double diplocci, pairs of spherical cells
dys- bad, faulty, painful dysentery, a disease of the enteric system
ec-, ecto-, ex outside, outer Ectoparasite, found on the outside of the body
em-, en- in, inside encapsulated, inside a capsule
-emia of the blood pyemia, pus in the blood
endo- inside endospore, spore found inside a cell
entero- intestine enteric, bacteria found in the intestine
epi- atop, over epidemic, a disease spreading over an entire population at one time
erythro- red lupus Erythematosus, disease with a red rash
etio- cause etiology, study of the causes of disease
eu- true, good, normal eukaryote cell with a true nucleus
exo- outside exotoxion, toxin released outside of a cell
extra- outside, beyond Extracellular, outside of a cell
fil- thread filament, thin chain of cells
flav- yellow flavivirus, cause of yellow fever
-fy to become, make solidify, to become solid
galacto- milk galactose, monosaccharide from milk sugar
gamet- marriage gamete, a reproductive cell, such as egg or sperm
gastro- stomach gastroenteritis, inflammation of the stomach and intestines
gel- to stiffen, congeal gelatinous, jellylike
gen-, -gen to give rise to pathogen, microbe that causes disease
-genesis origin, development pathogenesis, development of disease
germ, germin- bud germination, process of growing from a spore
-globulin protein immunoglobulins, proteins of the immune system
haem-, hem- blood hemmagglutinatin, clumping of blood cells
halo- salt halophilic, organisms that thrive in salty environments
hepat- liver hepatitis, inflammation of the liver
hetero- different, other heterotroph, organism deriving nutrition from other sources
histo- tissue histology, the study of tissues
homo- same homologous, having the same structure
hydro- water hydrologic cycle, water cycle
hyper- over, above hyperbaric oxygen, higher than atmospheric pressure oxygen
hypo- under, below hypodermic, going beneath the skin
im-, in- not insoluble, cannot be dissolved
inter- between intercellular, between cells
intra- inside intracellular, inside a cell
io- violet iodine, element that is purple in gaseous state
iso- same, equal isotonic, having the same osmotic pressure
-it is inflammation of meningitis, inflammation of the meninges
kin- moving kinetic energy, energy of movement
leuko- white leukocyte, white blood cell
lip-. lipo- fat, lipid lipoprotein, molecule having both fatty and proteinaceous parts
-logy, -ology study of microbiology, study of microbes
lopho- tuft lophotrihous, having a tuft or group of flagella
luc-, luci- light luciferase, enzyme that catalyzes a light-producing reaction
luteo- yellow *Micrococcus luteus*, bacterium producing yellow colonies
lys-, lysis slitting cytolysis, rupture of a cell
macro- large macroconidia, large spores
meningo- membrane meninges, membranes of the brain
meso- middle Mesophile, organism growing best a medium temperatures
micro- small, tiny microbiology, study of tiny forms of life
mono- one, single monosaccharide, a single sugar unit
morph- shape, form pleionorphic, having many different shapes

multi- many mutticellular, having many cells
mur- wall muramic acid, a component of cell walls
muri-, mus- mouse murine, in or of mice
mut-, mute to change mutagen, agent that causes genetic change
myc-, myces fungus *Actinomyces*, a bacterium that resembles a fungus
myxo- slime, mucus myxomycetes, slime molds
necro- dead, corpse necrotizing toxin, causes death of tissue
nema-, -nema thread *Treponema*, nematode, threadlike organisms
nigr- black *Rhizopus nigricans*, a black mold
oculo- eye binocular, microscope with two eyepieces
-oid like, resembling Toxoid, harmless molecule that resembles a toxin
-oma tumor carcinoma, tumor of opithelial cells
onco- mass, tumor oncogenes, genes that cause tumors
-osis condition of brucellosis, condition of being infected with Brucella
patho- abnormal pathology, study of abnormal diseased states
peri- around peritrichous flagella located all around an organism
phago- eating phagocytosis, cell eating by engulfing
philo-, -phil, -phile loving, preferring Capnophile, organism needing higher than normal levels of carbon dioxide
-phob, -phobe hating, fearing hydrophobic, water-repelling
-phore bearing, carrying electrophoresis, technique in which ions are carried by an electric current
-phyte plant Dermatophyte, fungus that attacks skin
pil- hair pilus, hairlike tube on bacterial surface
-plast formed part chloroplast, green body inside plant cell
pod-, -pod foot podocyte, foot cell of kidney
poly- many polyribosomes, many ribosomes on the same piece of messenger RNA
post- afterward, behind post-streptoccal glomerulonephritis, kidney damage following a streptococcal infection
pre-, pro- before, toward prepubertal, before puberty
pseudo- false pseudopod, projection resembling a foot, false foot
psychro- cold psychrophilic, preferring extreme cold
pyo- pus pyogenic, producing pus
pyro- fire, heat pyrogen, fever-producing compound
rhin- nose rhinitis, inflammation of nasal membranes
rhizo- root mycorrhiza, symbiotic growth of fungi and roots
rhodo- red *Rhodospirillum*, a large red spiral bacterium
-rrhea flow diarrhea, abnormal flow of liquid feces
rubric- red *Rhodospirillum rubrum*, a large red spiral bacterium
saccharo- sugar polysaccharide, many sugar units linked together
sapro- rotten, decaying saprophyte, organism living on dead matter
sarco- flesh sarcoma, tumor made up of muscle or connective tissue
schizo- to split schizogony, a type of fission in malarial parasites
-scope, -scopy to see, examine microscopy, use of the microscope to examine small things
sept-, septo- partition, wall septum, wall between cells
septi- rotting septic, exhibiting decomposition due to bacteria
soma-, -some body chromosome, colored body (when stained)
spiro- coil spirochete, spiral-shaped bacterium
sporo- spore sporocidal, spore killing
staphylo- in bunches, like grapes staphylococci, spherical bacteria growing in clusters
-stasism stat- stopping, not changing bacteriostatic, able to stop the growth of bacteria
strepto- twisted *Streptobacillus*, twisted chains of bacilli
sub- under, below Subclinical, signs and symptoms not clinically apparent
super- above, more than superficial mycosis, fungal infection of the surface tissues
sym-, syn- together symbiosis, living together
tact-, -taxis touch chemotaxis, orientation or movement in response to chemicals
tax-, taxon- arrangement taxonomy, the classification of organisms
thermo- heat thermophile, organism preferring or needing high temperatures
thio- sulfur *Thiobacillus*, organism that oxidizes hydrogen sulfide to sulfates
tox- poison toxin, a harmful compound
trans- through, across transduction, movement of genetic information from one cell to another
trich- hair Monotrichous, having a single, hairlike flagellum
-troph feeding, nutrition phototroph, organism that makes its on food, using energy from light
uni- one, singular unicellular, composed of one cell
undul- waving undulant fever disease in which fever rises and falls
vac-, vaccine- cow vaccine, disease-preventing product originally produced by inoculating it onto skin of calves
vacu- empty vacuole, empty-appearing structure in cytoplasm
vesic- blister, bladder vesicle, small blisterlike lesions

vitr- glass *in vitro*, grown in laboratory glassware
xantho- yellow *Xanthomonas oryzae*, bacterium producing yellow colonies
zeno- strange, foreign xenograft, graft from a different species
zoo- animal protozoan, first animal
zygo- yoke, joining zygote, fertilized egg
-zyme ferment enzymes, biological catalysts, some of which are involved in fermentation

Garrity, George M, Winters, Matthew and Searles, Denise B, Taxonoic Outline of the Procaryotic Genera, Bergey's Manual of Systematic Bacteriology, 2nd edn., April 2001, Sprinter-Verlag, New York.

Appendix D3

Fungi, Mildew and Yeasts

The classification below is based on that in the 8th edn. of Ainsworth and Bisby's Dictionary of the Fungi (Hawksworth, Kirk, Pegler, Sutton, Ainsworth 1995; republished in The Fungi, 2nd edn., Carlile MJ, Watkinson SC, Gooday GW Academic, New York, 2001). Species in the text are listed, but where more than one member of a genus has been mentioned in the text, only one, or sometimes a few, are given here. Where the names for both an anamorphic and a telemorphic state, or a synonym, are widely used, these are given (ana, tel., syn.). Yeasts are assigned to Ascomycetes, Basidiomycetes and mitosporic fungi, but with their yeast status indicated (Y).

The organisms studies by mycologists occur in three kingdoms, the Protozoa, Chromista and Fungi, in the domain Eukaryota.

The Kingdom Protoza

Four protozoan phyla have been studies by mycologists. Two of these, the Dictyosteliomycetes and Acrasiomycetes (not mentioned further) lack plasmodia and are informally known as cellular slime moulds and two, the Myxomycota and Plasmodiophoromycota, form plasmodia and hence are plasmodial slimie moulds.

I. Phylum Dictyosteliomycota

Cellular slime molds (Brit., moulds), with an amoeboid trophic (feeding) phase, and multicellular fruit bodies. Contain a single order, the Dictyosteliales.

Dictyostelium discoideum

II. Phylum Myxomycota

Plasmodial slime moulds, with an amoeboid trophic (feeding) phase, a plasmodial phase also able to ingest fod particles, and multicellular fruit bodies. Contains two classes, the Myxomycetes and Protosteliomycetes, only the former being dealt with in the present work.

Didymium iridis
Physarum polycephalum

III. Phylum Plasmodiophoromycota

Contains a single class, the Plasmodiophoromycetes, obligate parasites of plants having minute intracellular plasmodia.

Plasmodiophora brassicae

The Kingdom Chromista

Three Chromistan phyla, the Oomycota, the Hyphochytriomycoat and the Labyrinthulomycoat have been studied by mycologists, but only the Oomycota is considered in detail in the present work.

I. Phylum OOMCOTA, the Oomycetes

Most species have a mycelial trophic phase resembling that of fungi, and many produce zoospores with an anterior tinsel and a posterior smooth flagellum. Nine orders are recognizedm with members of four being mentioned in the present work.

(a) Order Sprolegniales. The water moulds
  Achlya bisexualis
  Aphanomyces eutiches
  Dictyuchus
  Nematophthora gynophila
  Pythiopsis
  Saprolegnia
  Thraustotheca
(b) Order Leptomitales
  Aqualinderella fermentans
  Leptomitus
(c) Order Pythiales
  Lagenidium
  Phytophthora infestans
  Pythium ultimum
(d) Order Peronosporales
  (i) Family Peronosporaceae. The downy mildews
    Bremia lactucae
    Peronospora tabacina
    Plasmopora viticola
    Pseudoperonospora humuli
  (ii) Family Albuginaceae. The white rusts
    Albugo candida

II. Phylum LABYRINTHULOMYCOTA

Two orders of freshwater and marine organisms, only one of which is mentioned in the present text.

(a) Order Thraustochytriales
Thraustochytrium
The Kingdom Fungi

Heterotrophic, absorptive organisms, typically mycelial, but sometimes unicellular, as in yeasts. There are four phyla, the Chytridiomycota, the Zygomycota, the Ascomycota and the Basidiomycota, and an information category, the mitosporic fungi, members of which lack a sexual phase and have not yet been assigned to taxa in the formal system of classification.

I. Phylum CHYTRIDIOMYCOTA, the Chytridiomycetes

Zoospores, and gametes if motile, have posterior smooth flagella – usually one, occasionally more. Five orders are recognized, and four of these are mentioned in the present work.

(a) Order Blastocladiales
Allomyces macrogynus
Blastocladia
Blastocladiella emersonii
Catenaria anguillulae
Coelomomyces
(b) Order Chytridiales. The chytrids
Batrachochytrum dendrobatides
Chytriomyces
Rhizophylyctis rosea
Rhizophydium
Synchytrium endobioticum
(c) Order Monoblepharidales
Monoblepharis
(d) Order Neocallimastigales. The anaerobic rumen fungi
Neocallimastix frontalis
Piromonas
Sphaeromonas
(e) Order Spizellomycetales
Karlingia asterocyta

II. Phylum ZYGOMYCOTA

Zygospores, large resting spores, result from the sexual process.

Class Trichomycetes. Obligate parasites of arthropods.

Class Zygomycetes. The Zygomycetes. Seven orders are recognized, and five are mentioned in the present work.

(a) Order Mucorales
Absidia glauca
Actinomucor elegans (syn. A. repens).
Amylomyces rouxii
Blakeslea trispora
Chaetocladium jonesii
Cunninghamella
Dicranophora fulva
Mortierella
Mucor mucedo
Mycotypha
Phycomyes blakesleeanus
Pilaira
Pilobolus kleinii
Rhizopus oryzae
Spinellus
Syzygites
Zygorhynchus
(b) Order Dimargaritales. Obligate parasites, mainly of Mucorales
Dimargaris
Dispira
Tieghemomyces
(c) Order Zoopagales. Parasites of fungi (mycoparasites) and of small animals such as amoebae and nematodes
Piptocephalis
(d) Order Entomophthorales. Parasites of insects and other arthropods
Basidiobolus ranarum
Conidiobolus coronatus
Entomophaga grylli
Entomo9phthora muscae
Erynia
Massospora
(e) Order Glomales. Mycorrhiza-forming fungi, the vesicular-arbuscular endophytes (VAMs)
Acaulospora
Entrophospora
Gigaspora
Glomus macrocarpum
Sclerocystis
Scutellospora

III. Phylum ASCOMYCOTA, the Ascomycetes

Asci, containing ascospores, result from the sexual process. Forty-six orders are recognized, with members of nineteen being mentioned in the present work. Lichenized fungi are indicated (L).

(a) Order Saccharomycetales. The asci of members of this order and of the Schizosaccharomycetales are solitary, not in ascocarps. Hence at tone time such fungi were included in the class Hemiascomycetes, "half Ascomycetes." In contrast, other orders have the asci in ascocarps and hence were regarded as Euascomycetes, "true Ascomycetes." Most of the Saccharomycetales are yeasts (Y), but a few do not have a yeast phase.

Ashbya gossypyii (Y)
Debaromyces hansenii
Dipodascus geotrichum (ana. Geotrichum candidum)
Endomycopsis fibuligera (syn. Saccharomycopsis fibuligera) (Y)
Hansenula anomala (Y)
Kluyveromyces lactis (Y)
Pichia saitoi (Y)
Saccharomyces carlsbergensis (lager strains of S. cerevisiae) (Y)
Saccharomyces cerevisiae (Y)
Saccharomyces ellipsoidum (wind strains of S. cerevisiae) (Y)
Saccharomyces rouxii (syn. Zygosaccharomyces rouxii) (Y)
Saccharomycodes ludwigii (Y)
Saccharomycopsis fibuligera (syn. Endomycopsis fibuligera) (Y)
Saccharomycopsis lipolytica (ana. Candida lipolytica) (Y)
Zygosaccharomyces rouxii (syn. Saccharomyces rouxii) (Y)

(b) Order Schizosaccharomycetales
Schizosaccharomyces pombe (Y)

(c) Order Diaporthales
Cryphonectria parasitica (syn. Endothia parasitica)
Diaporthe (ana. Phomopsis)
Endothia parasitica (syn. Cryphonectria parasitica)
Gaeumannomyces graminis (syn. Ophiobolus graminis)
Ophiobolus graminis (syn. Gaeumannomyces graminis)

(d) Order Dothidiales
Cochliobolus miyabeanus (ana. Drechslera oryzae, Helminthosporium oryzae)
Didymella (ana. Ascochyta)
Eudarluca caricis (ana. Sphaerellopsis filum)
Leptosphaeria
Leptosphaerulina trifolii
Mycosphaerella tassiana (ana. Cladosporium herbarum)
Pleospora herbarum (ana. Stemphylium botryosum)
Venturia inaequalis (ana. Spilocaea pomi)

(e) Order Erysiphales. The powdery mildews
Blumeria graminis (syn. Erysiphe graminis)
Erysiphe graminis (syn. Blumeria graminis)

(f) Order Eurotiales
Emericella nidulans (ana. Aspergillus nidulans)
Eupenicillium brefeldianum (ana. Penicillium dodgei)
Eurotium herbariorum (ana. Aspergilus glaucus)
Talaromyces (ana. Penicillium)

(g) Order Halospheriales
Chadefaudia corallinum
Remispora maritime

(h) Order Hypocreales
Atkinsonella hypoxylon
Balansia cyperi
Claviceps purpurea
Cordyceps
Epichoë typhina
Gibberella zeae (ana. Fusarium graminearum)
Hypocrella
Nectria haematococca (ana. Fusarium solani)
Neocosmospora
Neotyphodium lolii

(i) Order Lecanorales
Cladonia arbuscula (L)
Rhizocarpon concentricum (L)

(j) Order Leotiales (formerly Helotiales)
Amorphotheca resinae (ana. Hormoconis resinae, Cladosporium resinae)
Botryotinia fuckelina (ana. Botrytis cinerea)
Bulgaris inquinans
Helotium chloropodium
Hymenoscyphus ericae
Pezicula aurantiaca

(k) Order Onygenales (formerly Gymnoascales)
Ajellomyces capsulate (syn. Emmonsiella capsulate, ana. Histoplasma apsulata) (Y)
Ajellomyces dermatitidis (ana. Blastomyces dermatitidis) (Y)
Emmonsiella capsulate (sny. Ajellomyces capsulate, ana. Histoplasma capsulate)
Gymnoascus

(l) Order Ophiostomatales
Ceratocystis fagacearum
Ophiostoma novo-ulmi and Ophiostoma ulmi (syn. Ceratocystis ulmi)

(m) Order Peltigerales
Peltigera canina (L)
Pseudocyphellaria rufovirescens
Sticta (L)

(n) Order Pezizales
Aleuria vesiculosa
Ascobolus immerses
Gyromitra esculenta
Monascus bisporus (syn. Xeromyces bisporus)
Morchella hortensis
Pyronema confluens (syn. omphalodes)
Sclerotinia fructigena
Tuber melanosporum
Xeromyces bisporus (syn. Monascus bisporus)
(o) Order Phyllachorales
Glomerella cingulata (ana. Colletotrichum gloeosporoides)
Magnaporthe grisea (sny. Pyricularia oryzae)
(p) Order Rhytismatales
Rhytisma acerinum
(q) Order Sordariales
Chaetomium globosum
Neurospora crassa
Podospora anserine
Sordaria fimicola
(r) Order Teloschistales
Xanthoria parietina (L)
(s) Order Xylariales
Biscogniauxia nummularia
Daldinia concentrica
Pseudomassaria
Xylaria hypoxylon

IV. Phylum BASIDIOMYCOTA, the Basidiomycetes

A basidum bearing basidiospores results from the sexual process. Three classes are recognized.

1. Class Basidiomycetes. The Basidiomycetes that have fruit bodies (basidiocarps), plus a few yeasts. The class replaces two classes that formerly accommodated Basidiomycete macrofungi, the Hymenomycetes and the Gasteromycetes, the latter a very artificial assemblage. Thirty-two orders are recognized, with members of fourteen being mentioned in the present work.
   (a) Order Agaricales. Most mushrooms and toadstools.
   Agaricus bisporus
   Agrocybe gibberosa
   Amanita phalloides
   Armillaria mellae
   Bolbitius
   Clitocybe
   Collybia velutipes (sny. Flammulina velutipes)
   Coprinus cinereus
   Flammulina velutipes (syn. Collybia velutipes)
   Hypsizygus marmoreus
   Laccaria
   Lepiota
   Marasmius androsaceus
   Mycena galopus
   Oudemansiella mucida
   Panellus
   Pholiota nameko
   Pluteus
   Podaxis
   Psilocybe cubensis
   Stobilurus tenasellus
   Termitomyces titanicus
   Volvaria volvacia (syn. Volvariella diplasia)
   (b) Order Auriculariales
   Auricularia polytricha
   (c) Order Boletales
   Boletus subtomentosus
   Coniophora putena
   Omphalotus
   Rhizopogon roseolus
   Serpula lachrymans
   Suillus bovines
   (d) Order Cantharellales
   Cantharellus cibarius
   Clavaria
   (e) Order Ceretobasidiales
   Thanatephorus cucumeris (ana. Rhizoctonia solani)
   (f) Order Cortinariales
   Cortinarius speciosissimus
   Inocybe
   (g) Order Ganodermatales
   Ganoderma adspersum
   (h) Order Halospheriales
   Remispora maritime
   (i) Order Hericiales
   Hericium erinaceus
   (j) Order Lycoperdales, Puffballs
   Geastrum triplex
   Lycoperdon perlatum
   (k) Order Nidulariales, Birds' nest fungi
   Crucibulum leave
   Sphaerobolus
   (l) Order Phallales. The stinkhorns
   Phallus impudicus
   (m) Order Poriales
   Bjerkandera adusta

Coriolus versicolor
Fomes fomentarius
Gloeophyllum trabeum
Grifola frondosa
Heterobasidion annosum
Lentinus edodes
Phaeolus schweinitzii
Pleurotus ostreatus
Polyporus brumalis
Poria vaillantii
Trametes gibbosa

(n) Order Russulales
Russula

(o) Order Schizophyllales
Schizophyllum commune

(p) Order Sclerodermatales
Pisolithus tinctorius

(q) Order Stereales
Phanerochaete chrysosporium (ana. Sporotrichum pulverulentum)
Phlebia gigantean
Phlebiopsis
Stereum sanguinolentum

(r) Order Tremellales
Silobasidiella neoformans (ana. Cryptococcus neoformans) (Y)
Tremella mesenterica

2. Class Ustomycetes. Seven orders are recognized, with members of two being mentioned in the present work.
   (a) Order Ustilaginales
   Tilletia caries
   Ustilago maydis
   (b) Order Sporidiales
   Rhodosporidium toruloides (ana. Rhodotorula rubrum)

3. Class Teliomycetes. Two orders are recognized
   (a) Order Uredinales. The rusts
   Hemileia vastatrix
   Melampsora lini
   Puccinia graminis
   Uromyces phaseoli
   (b) Class Septobasidials. Parasites of scale insects

Informal group MITOSPORIC FUNGI

Previously known as Deuteromycetes, Fungi Imperfecti or Imperfect Fungi, are classified on the basis of asexual morphology, although with some a sexual, telemorphic phase is known. This is indicated (tel.) where the name is widely used.

1. Coelomycetes. Conidia are formed in a cavity composed of fungal or mixed host and fungal tissue.
   Ascochyta (tel. Didymella)
   Colletotrichum gloeosporoides (tel. Glomerella cingulata)
   Coniothyrium minitans
   Microsphaeropsis
   Phoma (tel. Pleospora)
   Phomopsis (tel. Diaporthe)
   Sphaeropsis filum (tel. Eudarluca caricis)
2. Hyphomycetes. Conidia not within cavities. Includes yeasts (Y), sometimes separated as Blastomycetes.
   Acremonium strictum
   Akanthomyces
   Alternaria tenuis
   Aphanocladium album
   Articulospora tetracladia
   Arthrobotrys oligospora
   Aspergillus glaucus (tel. Eurotium herbariorum)
   Aspergillus nidulans (tel. Emericella nidulans)
   Aspergillus niger
   Aureobasidium pullulans
   Beauveria bassiana
   Bipolaris (tel. Cochliobolus)
   Blastomyces dermatitidis (tel. Ajellomyces dermatitidis) (Y)
   Botrytis cinerea (tel. Botryotinia fuckeliana)
   Candidaalbicans (Y)
   Candida lipolytica (tel. Saccharomycopsis lipolytica) (Y)
   Candida utilis (Y)
   Cephalosporium
   Cercospora (tel. Mycosphaerella)
   Cercosporidium personathum
   Cladosporium herbarum (tel. Mycosphaerella tassiana)
   Cladosporium resinae (sny. Hormoconis resinae, tel., Amorphotheca resinae)
   Coccidioides immitis (Y)
   ryptococcus neoformans (tel. Filobasidiella neoformans) (Y)
   Curvularia lunata
   Cylindrocarpon tonkinense
   Deightoniella
   Dendrypheilla salina
   Drechslera oryzae (syn. Helmithosporium oryzea, tel. Cochliobolus miyabeanus)
   Epicoccum

Epidermophyton
Fusarium culmorum
Fusarium graminearum (tel. Gibberella zeae)
Fusarium solani (tel. Nectria haematococca)
Fusidium coccineum
Geotrichum candidum (tel. Dipodascus geotrichum)
Glioxastix
Graphium
Helicodendron glomeratum
Helminthosporium oryzea (ana. Drechslera oryzea, tel. Cochliobolus miyabeanus)
Hirsutella
Histoplasma capsulatum (tel. Ajellomyces capsulate, Emmonsiella capsulate) (Y)
Hormoconis resinae (syn. Cladosporium resinae, tel. Amorphotheca resinae)
Humicola lanuginose
Malassezia furfur (Y)
Memnoniella
Meria coniospora
Metarhizium anisopliae
Myrothecium
Nigrospora
Oidiodendron oryzae
Paecilomyces lilacinus
Paracoccidioides brasiliensis (Y)
Penicillium chrysogenum
Penicillium dodgeii (tel. Eupenicillium brefeldianum)
Periconia abyssa
Phialophora fastigiata
Pithomyces (tel. Leptosphaerulina)
Pityrosporon (Y)
Pyricularia oryzae (tel. Magnaporthe grisea)
Retiarius
Rhizoctonia solani (tel. Thanatephorus cucumeris)
Rhodotorula rubra (tel. Rhodosporidium toruloides) (Y)
Rhynchosporium secalis
Spilocaea pomi (tel. Venturia inaequalis)
Sporobolomyces (Y)
Sporodesmium schlerotivorum
Sporotrichum pulverulentum (tel. Phanerochaete chrysosporium)
Stachybotrys
Stemphylium botryosum (tel. Pleospora herbarum)
Thermomyces lanuginosus
Tilletiopsis minor (Y)
Tolypocaladium inflation
Torulopsis (Y)
Trichoderma viride
Trichophyton
Trichosphoron
Verticillium albo-atrum

3. Agonomycetes. Also known as Mycelia Sterilia. Lacking spores, they are not strictly speaking mitosporic, but are included with mitosporic fungi for convenience. Some have morphological features such as sclerotia or traps for eelworms that facilitate identification.
   Sclerotium cepivorum

Appendix E

A

Abbreviations and Symbols

Terms	Symbol
A member of	ϵ
Addition	$+$
Alternative hypothesis	H_1
Amalgamation	$\coprod$
And, conjunction	$\wedge$
Angle	$\angle$
Approaches a limit, definition	$\doteq$
Arithmetic mean operator	$< >$
Assertion	$\vdash$
Assignment	$:=$
Autocorrelation	$R_{xx}()$
Autocorrelation coefficient	$\rho_{xx}()$
Autocovariance	$C_{xx}()$
Base of natural logarithms ($\approx$ 2.71828183)	e
Because	$\because$
Binomial coefficient	C_k^n
Composition	$\circ$
Conditional operator	$\mid$
Contour integral	$\oint$
Convolution	$*$
Correlation coefficient	ρ_{xy}
Covariance	$C_{ov}()$
Cross-correlation	$R_{xy}()$
Cross-correlation coefficient	$\rho_{xy()}$
Cross-covariance	$C_{xy}()$
Cube root	$\sqrt[3]{}$
d'Alembertian operator	$\Box^2$
Degree	$\circ$
Delta	Δ
Direct sum, various	$\oplus$
Divides, divisible by	$\mid$
Division	$\div$ or /
Double factorial	!!

Abbreviations and Symbols (Continued)

Terms	Symbol
Double prime	$''$
Empty set, null set	$\emptyset$
Equal angles	$\vee$
Event	E
Equals approximately, isomorphic to	$\cong$
Equal to	$=$
Equivalent to, congruent to	$\equiv$
Existential quantifier	$\exists$
Expectation value operator	$\mathbf{E}\{\}$
Factorial	!
Gamma function	$\Gamma()$
Greater than	$>$
Greater than or equal to	$\geq$
Hence, therefore	$\therefore$
If and only if	$\Leftrightarrow$ or $\leftrightarrow$
Implies	$\Rightarrow$ or $\rightarrow$
Infinity	∞
Integral	$\int$
Intersection	$\cap$
Kurtosis	κ or γ_2
Laplacian operator	∇^2, Δ
Less than	$<$
Less than or equal to	$\leq$
Maps from	$\leftarrow$
Maps into	$\hookrightarrow$ or $\hookleftarrow$
Maps to	$\mapsto$
Mean	μ
Minute	$'$
Minus or plus	$\mp$
Moment operator	$M\{\}$
Much greater than	$\gg$
Much less than	$\ll$
Multiplication	X or $\bullet$
Nabla, del	∇
Nearly equal to	$\approx$
Negation	$\neg$
nth root	$\sqrt[n]{}$
Not a member of	$\ni$

Abbreviations and Symbols (Continued)

Terms	Symbol
Not equal to	$\neq$
Not equivalent to, not congruent to	$\not\equiv$
Not parallel	$\nparallel$
Null hypothesis	H_0
Or, disjunction	$\vee$
Parallel	$\parallel$
Partial differential	∂
Perpendicular	$\perp$
Pi = 3.14159265	π
Plus or minus	$\pm$
Prime	$'$
Probability or probability distribution	P
Product	Π
Ratio	:
Sample space	S or Ω
Second	$''$
Similar to, asymptotically to	$\sim$
Skewness	γ or γ_1
Spherical angle	$\sphericalangle$
Square root, radical	$\sqrt{}$
Square root of −1	i
Standard deviation	σ
Statistical weights	W
Subset of	$\subset$
Subset of or equal to	$\subseteq$
Subtraction	$-$
Summation	Σ
Superset of	$\supset$
Superset of or equal to	$\supseteq$
Tends to, maps to	$\rightarrow$
Triple prime	$'''$
Union	$\cup$
Universal quantifier	$\forall$
Variance	σ^2 or Var()
Varies as, proportional to	$\propto$
Various	$\ominus, \otimes, \odot$

Acceptance Region *n.* The range for the test statistic within which the null hypothesis is accepted. The limiting value(s) of the acceptance region are known as the critical values. See also ▶ Critical Value(s).

Addition Rule *n.* Also called the *additive law of probability,* one of the primary rules of probability. The addition rule states that the probability that event 1, E_1, occurs, event 2, E_2, occurs, or both events 1 and 2 occur is given by:

$$\mathbf{P}(E_\downarrow 1 \cup E_\downarrow 2) = \mathbf{P}(E_\downarrow 1) + \mathbf{P}(E_\downarrow 2) - \mathbf{P}(E_\downarrow 1 \cap E_\downarrow 2)$$

where: $\mathbf{P}(E_1)$ is the probability that E1 occurs,
$\mathbf{P}(E_2)$ is the probability that E2 occurs, and
$\mathbf{P}(E_1 \cap E_2)$ is the probability that both E_1 and E_2 occur.
$\mathbf{P}(E_\downarrow 1 \cap E_\downarrow 2)$ can be calculated using the multiplication rule. If the two events are independent then $\mathbf{P}(E_\downarrow 1 \cap E_\downarrow 2)$ is equal to $\mathbf{P}(E1)\mathbf{P}(E2)$ and the addition rule becomes:

$$\mathbf{P}(E_1 \cup E_2) = \mathbf{P}(E_1) + \mathbf{P}(E_2) - \mathbf{P}(E_1)\mathbf{P}(E_2)$$

If the two events 1 and 2 are mutually exclusive then the probability that both events 1 and 2 occur, $\mathbf{P}(E_1 \cap E_2)$, is zero and the addition rule reduces to:

$$\mathbf{P}(E_1 \cup E_2) = \mathbf{P}(E_1) + \mathbf{P}(E_2)$$

Additive Law of Probability *n* See ▶ Additions Rule.

Alpha Level *n* The alternate name for the *significance level.*

Alternative Hypothesis *n* The alternative or negation of the null hypothesis in hypothesis testing. It is normally symbolized by H_1.

Arithmetic Mean *n* Also referred to as *mean* is the simple average of a set of numbers, i.e., the sum of all numbers in the set divided by the number of numbers in the set. The arithmetic mean of the set X, <X>, is given by:

$$<\mathbf{X}> = \frac{\sum_i x_i}{N}$$

where: x_i are the elements of the set **X** and
N is the number of elements in **X**.
The arithmetic mean is one of several types of *means* which are measures of the central tendencies of a set, sample, population, or probability distribution.

Autocorrelation *n* A measure of the correlation between two sections or sub-sequences of a sequence of a single random variable. Each sub-sequence has the same number of elements, ordering, and spacings. The autocorrelation can be defined as the expectation value of the term-by-term product of a continuous section of a sequence with another continuous section of the same sequence with the same length. The autocorrelation is a function of three values, the starting element of the first section, the starting element of the second section, and the length of the sections. The autocorrelation, $\mathbf{R}_{xx}(i, j, \mathrm{T})$, of a sequence, **X**, is given by:

$$\mathbf{R}_{xx}(i, j, \mathrm{T}) = \mathbf{E}\{\mathrm{X}_{\mathrm{T}i}\mathrm{X}_{\mathrm{T}j}\}$$

where: $\mathbf{X}_{\mathrm{T}i}$ is the first sub-sequence of **X** with starting element, i, and length, T,
$\mathbf{X}_{\mathrm{T}j}$ is the second sub-sequence with starting element, j, and length, T, and

$\mathbf{E}\{\}$ is the expectation value operator.

For a continuous random variable, $\mathbf{X}$, over the sub-space, $\mathbf{S}_T$, of the sample space, $\mathbf{S}$, this becomes:

$$R_{xx}(s,t,\mathbf{S}_T) = \int_{S_T} \mathbf{x}(s)\mathbf{x}(t)f(s,t)d\mathbf{S}_T$$

for a continuous random variable with probability density function, $f(s,t)$. The autocorrelation is given by:

$$\mathbf{R}_{xx}(i,j,\mathrm{T}) = \sum_k x_k x_{k+j-i}\mathbf{P}_{ijk}$$

for a discrete random variable, with discrete values, x_k and x_{k+j-i}, with associated probabilities, $\mathbf{P}_{ijk}$, *and by:*

$$\mathbf{R}_{xx}(i,j,\mathrm{T}) = \frac{\sum_k xkxk+j-i\,\mathbf{W}_{ijk}}{\sum_k \mathbf{W}_{ijk}}$$

for statistical samples, x_k and x_{k+j-i}, with weights, $\mathbf{W}_{ijk}$. The autocorrelation of a random variable is often used interchangeably with and confused with the *autocorrelation coefficient* which is the *autocovariance* normalized by the product of the *standard deviations* of the two sections in the *autocovariance*, which is often referred to as simply the autocorrelation. The *autocorrelation coefficient* is generally used more often than the true autocorrelation for comparing two sections of a sequence.

Autocorrelation Coefficient *n* The autocovariance normalized by the product of the standard deviations of the two sections from the single random variable sequence used to calculate the autocovariance. In other words the autocorrelation coefficient is the cross-correlation coefficient of two sub-sequences of the same random variable. It is probably the most commonly used measure of the correlation between two sections of a single random variable sequence. It is often simply but incorrectly referred to as the *autocorrelation*, which is the un-normalized expectation value of the product of the two sequence sections. The autocorrelation coefficient of the two sub-sequences of random variable, $\mathbf{X}$, is often denoted by $\rho_{XX}(i,j,\mathrm{T})$; where i is the starting index of the second section, and T is the length of the sections. The precise mathematical definition of autocorrelation coefficient of two random variable sequence sections is given by:

$$\rho_{XX}(i,j,\mathbf{T}) = \frac{\mathbf{Cov}(\mathbf{X}_{\mathrm{T}i},\mathbf{X}_{\mathrm{T}j})}{\boldsymbol{\sigma}_{XTi}\boldsymbol{\sigma}_{XTj}} = \frac{\mathbf{E}\left\{(\mathbf{X}_{\mathrm{T}i}-\mu_{\mathrm{XT}i})(\mathbf{X}_{\mathrm{T}j}-\mu_{\mathrm{XT}j})\right\}}{\boldsymbol{\sigma}_{XTi}\boldsymbol{\sigma}_{XTj}}$$

where: $\mathbf{X}_{\mathrm{T}i}$ is the first sub-sequence of $\mathbf{X}$,

$\mu_{\mathrm{XT}i}$ is the mean over $\mathbf{X}_{\mathrm{T}i}$,

$\sigma_{\mathrm{XT}i}$ is the standard deviation over $\mathbf{X}_{\mathrm{T}i}$,

$\mathbf{X}_{\mathrm{T}j}$ is the second sub-sequence of $\mathbf{X}$,

$\mu_{\mathrm{XT}j}$ is the mean over $\mathbf{X}_{\mathrm{T}j}$,

$\sigma_{\mathrm{XT}i}$ is the standard deviation over $\mathbf{X}_{\mathrm{T}j}$, and

$\mathbf{E}\{\}$ is the expectation value operator.

For continuous random variables over the sub-space, $\mathbf{S}_T$, of the sample space, $\mathbf{S}$, this becomes:

$$\rho_{XX}(s,t,\mathbf{S}_T) = \frac{\int_{S_T}[\mathbf{X}(s)-\mu_{\mathrm{XT}s}][\mathbf{X}(t)-\mu_{\mathrm{XT}t}]f(s,t)d\mathbf{S}_T}{\boldsymbol{\sigma}_{\mathrm{XT}s}\boldsymbol{\sigma}_{\mathrm{XT}t}}$$

for continuous random variables with probability density function, $f(s,t)$, sub-space means, $\mu_{\mathrm{XT}s}$ and $\mu_{\mathrm{XT}t}$, and sub-space standard deviations, $\sigma_{\mathrm{XT}s}$ and $\sigma_{\mathrm{XT}t}$ by:

$$\rho_{\mathbf{xx}}(i,j,\mathbf{T}) = \frac{\sum_k (x_k-\mu_{\mathrm{XT}i})(x_{k+j-i}-\mu_{\mathrm{XT}j})\mathbf{P}_{ijk}}{\boldsymbol{\sigma}_{\mathrm{XT}i}\boldsymbol{\sigma}_{\mathrm{XT}j}}$$

for discrete random variables, with discrete values, x_k and x_{k+j-i}, with associated probabilities, $\mathbf{P}_{ijk}$, and by:

$$\rho_{\mathbf{xx}}(i,j,\mathbf{T}) = \frac{\sum_k (x_k-\mu_{\mathrm{XT}i})(x_{k+j-i}-\mu_{\mathrm{XT}j})\mathbf{W}_{ijk}}{\boldsymbol{\sigma}_{\mathrm{XT}i}\boldsymbol{\sigma}_{\mathrm{XT}j}\sum_k \mathbf{W}_{ijk}}$$

for statistical samples, x_k and x_{k+j-i}, with weights, $\mathbf{W}^{ijk}$. See also ► Cross-correlation Coefficient and ► Autocorrelation.

Autocovariance *n* The cross-covariance of two regions or sequences of a single random variable. It is a quantification of how much the two sub-sequences change together. The autocovariance is a function of three variables, the starting element of the first section, the starting element of the second section, and the length of the sections. The autocovariance of two sequences of a random variable, $\mathbf{X}$, is often denoted by $\mathrm{C}_{xx}(i,j,\mathrm{T})$; where i is the starting index of the first section or region of the sequence, $\mathbf{X}$, j is the starting index of the second, and T is the length of the sections. The precise mathematical definition of autocovariance of two regions of a random variable is the covariance of the two sections which is given by:

$$\mathrm{C}_{xx}(i,j,\mathrm{T}) = \mathrm{Cov}(\mathrm{X}_{\mathrm{T}i}\mathrm{X}_{\mathrm{T}j}) = \mathbf{E}\left\{(\mathrm{X}_{\mathrm{T}i}-\mu_{\mathrm{XT}i})(\mathrm{X}_{\mathrm{T}j}-\mu_{\mathrm{XT}j})\right\}$$

where: $\mathbf{X}_{\mathrm{T}i}$ is the sub-sequence of $\mathbf{X}$ with a mean over the section of $\mu_{\mathrm{XT}i}$,

$\mathbf{X}_{\mathrm{T}j}$ is the sub-sequence of $\mathbf{X}$ with a mean over the section of μ_{XTj}, and

$\mathbf{E}\{\}$ is the expectation value operator.

For a continuous random variable, **X**, over the sub-space, $\mathbf{S}_T$, of the sample space, **S**, this becomes:

$$C_{XX}(s,t,\mathbf{S}_T) = \int \mathbf{S}_T[X(s) - \mu_{XTs}][X(t) - \mu_{XTt}]f(s,t)d\mathbf{S}_T$$

for a continuous random variable with probability density function, *f(s, t)* and sub-space means, μ_{XTs} and μ_{XTt}, by:

$$C_{XX}(i,j,\mathbf{T}) = \sum_k (x_k - \mu_{XTi})(x_{k+j-i} - \mu_{XTj})\mathbf{P}_{ijk}$$

for a discrete random variable, with discrete values, x_k and x_{k+j-i}, with associated probabilities, $\mathbf{P}_{ijk}$, and by:

$$C_{XX}(i,j,\mathbf{T}) = \frac{\sum_k (x_k - \mu_{XTi})(x_{k+j-i} - \mu_{XTj})\mathbf{W}_{ijk}}{\sum_k \mathbf{W}_{ijk}}$$

for statistical samples, x_k and x_{k+j-i}, with weights, $\mathbf{W}_{ijk}$. The autocovariance of a random variable is often used interchangeably with and confused with the *autocorrelation* which is actually the *cross-correlation* of a random variable with itself. The cross-covariance is related to and often confused with the *cross-correlation.*

The autocovariance normalized by the product of the standard deviations of the two sections is the *autocorrelation coefficient,* which is often referred to as simply the autocorrelation.

Axioms of Probability *n* The three axioms from which the probability model is derived. The axioms of probability are as follows:

1. $\mathbf{P}(E_i)$ is greater than or equal to zero, for all events E_i in **S**,
2. $\mathbf{P}(\mathbf{S}) = 1$, and
3. **P**(union of any two events, E_j and E_k) equals $\mathbf{P}(E_j) + \mathbf{P}(E_k)$, whenever the intersection of the two events is the empty set, i.e., $E_j \cap E_k = \theta$

Where, and event, E_i, is a subset of the sample space, **S**, of a random experiment and $\mathbf{P}(E_i)$ is the probability of event E_i.

B

Bar Chart *n* A type of graph where quantities are represented by rectangular bars whose lengths are proportional to the quantities they represent. A bar chart also known as a *bar graph* can be oriented either vertically or horizontally. The bars are often segmented to show the quantities that make up the total group represented by the single bar or individual bars can be grouped to show the composition of a whole.

Bar Graph *n* See ▶ Bar Chart.

Bayes' Rule *n* See ▶ Bayes' Theorem.

Bayes' Theorem *n* Also known as *Bayes' Rule.* For two events, E_1 and E_2, in the sample space, **S**, of a random experiment, Bayes' Theorem relates the conditional probability of E_1 given E_2, $\mathbf{P}(E_1|E_2)$, to the conditional probability of E_2 given E_1, $\mathbf{P}(E_2|E_1)$. Bayes' Theorem states that:

$$\mathbf{P}(E_1|E_2) = \frac{\mathbf{P}(E_2|E_1)\mathbf{P}(E_1)}{\mathbf{P}(E_2)}$$

where: $\mathbf{P}(E_1)$ is the probability of event E_1 and $\mathbf{P}(E_2)$ is the probability of event E_2.

If the sample space, **S**, is composed of a set of mutually exclusive events, e_i, that completely cover **S**, then Bayes' theorem states that the conditional probability of event, e_k given an event, *E*, is given by:

$$\mathbf{P}(e_k|E) = \frac{\mathbf{P}(\mathbf{E} + \mathbf{e_k})\mathbf{P}(\mathbf{e_k})}{\sum_i \mathbf{P}(\mathbf{E}|\mathbf{e_i})\mathbf{P}(\mathbf{e_i})}$$

where: $\mathbf{P}(e_k|E)$ is he conditional probability of e_k given *E*, $\mathbf{P}(E|e_k)$ is the conditional probability of *E* given e_k, $\mathbf{P}(e_k)$ is the probability of e_k, $\mathbf{P}(E|e_i)$ is he conditional probability of *E* given e_i, and $\mathbf{P}(e_i)$ is the probability of e_i.

Before-after Design *n* A type of study or experiment design that employs *correlated samples,* where the same element is measured twice, once before and once after the modification of a variable (before and after a test or change in condition). See also ▶ Correlated Samples.

Bernoulli Distribution *n* See ▶ Binomial Distribution.

Bernoulli Trial *n* A set of independent trials where each trial will result in one of two results whose probabilities do not change over the set of trials are referred to as *Bernoulli Trials.* Since there are only two possible outcomes for each trial, the probability for the second outcome is equal to one minus the probability of the first outcome. The *binomial distribution* describes the probability of getting exactly *k* results of the first outcome in *n* trials.

Bias *n* The difference between the expectation value of the estimate of the value of a parameter of a population and its actual value. An estimator is referred to as *biased* if its estimate of a parameter's value is not equal to its true value.

Binary Data *n* A type of *categorical data* in which the property of the *population* has only two states, most often it is the presence or lack of a property.

Binomial Coefficient *n* The numerical coefficients of the terms of a binomial multiplied by itself an integer number of times. For example, the coefficient of the $a^{n-k}b^k$ term of $(a+b)^n$ is the binomial coefficient C_k^n. It can be shown that C_k^n is given by:

$$C_k{}^n = \frac{n!}{k!(n-k)!}$$

Binomial Distribution *n* Also known as the *Bernoulli Distribution.* It is a discrete probability distribution that describes the probability of getting exactly *k* outcomes of one type in a set of *n Bernoulli Trials.* If the *p* is the probability of the desired outcome in a single trial then the probability of getting exactly *k* desired outcomes in *n* trials is given by the binomial distribution as:

$$\mathbf{P}(k, n, p) = C_k{}^n p^k (1-p)^{(n-k)}$$

where: C_k^n is the binomial coefficient.
The mean, μ, is given by: $\mu = np$
The variance, σ^2, is given by: $\sigma^2 = np(1-p)$
and the standard deviation, σ, by: $\sigma = \sqrt{np(1-p)}$
The skewness, γ, is given by: $\gamma = \dfrac{1-2p}{\sqrt{np(1-p)}}$
The kurtosis, *k*, is given by: $k = \dfrac{1+3p(n-2)(1-p)}{np(1-p)}$

Blinding *n* A type of study or experiment design which attempts to eliminate *bias* induced by human observers or subjects involved with the experiment. Do to human psychology and the qualitative nature of many experiments, especially those involving medical trials, it is often the case that the results are biased by the observer's or subject's feelings or opinions. The design of *single-blind* or *double-blind* experiments are often used especially in clinical trials as an attempt to minimize these biases. In a *double-blind* study neither the observers nor the subjects are aware of which group is associated with which situation. For example in a clinical trial for a new treatment, one group would receive the treatment and a second group would receive a *placebo,* but neither the researchers nor the subjects are informed which individuals are given the treatment. The double-blind study is considered the least affected by *bias.* There are many situations in studies where the observers or researchers must know which individuals are in each group, due to the logistics of the study. In this case only the subjects are not aware into which group they fall. This type of study is referred to as a *single-blind* study. There can also be single-blind studies in which the subjects but not the observers are aware of the treatment. Single-blind studies are, in general, more susceptible to bias than double-blind studies.

Block *n* A *block* is a subset of similar experimental units within the whole experimental space. Experimental space is divided into blocks whose elements are homogeneous in an attempt to reduce the experimental error due to uncontrollable factors within the whole experimental space. See ▶ Blocking.

Blocking *n* The grouping of experimental units into sets where the units are as similar as possible. The groups are referred to as *blocks.* The experimental error will generally be smaller if the proportions of the experiment are randomly distributed within each block. This technique is used to experiment by lowering the effects due to uncontrollable factors within the whole experimental space.

Box-and-whisker plot *n* Also referred to as a *box plot* is a type of plot that is used to represent the distribution of a set of data, by indicating five values associated with the data. The values are the lower limit, the *first quartile,* the *median,* the *third quartile,* and the upper limit.
An element of the box-and-whisker plot consists of a central rectangle, whose first edge is located at the first quartile and whose last edge is located at the third quartile. The rectangle is divided by a line along the median. A line extends from the center of the first edge (first quartile) to the lower limit of the data and a second line extends from the center of the last edge (third quartile) to the upper limit of the data. A box-and-whisker plot contains one or more elements.
The box-and whisker plots are often used to display a comparison between different segments of a population such as male and female.

Boxplot *n* An alternate name for the *box-and whisker plot.*

C

Categorical Data *n* Data or observations that can be separated into non-overlapping classes or sets. These sets are called *categories* and are usually related to some *parameter* of the *population.*

Centile *n* An alternate term for *percentile.*

Central Limit Theorem *n* This theorem states that for any set of independent, identically distributed *random variables,* $X_1, X_2, \ldots, X_n$, with a finite *mean,* μ, and finite *variance,* σ^2, the distribution of the set approaches the *normal distribution* as *n* goes to infinity; furthermore, the value of the set mean, μ_n, will approach the random variable mean, μ, and the set variance, σ_n^2 will approach σ^2/n where, σ^2 is the random variable variance. This will be true no matter what the configuration of the random variable distribution happens to be.

Chi Distribution *n* A probability distribution that describes the variation from the mean value of the normalized distance of a set of independent random variables that have a normal distribution. For a set of random variables, such as set of *x*, *y*, and *z* coordinates of a distribution of random particles in space, common metrics such as average position and higher moments can be used to describe the distribution of these variables. For three dimensional space, the most often used metric is the average position, (x_0, y_0, z_0), which is found by summing the individual types of coordinates and dividing by the number of positions. For example x_0 would equal the sum of all of the *x*-coordinates in the set divided by the number of particles. This is equivalent to finding the mean of the x_is. If one were interested in how tightly the particles were clustered, the rms (root-mean-square) distance would be an appropriate measure. The rms distance from the mean position is calculated by taking the square root of the sum of the square distances from the mean position and dividing by the number of particles. In many cases the variables are scaled or scattered differently due to the particular nature of the sets being described. Since the standard deviation of each variable gives a measure of its spread, an appropriate scaling factor for each variable would be the standard deviation of that variable. This scenario can be generalized to any set of *N* independent variables, $\mathbf{X}_i$, in which case the normalized distance, *d*, of any *N*-dimensional point from the mean point is given by:

$$d = \left[\sum_{i=1}^{N}\left(\frac{\mathbf{X}_i - \mu_i}{\sigma_i}\right)^2\right]^{1/2}$$

where: μ_i and σ_i are the mean and standard deviation of the ith independent random variable.

This formulation for the *N*- dimensional distance is directly related to the chi distribution. More formally if $\mathbf{X}_1, \mathbf{X}_2, \ldots \mathbf{X}_N$ are a set of *N* independent random variables, where each $\mathbf{X}_i$ has a normal distribution, then a random variable, **Y**, can be defined as:

$$\mathbf{Y} = \left[\sum_{i=1}^{N}\left(\frac{\mathbf{X}_i - \mu_i}{\sigma_i}\right)^2\right]^{1/2}$$

where: μ_i and σ_i are the mean and standard deviation of the i^{th}independent random variable.

The random variable, **Y**, as defined above has the *chi distribution* which is described by the following probability density function, *f*(*x*, *N*):

$$f(x,N) = \frac{x^{N-1}e^{\frac{-x^2}{2}}}{2^{N/2-1}\Gamma\left(\frac{N}{2}\right)}$$

where: Γ () is the *Gamma function* and *x* is greater than or equal to 0 and $f(x,N) = 0$ when *x* is less than 0.

N is referred to as the degrees of freedom of the chi distribution. The chi distribution for an *N* of one is the half-normal distribution centered on zero with a standard deviation of $(\pi/2)^{1/2}$. For larger *N*, the distribution has a mean, μ, given by:

$$\mu = \frac{2^{1/2}\Gamma\left[\frac{(N+1)}{2}\right]}{\Gamma\left(\frac{N}{2}\right)}$$

and a standard deviation, $\sigma = (N - \mu^2)^{1/2}$

The term *chi distribution* is often confused with and used to describe the *chi-square distribution* which is the probability distribution that describes $\mathbf{Y}^2$, which is given by:

$$\mathbf{Y}^2 = \sum_{i=1}^{N}\left(\frac{\mathbf{X}_i - \mu_i}{\sigma_i}\right)^2$$

The chi-square distribution has the probability density function, *f*(*x*,*N*):

$$f(n,N) = \frac{x^{N/2}e^{-x/2}}{2^{N/2}\Gamma(N/2)}$$

where: *x* is greater than or equal to 0 and $f(x,N) = 0$ when *x* is less than 0.

Chi-square Distribution *n* A probability distribution that describes the variation from the mean value of the normalized sum of the squares of a set of independent random variables that have a normal distribution. More formally, if $\mathbf{X}_1, \mathbf{X}_2, \ldots \mathbf{X}_N$ are a set of *N* independent random variables, where each $\mathbf{X}_i$ has a normal distribution, then a random variable, $\mathbf{Y}^2$, can be defined as:

$$\mathbf{Y}^2 = \sum_{i=1}^{N}\left(\frac{\mathbf{X}_i - \mu_i}{\sigma_i}\right)^2$$

where: μ_i and σ_i are the mean and standard deviation of the ith independent random variable.

$\mathbf{Y}^2$ as defined above has the *chi-square distribution* or χ^2 *distribution* which is described by the following probability density function, $f(x,N)$:

$$f(x,N) = \frac{x^{N/2-1}e^{-x/2}}{2^{N/2}\Gamma(N/2)}$$

where: Γ () is the *Gamma function* and

x is greater than or equal to 0 and $f(x,N,) = 0$ when *x* is less than 0.

N is referred to as the degrees of freedom of the distribution. The chi-square distribution has a mean that is equal to the degrees of freedom of the distribution, N, and a standard deviation, σ, given by:

$$\sigma = (2N)^{1/2}$$

As the distribution's degrees of freedom, N, increases the chi-square distribution approaches the normal distribution with the same mean and standard deviation. The term *chi-square distribution* is often confused with and used to describe the *chi distribution* which is the probability distribution that describes **Y**, which is given by:

$$\mathbf{Y} = \left[\sum_{i=1}^{N}\left(\frac{X_i - \mu_i}{\sigma_i}\right)^2\right]^{1/2}$$

The chi-square distribution has the probability density function, $f(x,N)$:

$$f(x, N) = \frac{x^{N-1}e^{-x^2/2}}{2^{N/2-1}\Gamma(N/2)}$$

where: x is greater than or equal to 0 and if $f(x, N) = 0$ when x is less than 0.

Chi-square Goodness of Fit Test *n* The *chi-square goodness of fit test* is a type of *chi-square test* for quantifying how well a model predicts the observed data for a sample. The test uses the chi-square distribution to calculate the probability that the difference between the model and the observed data was due to chance alone. If the probability of having a difference that large is small, below a predetermined *significance level*, the model is rejected. The model is often a theoretical probability distribution. See ▶ Chi-Square Tests.

Chi-square Test of Association *n* An alternate name for *chi-square test of independence.*

Chi-square Test of Homogeneity *n* A type of *chi-square test* for quantifying how consistent a proportion, result, or outcome is across more than one sample of a population or across more than one population. One would not expect the proportions, etc., to be identical across different samples even from the same population; however, one would expect the proportions to be closer than that predicted by chance alone if the proportions were not due to random chance alone. The chi-square test of homogeneity uses the *null hypothesis* that the samples are homogenous with respect to the proportion, etc., under consideration. The *chi-square test* to test whether the differences between the two samples were less than that expected by chance alone.

Chi-square Test of Independence *n* Sometimes referred to as the *chi-square test of association,* it is a type of *chi-square test* that quantifies the relationship between two variables, attributes, or outcomes of an experiment. In its most widely used form it compares the outcome of an experiment with and without the assertion of a change or added input. This differentiates it from the *chi-square goodness of fit test,* which is concerned with the comparison of one variable with a model.

In the primary use of the *chi-square test of independence,* there are usually two variables, the type of procedure and the outcome. An example would be a study to determine whether a treatment changes the survival rate of rats with a specific affliction. The experiment is designed with two groups, the treated group and the control group. Each group would then have two outcomes, survival and did not survive. At the end of the experiment the data are organized into a *contingency table.* In this scenario, the contingency table is a row column matrix type table whose rows show the segregation of the population into types, e.g., treated and not treated. The contingency table columns show the range groups of the outcome, e.g., survived, and did not survive. There is usually an extra row showing the outcome range group totals and an extra column showing the population type totals. These are referred to as the *marginal totals.* A possible contingency table for the example study described above would be as follows:

Study Groups	Survived	Did Not Survive	Group Totals
Treated Group	25	35	60
Control Group	8	37	45
Outcome Totals	33	72	105

The table above shows that there were only 105 total individuals of which 60 were treated and 45 were not treated. Of the total group 33 individuals survived, 25 in the treated group and 8 in the untreated group. In the example, the null hypothesis to be verified or rejected is that the treatment had no effect on the number of surviving individuals. The verification is made using the chi-square test procedure described under *chi-square tests,* where the control group is treated as the model and the treated group is treated as the actual outcome, using the equation:

$$\mathbf{Y}^2 = \sum_{i=1}^{k} \frac{(\mathbf{x}_i - \mathbf{n}_i)2}{\mathbf{n}_i}$$

Chi-square Tests *n* The *chi-square* (chi, the Greek letter) statistic is a nonparametric statistical technique used to determine if a distribution of observed frequencies differs from the theoretical expected frequencies. Chi-square statistics use nominal (categorical) or ordinal level data, thus instead of using means and variances, this test uses frequencies.

The chi-square distribution is extremely useful in determining whether a hypothesis is valid for large populations. For example if a measurement is made on a group of randomly selected individuals from a large population and the measurements are divided into k ranges, the sum of the square of the number of measurements falling into each range minus the expected number in that range normalized by the expected number in each range will exhibit a discrete probability distribution that can be very closely approximated in most instances by the continuous chi-square probability distribution, which as stated above has well known and well understood properties and readily available values. The value of this sum, $\mathbf{Y}^2$, for a single measurement is given by:

$$\mathbf{Y}^2 = \sum_{i=1}^{k} \frac{(x_i - n_i)2}{n_i}$$

where: x_i is the measured number in range i and

n_i is the expected number in range i.

Note that the x_is and n_is are actual measured quantities or numbers and not percentages.

The chi-square distribution for the above example will have $N = k - 1$ degrees of freedom and not k. This is due to the fact that the total number of measurements is a constant and thus only $k - 1$ of the measured numbers, x_i, are independent; since the last measured x_i is the difference of the total number of measurements minus the sum of the rest of the x_is. The probability for having $\mathbf{Y}^2$s greater than or equal to the $\mathbf{Y}^2$ calculated from the actual measured data can be calculated from the actual measured data can be calculated from the chi-square distribution with $N = k - 1$. This value is often referred to as the *P-value* or probability value. This chi-square probability calculation is usually performed using interpolation from precalculated tables or by using calculators or computer analysis programs designed for this purpose. If the probability calculated is less than that expected from chance alone then the measurement is significant in that the hypothesis that it occurred by chance is improbable.

If an event occurs and the calculated probability of the event's occurring is so small that it is unlikely that the event occurred due to chance alone, the hypothesis, from which the probability that indicated the event would not occur was calculated, should be considered invalid. This qualitative decision based on "unlikely" is insufficient for comparison; therefore, the decision is based on a specific value that should always be specified along with the hypothesis. The limiting value chosen for this decision is termed the *significance level* and is usually represented by α. Significance levels of 0.01 (1%) and 0.05(5%) are common. Some researchers use values as low as 0.001. One would generally believe that events whose probabilities of occurrence are less than 1% or 5% did not occur through chance alone. The significance level commonly used in medical and biological research is $\alpha = 0.05$, which means a calculated probability, P-value <0.05. At this level there is only a one in twenty chance that the outcome was due to a random fluctuation in the variables. On the other end of the probability distribution range, if the probability is too high, greater than 95% or 99% that the hypothesis is true, one might expect that things are too good to be true, since it is statistically improbable that random measured data would agree so well even with a distribution that correctly describes it. This often implies that the measurements were not statistically independent or that there was some error in the measurement. The steps in using this test are summarized as follow:

1. Determine the hypothesis to be evaluated and the method to calculate the expected results.
2. Gather the data by conducting the proposed experiment.
3. Determine the expected numbers, n_i, for each observational class.
4. Calculate $\mathbf{Y}^2$ using the formula above.
5. Determine degrees of freedom, N, from the number of ranges, k, and the number of parameters estimated from the data, m.
6. Determine a significance level, α, to serve as the basis for accepting or rejecting the hypothesis, 0.05, 0.01, etc.
7. Use the chi-square distribution table, program, etc., to determine $\mathbf{Y}_\alpha{}^2$ value associated with the chosen significance value and degrees of freedom.
8. Accept (if $\mathbf{Y2} < \mathbf{Y}_\alpha{}^2$) or reject (if $\mathbf{Y2} \geq \mathbf{Y}_\alpha{}^2$) your hypothesis.

9. Repeat steps 6 through 8 choosing a significance level for the lower end, α, 0.95, 0.99, etc.

The above analysis is referred to as the *chi-square goodness of fit test.* It is also known as *Pearson's chi-square test.* The freedom in choosing the method of generating the expected value, n_i, for the measurement ranges allows for a wide range of uses of this test. For example, this test can be used to determine whether or not a measured variable can be described by a particular probability distribution by using that distribution to calculate the expected values, n_i. This can be accomplished even if the particulars of the distribution are not known. The unknown probability parameters such as mean and standard deviation can be calculated using the measured data. There is, however, an adjustment that needs to be made for this case. The degrees of freedom, *N*, of the chi-square distribution needs to be reduced by the number of independent parameters that are estimated from the measurements; that is $N = k-1- m$, where: *m* is the number of independent parameters estimated from the measurements. It should be noted that the standard deviation and the variance are not independent since the variance is equal to the square of the standard deviation and would only decrease the degrees of freedom, *N*, by one. In addition to the freedom of choosing the expected values, the ease of calculation of, by one. In addition to the freedom of choosing the expected values, the ease of calculation of $\mathbf{Y}^2$ and the broad availability of tables and programs for calculating the chi-square probabilities have contributed to the widespread use of chi-square tests.

Cluster *n* A subset of a population that is, theoretically, statistically similar or representative of the whole population. In *cluster sampling* the whole population or the whole population that is known is divided into clusters. See ► Cluster Sampling.

Cluster Sampling *n* A sampling technique that is applied to large populations, by dividing the total population into groups or subsets referred to as *clusters*, then randomly selecting a subset of the clusters, and sampling those clusters. Each cluster is theoretically a representative sample of the population. The selected clusters are usually but not always completely sampled.

Coefficient of Determination *n* Also known as *regression correlation coefficient* or *multiple regression correlation coefficient*, it is the ratio of the variation explained by the linear *regression analysis* (either *simple* or *multiple*) to the total variation of the data set. The *explained variation*, **REGSS**, is the sum of the squares of the regression estimated data points, y_{Ri}, minus the *sample mean*, μ_Y, which is given by:

$$\mathbf{REGSS} = \sum_i (y_{Ri} - \mu_Y)2$$

The total variation of the data set, **TSS**, is the sum of the squares of the observed data points, y_i, minus the sample mean, which is given by:

$$\mathbf{TSS} = \sum_i (y_i - \mu_Y)2$$

The *coefficient of determination,* which is often denoted by R^2, is equal to **REGSS/TSS** and is given by:

$$R^2 = \frac{\sum_i (y_{Ri} - \mu_Y)2}{\sum_i (y_i - \mu_Y)2}$$

A value for the coefficient of determination close to one indicates a good fit of the linear regression model and a value close to zero indicates a poor fit.

Coefficient of Variation *n* The ratio of the *standard deviation*, of a sample or population. It is sometimes defined as the percentage of the ratio as opposed to the fraction. The *coefficient of variation* is a *measure of relative dispersion.*

Completely Randomized Block Design *n* A type of *randomized block design* in which each *treatment* occurs at least once within a *block.*

Completely Randomized Design *n* A type of *experimental design* in which *units* are assigned to groups randomly. The most common case in the comparison of *treatments*, either multiple treatments or a single treatment and a control, where the units are randomly chosen for each type of treatment.

Composite Hypothesis *n* A hypothesis that partially but not completely describes the distribution of the *outcomes* of a *random experiment.* For example, if the hypothesis states only the mean and not the variance of the normal distribution that describes the distribution. If the hypothesis completely describes the distribution, it is referred to as a *simple hypothesis.*

Conditional Probability *n* For two events, E_1 and E_2, in the same space, **S**, of a random experiment, the conditional probability of E_1 given E_2 is defined as the probability that both E_1 *and* E_2 occurred, $\mathbf{P}(E_1 \cap E_2)$, divided by the probability that E_2 occurred, $\mathbf{P}(E_2)$. The conditional probability of E_1 given E_2 is usually denoted by $\mathbf{P}(E_1|E_2)$, thus:

$$\mathbf{P}(E_1|E_2) = \frac{\mathbf{P}(E_1 \cap E_2)}{\mathbf{P}(E_2)}$$

The conditional probability is, therefore, the probability that an event will occur given that another event is known to have occurred.

Confidence Interval *n* An interval of a population parameter that probably includes the actual value of the parameter. Associated with the confidence interval is the confidence level which is the probability or fraction that the true value of the parameter is contained in the interval. The confidence interval is generally constructed with the use of an estimator for the parameter that will give the *confidence limits* which are the boundary values for the confidence interval. The probability that a parameter is within the confidence interval is usually denoted by $1-\alpha$, where α is a small number which identifies the error or the probability that the parameter resides outside the confidence interval.

Confidence Interval for a Mean *n* An interval estimate for the *mean* of a population parameter. See ► Confidence Interval.

Confidence Interval for a Proportion *n* The confidence interval for a population proportion which provides a range of values within which the true proportion of the population will likely lie. See ► Confidence Interval.

Confidence Interval for the Difference Between Two Means *n* A *confidence interval* that defines the range of values within which the difference between the *means* of two population *parameters* will probably lie.

Confidence Interval for Difference Between Two Proportions *n* A *confidence interval* that defines the range of values within which the difference between the values of two different population proportions will probably lie.

Confidence Level *n* The probability or fraction that the actual value of a population parameter resides within the confidence interval. The confidence interval is generally constructed with the use of an estimator for the parameter that will define or generate the *confidence limits* which are the boundary values for the confidence interval. The probability that a parameter is within the confidence interval is usually denoted by $1-\alpha$, where α is a small number which identifies the error or the probability that the parameter resides outside the confidence interval.

Confidence Limits *n* The boundary values for the *confidence interval* of a population *parameter*. The confidence limits are usually generated with the use of an *estimator*. See ► Confidence Interval.

Contingency Table *n* A two or more dimensional table whose cells contain a measure of the relationship of the categories of the variables of the population used to index the cell. That is, each variable is divided into categories and each cell quantifies the relationship among the specific variable categories associated with the cell. Most often this relationship is the frequency of occurrence of the indicated category of variable 1 with the indicated category of variable 2, etc. The contingency table, thus, summarizes the relationships among the categorized variables. A 2-dimensional contingency table is often referred to as a *crossed table* and the two variables referred to as *cross classified.* A 2-dimensional contingency table is usually a row column matrix type table whose rows show the segregation of the population into types, e.g., treated and not treated. The contingency table columns show the range groups of the outcome, e.g., survived, did not survive. There is usually an extra row showing the outcome range group totals and an extra column showing the population type totals. These are referred to as the marginal totals. A possible contingency table for the example study described above would be as follows:

Study Groups	Survived	Did Not Survive	Group Totals
Treated Group	25	35	60
Control Group	8	37	45
Outcome Totals	33	72	105

The table above shows that there were 105 total individuals of which 60 were treated and 45 were not treated. Of the total group 33 individuals survived, 25 in the treated group and 8 in the untreated group. In the example, the null hypothesis to be verified or rejected is that the treatment had no effect on the number of surviving individuals. The verification is made using the chi-square test of independence. The contingency table above is referred to as a 2 by 2 contingency table. This concept can be expanded to any number of rows and columns, i.e., an N by M contingency table and two any number of dimensions, i.e., an N_1 by N_2 by ... by N_k contingency table. The general form of an N by M contingency table would be:

Study Groups	Range 1	Range 2	...	Range M	Group Totals
Study Group 1	x_{11}	x_{12}	...	x_{1M}	Group 1 Total
Study Group 2	x_{21}	x_{22}	...	x_{2M}	Group 2 Total
...	...	...	...	...	...
Study Group N	x_{N1}	x_{N2}	...	x_{NM}	Group N Total
Range Totals	R1 Total	R2 Total	...	RM Total	Study Total

where x_{ij} is the number of measurements that fall in both group i and range j.

Continuous Data *n* Data whose possible values are uncountably infinite (continuous or piecewise continuous). *Uncountably infinite* means that the set of possible values **cannot** be mapped to the set of integers, for example the real numbers, the irrational numbers, the real numbers between 0 and 1, or the real numbers greater than 3 and less than 5 along with the real numbers greater than 11 and less than 21. **Note:** the set of rational numbers is *countably infinite* and would, therefore, not be continuous. Continuous data results from the measurements of a *continuous random variable.* See also ► Discrete Data.

Continuous Distribution Function *n* The *distribution function* for a *continuous random variable.* See ► Distribution Function.

Continuous Random Variable *n* A *random variable* whose possible values are uncountably infinite (continuous or piecewise continuous). *Uncountably infinite* means that the set of possible values **cannot** be mapped to the set of integers, for example the real numbers, the irrational numbers, the real numbers between 0 and 1, or the real numbers greater than 3 and less than 5 along with the real numbers greater than 11 and less than 21. **Note:** the set of rational numbers is *countably infinite* and would, therefore, not be continuous. See also ► Discrete Random Variable.

Continuous Uniform Distribution *n* The form of the *uniform distribution* over a *continuous random variable.*

Correlated Groups *n* An alternate term for *correlated samples.*

Correlated Samples *n* Referred to as *correlated groups, matched samples,* and *matched groups,* are *samples* where the *elements* are purposely paired by the researcher or observer so that one or more features, or variables associated with one or more *statistics* are matched or closely matched between the elements.

This pairing is performed in two primary methods:

1. *Before-after design,* where the same element is measured twice, once before and once after the modification of a variable (before and after a test or change in condition).
2. *Matched group design,* where the elements in the two groups having different modifications of the variable under study are matched. The simplest and most often group division is the control group and the test group. Often the groups are not explicitly paired by elements but paired by subsets of equal numbered elements. For example, two groups each containing *M* females and *N* males.

Correlation Coefficient *n* The covariance normalized by the product of the standard deviations of the two random variable sequences used to calculate the covariance. The correlation coefficient is a widely used measure of the correlation between two different random variables. If the correlation coefficient is calculated over subsequences of the random variables instead of the whole sequences it is referred to as the *cross-correlation coefficient.* The correlation coefficient of two sequences of random variables, **X** and **Y**, is often denoted by ρ_{XY}. The precise mathematical definition of cross-correlation coefficient of two random variable sequence sections is given by:

$$\rho_{XY} = \frac{\mathbf{Cov(X, Y)}}{\sigma_X \sigma_Y} = \frac{\mathbf{E}\{(\mathbf{X} - \mu_X)(\mathbf{Y} - \mu_Y)\}}{\sigma_X \sigma_Y}$$

where: **X** is a random variable,
μ_X is the mean of **X**,
σ_X is the standard deviation of **X**,
Y is a second random variable,
μ_Y is the mean of **Y**,
σ_Y is the standard deviation of **Y**, and
E {} is the expectation value operator.
For continuous random variables over the sample space, **S**, this becomes:

$$\rho_{XY} \frac{\int_S [\mathbf{X} - \mu_X][\mathbf{Y} - \mu_Y] f(\mathbf{X}, \mathbf{Y}) d\mathbf{X} d\mathbf{Y}}{\sigma_X \sigma_Y}$$

for continuous random variables with probability density function, f(**X**,**Y**), means, μ_X and μ_Y, and standard deviations, σ_X and σ_Y. The correlation coefficient is given by:

$$\rho_{XY} = \frac{\sum_k (x_k - \mu_X)(y_k - \mu_Y)\mathbf{P}_k}{\sigma_X \sigma_Y}$$

for discreet random variables, with discrete values, x_k and y_k, with associated probabilities, $\mathbf{P}_k$, and by:

$$\rho_{XY} = \frac{\sum_k (x_k - \mu_X)(y_k - \mu_Y)\mathbf{W}_k}{\sigma_X \sigma_Y \sum_k \mathbf{W}_k}$$

for statistical samples, x_k and y_k, with weights, $\mathbf{W}_k$. See also ► Cross-correlation Coefficient and ► Autocorrelation Coefficient.

Covariance *n* A basic measure of the association between two random variables. It is a quantification of how much the two variables change together. The covariance of two random variables, $\mathbf{X}_1$ and $\mathbf{X}_2$, is often denoted by Cov($\mathbf{X}_1$, $\mathbf{X}_2$). The precise mathematical definition of covariance of two random variables, $\mathbf{X}_1$ and $\mathbf{X}_2$, defined on a probability space, **S**, is given by:

$$\mathrm{Cov}(\boldsymbol{X}_1, \boldsymbol{X}_2) = \mathbf{E}\{(\boldsymbol{X}_1 - \mathbf{E}\{\boldsymbol{X}_1\})(\boldsymbol{X}_2 - \mathbf{E}\{\boldsymbol{X}_2\})\}$$

where: **E** {} is the expectation value operator. Or by:

$$\mathrm{Cov}(\boldsymbol{X}_1, \boldsymbol{X}_2) = \mathrm{E}\{(\boldsymbol{X}_1 - \mu_1)(\boldsymbol{X}_2 - \mu_2)\}$$

where: μ_1 and μ_2 are the means of $\mathbf{X}_1$ and $\mathbf{X}_2$ respectively.

$$\mathrm{Cov}(\boldsymbol{X}_1,\boldsymbol{X}_2) = \int_S (\mathrm{X}_1 - \mu_1)(\mathrm{X}_2 - \mu_2) f(\mathrm{X}_1,\mathrm{X}_2) d\mathrm{X}_1 d\mathrm{X}_2$$

Therefore:
for continuous random variables with probability density function, $f(\mathbf{X}_1,\mathbf{X}_2)$ and means μ_1 and μ_2, by:

$$Cov(\boldsymbol{X}_1, \boldsymbol{X}_2) = \sum_i \sum_j (x_{1i} - \mu_1)(x_{2j} - \mu_2)\boldsymbol{P}_{ij}$$

for discrete random variables, with discrete values, x_{1i} and x_{2j}, with associated probabilities, $\mathbf{P}_{ij}$, and by:

$$\mathrm{Cov}(\boldsymbol{X}_1, \boldsymbol{X}_2) = \frac{\sum_i \sum_j (x_{1i} - \mu_1)(x_{2j} - \mu_2)\mathbf{W}_{ij}}{\sum_i \sum_j \mathbf{W}_{ij}}$$

for statistical samples, x_{1i} and x_{2j}, with weights, $\mathbf{W}_{ij}$.
If two random variables do not depend on each other, they are independent and their covariance is zero; however, if two random variables have a zero covariance, they are not necessarily independent.
For a single random variable, **X**, the Cov (**X**,**X**) is the variance of **X**.

Critical Region *n* Also known as the *rejection region.* The critical region is the range for the *test statistic* within which the *null hypothesis* is rejected. The limiting value(s) of the critical region are known as the *critical value(s).* See also ► Acceptance Region.

Critical Region *n* The range for the *test statistic* within which the *null hypothesis* is rejected. The limiting value(s) of the critical region are known as the *critical value(s).* Also known as *rejection region.* See also ► Acceptance Region.

Critical Value *n* The value of the *test statistic* beyond which the *null hypothesis* is rejected. There is one critical value in a *one-sided test* and two critical values in a *two-sided test.* The critical value separates the *critical region* (the region where the null hypothesis is rejected) from the *acceptance region* (the region where the null hypothesis is accepted).

Cross Correlation *n* A measure of the correlation between two sections or sub-sequences of sequences of two different random variables. Each sub-sequence has the same number of elements, ordering, and spacings. The cross-correlation can be defined as the expectation value of the term-by-term product of a continuous section of a sequence of one random variable with another continuous section of with the same length from a different sequence from a different random variable. The cross-correlation is a function of three values, the starting element of the first section, the starting element of the second section, and the length of the section. The cross-correlation,

$$\mathbf{R}_{\mathrm{XY}}(i, j, \mathrm{T}) = \mathbf{E}\{\mathbf{X}_{\mathrm{T}i}\mathbf{Y}_{\mathrm{T}j}\}$$

where: $\mathbf{X}_{\mathrm{T}i}$ is the sub-sequence of **X** with starting element, *i*, and length, T,
$\mathbf{Y}_{\mathrm{T}j}$ is the sub-sequence of **Y** with starting element, *j*, and length, T, and
E {} is the expectation value operator.
For a continuous random variables, **X** and **Y**, over the sub-space, $\mathbf{S}_\mathrm{T}$, of the sample space, **S**, this becomes:

$$\mathbf{R}_{\mathrm{XY}}(s, t, \mathbf{S}_\mathrm{T}) = \int_{S_T} \mathbf{X}(s)\mathbf{Y}(t) f(s, t) d\mathbf{S}_\mathrm{T}$$

for continuous random variable with probability density function, $f(s,t)$. The cross-correlation is given by:

$$\mathbf{R}_{\mathrm{XY}}(i, j, \mathrm{T}) = \sum_k x_k y_{k+j-i}\mathbf{P}_{ijk}$$

for two discrete random variables, with discrete values, x_k and y_{k+j-i}, with associated probabilities, $\mathbf{P}_{ijk}$, and by:

$$\mathbf{R}_{\mathrm{XY}}(i, j, \mathrm{T}) = \frac{\sum_k x_k y_{k+j-i}\mathbf{W}_{ijk}}{\sum_k \mathbf{W}_{ijk}}$$

for statistical samples, x_k and y_{k+j-i}, with weights, $\mathbf{W}_{ijk}$.
The cross-correlation of two random variables is often used interchangeably with and confused with the *cross-correlation coefficient* which is the *cross-covariance* normalized by the product of the *standard deviations* of the two sections in the *cross-covariance.* The *cross-covariance* and the *cross-correlation coefficient* are both often referred to as simply the cross-correlation.
The *autocorrelation* is the cross-correlation of a sub-sequence of a single random variable with another sub-sequence of the same random variable.

Cross-Correlation Coefficient *n* The cross-covariance normalized by the product of the standard deviations of the two sections from the two random variable sequences used to calculate the cross-covariance. The cross-correlation coefficient is a widely used measure of the correlation between two sections of two different but random variables. It is often simply but incorrectly referred to as the *cross-correlation*, which is the un-normalized expectation value of the product of the two sequence sections. The cross-correlation coefficient of two sequences of random variables, **X** and **Y**, is often denoted by $\rho_{XY}(i, j, T)$; where i is the starting index of the section of the sequence, **X**, j is the starting index of the section of the sequence, **Y**, and T is the length of the sections. The precise mathematical definition of cross-correlation coefficient of two random variable sequence sections is given by:

$$\rho_{XY}(i,j,T) = \frac{\mathrm{Cov}(\mathbf{X}_{Ti}, \mathbf{Y}_{Tj})}{\sigma_{XTi}\sigma_{YTj}} = \frac{\mathbf{E}\left\{(\mathbf{X}_{Ti} - \mu_{XTi})(\mathbf{Y}_{Tj} - \mu_{YTj})\right\}}{\sigma_{XTi}\sigma_{YTj}}$$

where: $\mathbf{X}_{Ti}$ is the sub-sequence of **X**,
μ_{XTi} is the mean over $\mathbf{X}_{Ti}$,
σ_{XTi} is the standard deviation over $\mathbf{X}_{Ti}$,
$\mathbf{Y}_{Tj}$ is the sub-sequence of **Y**,
μ_{YTj} is the mean over $\mathbf{Y}_{Tj}$,
σ_{YTj} is the standard deviation over $\mathbf{Y}_{Tj}$, and
E {} is the expectation value operator.
For continuous random variables over the sub-space, $\mathbf{S}_T$, of the sample space, **S**, this becomes:

$$\rho_{XY}(s,t,\mathbf{S}_T) = \frac{\int_{S_T}[\mathbf{X}(s) - \mu_{XTs}][\mathbf{Y}(t) - \mu_{YTt}]f(s,t)d\mathbf{S}_T}{\sigma_{XTs}\sigma_{YTt}}$$

for continuous random variables with probability density function, $f(s,t)$, sub-space means, μ_{XTs} and μ_{YTt}, and sub-space standard deviations, σ_{XTs} and σ_{YTt}, by:

$$\rho_{XY}(i,j,T) = \frac{\sum_k (x_k - \mu_{XTi})(y_{k+j-i} - \mu_{YT_j})\mathbf{P}_{ijk}}{\sigma_{XTi}\sigma_{YTj}}$$

for discrete random variables, with discrete values, x_k and y_k, with associated probabilities, $\mathbf{P}_{ijk}$, and by:

$$\rho_{XY}(i,j,T) = \frac{\sum_k (x_k - \mu_{XTi})(y_{k+j-i} - \mu_{YT_j})\mathbf{W}_{ijk}}{\sigma_{XTi}\sigma_{YTj}\sum_k \mathbf{W}_{ijk}}$$

for statistical samples, x_k and y_k, with weights, $\mathbf{W}_{ijk}$.
The cross-correlation coefficient of a random variable with itself is referred to as the *autocorrelation coefficient*, which is often used interchangeably and confused with the *autocorrelation* which is actually the *cross-correlation* of a random variable with itself.
See also ▶ Correlation Coefficient which is the covariance normalized by the product of the standard deviations.

Cross-Covariance *n* A measure of the correlation between two sections of sequences of random variables. It is a quantification of how much the two sub-sequences change together. The cross-covariance is a function of three variables, the starting element of the first section, the starting element of the second section, and the length of the sections. The cross-covariance of two sequences of random variables, **X**, and **Y**, is often denoted by $\mathbf{C}_{XY}(i,j,T)$; where i is the starting index of the section of the sequence, **X**, j is the starting index of the section of the sequence, **Y**, and T is the length of the sections. The precise mathematical definition of cross-covariance of two random variable sequences is the covariance of the two sections which is given by:

$$\mathbf{C}_{XY}(i,j,T) = \mathrm{Cov}(\mathbf{X}_{Ti}, \mathbf{Y}_{Tj}) = \mathbf{E}\left\{(\mathbf{X}_{Ti} - \mu_{XTi})(\mathbf{Y}_{Tj} - \mu_{YTj})\right\}$$

where: $\mathbf{X}_{Ti}$ is the sub-sequence of **X** with a mean over the section of μ_{XTi},
$\mathbf{Y}_{Tj}$ is the sub-sequence of **Y** with a mean over the section of μ_{YTj}, and
E {} is the expectation value operator.
For continuous random variables over the sub-space, $\mathbf{S}_T$, of the sample space, **S**, this becomes:

$$\mathbf{C}_{XY}(s,t,\mathbf{S}_T) = \int_{S_T}[\mathbf{X}(s) - \mu_{XTs}][\mathbf{Y}(t) - \mu_{YTt}]f(s,t)d\mathbf{S}_T$$

for continuous random variables with probability density function, $f(s,t)$ and sub-space means, μ_{XTs} and μ_{YTt}, by:

$$\mathbf{C}_{XY}(i,j,T) = \sum_k (x_k - \mu_{XTi})(y_{k+j-i} - \mu_{YTj})\mathbf{P}_{ijk}$$

for discrete random variables, with discrete values, x_k and y_k, with associated probabilities, $\mathbf{P}_{ijk}$, and by:

$$\mathbf{C}_{XY}(i,j,T) = \frac{\sum_k (x_k - \mu_{XTi})(y_{k+j-i} - \mu_{YTj})\mathbf{W}_{ijk}}{\sum_k \mathbf{W}_{ijk}}$$

for statistical samples, x_k and y_k, with weights, $\mathbf{W}_{ijk}$.
The cross-covariance of a random variable with itself is referred to as the *autocovariance*, which is often used interchangeably and confused with the *autocorrelation* which is actually the *cross-correlation* of a random variable with itself. The cross-covariance is related to and often confused with the *cross-correlation*.
The cross-covariance normalized by the product of the standard deviations of the two sections is the *cross-*

correlation coefficient, which is often referred to as simply the cross-correlation.

See also ▶ Correlation Coefficient which is the covariance normalized by the product of the standard deviations.

Cumulative Distribution Function *n* An alternative term for *distribution function*.

Cyclical Component *n* The component of the values of a *time series* that are due to *cyclical fluctuation*.

Cyclical Fluctuation *n* An identifiable periodic movement in a *time series* that is not related to seasonal changes. The component of the values of the time series associated with the cyclical fluctuation is referred to as the *cyclical component*. See also ▶ Seasonal Variation.

D

Data Transformation *n* An alternate term for *transformation*.

Decision Maker *n* An alternate term for *test statistic*.

Density Function *n* Also known as the *probability density function* is defined for a *continuous random variable* as the derivative of the *distribution function* of the variable. This means that for a *density function*, $f(x)$, the distribution function, $F(x)$, is given by:

$$F(x) = \int_{-\infty}^{x} f(t)dt$$

Also see ▶ Distribution Function.

Dependent Events *n* Two events, E_1 and E_2, are dependent if and only if they are not independent. That is the conditional probability of either event given the other is not merely equal to the probability of the event itself. That is $\mathbf{P}(E_1|E_2) \neq \mathbf{P}(E_1)$ and $\mathbf{P}(E_2|E_1) \neq \mathbf{P}(E_2)$. In other words, two events are dependent if the occurrence of one is affected by the occurrence of the other.

Differencing *n* The process of creating a new *time series* from the original time series by calculating each element in the new series to be the difference of the corresponding consecutive elements in the original series. That is for an original time series $x_1, x_2, \ldots$ the new series, $y_1, y_2, \ldots$ would be created with $y_k = x_{k+1} - x_k$, for all $k = 1, 2, \ldots$. *Differencing* is often used to remove the *trend* from a time series.

Discrete Data *n* Data whose possible values are countable (finite or countably infinite). *Countably infinite* means that the set of possible values **can** be mapped to the set of integers, for example the odd numbers, the prime numbers, or the integer multiples of one third. **Note:** the set of rational numbers is *countably infinite* and would, therefore, be discrete. Discrete data results from the measurements of a *discrete random variable*. See also ▶ Continuous Data.

Discrete Distribution Function *n* The *distribution function* for a *discrete random variable*. See ▶ Distribution Function.

Discrete Random Variable *n* A *random variable* whose possible values are countable (finite or countably infinite). *Countably infinite* means that the set of possible values can be mapped to the set of integers. See also ▶ Continuous Random Variable.

Discrete Uniform Distribution *n* The form of the *uniform distribution* over a *discrete random variable*.

Dispersion *n* The variation in the value of a variable within a set of observations. A small dispersion indicates closely spaced values; whereas a large dispersion indicates widely spaced values. There are several *measures of dispersion* which define quantitative values related to the spread of the values in a set of observations. These include: *range, interquartile range, standard deviation,* and *variance*. A *measure of dispersion* along with a *measure of central tendency* give a simple quantitative description of shape of the distribution of values within the set of observations.

Distribution Function *n* Also known as the *cumulative distribution function* or *cumulative probability distribution*. The *distribution function* is defined for a *random variable*, **X**, and a real value, *x*, to be the *probability* that **X** has a value less than or equal to *x*. For a *discrete random variable*, the distribution function is referred to as the *discrete distribution function* and for a *continuous random variable* the distribution function is referred to as the *continuous distribution function*. Also see ▶ Probability and ▶ Probability Density Function.

Dot Plot *n* A type of quantitative graphical display used for displaying the *frequency* of *categorical data* by placing dots along the horizontal axis above the category whose frequency they enumerate. Each dot represents a fixed number of occurrences and the number of dots stacked vertically represents the total number of occurrences for a category. Dot plots are often used to visualize *frequency tables*.

Double-blind Study *n* A *double-blind study* is a type of study or experiment which attempts to eliminate *bias* induced by human observers or subjects involved with the experiment, by allowing neither the observers or the subjects to be aware of which group is associated with which situation. For example in a clinical trial for a new treatment, one group would receive the treatment and a second group would receive a *placebo*, but neither the researchers nor the subjects are informed which individuals are given the treatment. The double-blind study is considered the least affected by *bias*. See ► Blinding.

Dummy Variable *n* Also known as an *indicator variable* is a variable generated from a non-quantitative or non-numeric categorical feature of the data in order to associate a quantitative value with the feature so that its effects can be used in *regression analysis*. As a standard, the variable is given a range of zero to one. A dummy variable is often used to indicate the presence of a feature such as left-handedness. Multiple dummy variables can be used to represent independent but related categories such as the days of the week.

E

Element *n* An individual unit of a population.

Estimate *n* The value of the parameter of a population obtained from an estimator. It is referred to as an *estimate* of the parameter.

Estimation *n* The process of or procedure using an estimator to calculate the value of a parameter of a population.

Estimator *n* A rule or function applied to a sample in order to guess or estimate a parameter of a population. The value of the parameter obtained from the estimator is called an *estimate* of the parameter. A process of or procedure using an estimator to calculate the value of a parameter is referred to as *estimation*.

Event *n* An event is a subset of the sample space of a random experiment.

Excess *n* See ► Kurtosis.

Excess Kurtosis *n* See ► Kurtosis.

Exhaustive *n* If a set of events of a sample space, **S**, represents all of the possible elements of **S**, then the set of events is said to be *exhaustive*.

Expectation *n* See ► Expectation Value.

Expectation Value *n* Also referred to as *expectation*, *expected value*, or *mathematical expectation*. In simple terms the expectation value for a function of a *random variable* is the expected average value of the function over a large number of samples. The precise mathematical definition is: for a continuous random variable, $\mathbf{X}$, defined on a probability space, $\mathbf{S}$, with a probability density function, $f(\mathbf{X})$, the expectation value of a function, $g(\mathbf{X})$, is denoted by $\mathbf{E}\{g(\mathbf{X})\}$ and defined to be the integral over $\mathbf{S}$ of the product $g(X)f(\mathbf{X})$, i.e.:

$$\mathbf{E}\{g(\mathbf{X})\} = \int_{\mathrm{S}} g(\mathbf{X})f(\mathbf{X})d\mathbf{X}$$

For a discrete random variable, $\mathbf{X}$, assuming the values x_i, defined on a probability space, $\mathbf{S}$, with discrete probabilities, $\mathbf{P}_i$, the expectation value of a function, $g(\mathbf{X})$, is defined to be the sum over all x_i in $\mathbf{S}$ of the product $g(x_i)\mathbf{P}_i$, i.e.:

$$\mathbf{E}\{g(\mathbf{X})\} = \sum_i g(x_i)\mathbf{P}_i$$

The expectation value exists in the continuous case only if the integral is absolutely integrable and in the discrete case only if the sum converges absolutely. For statistical samples, x_i, with weights, $\mathbf{W}_i$, the expectation of g is given by:

$$\mathbf{E}\{g(\mathbf{X})\} = \frac{\sum_i g(x_i)\mathbf{W}_i}{\sum_i \mathbf{W}_i}$$

For random statistical samples where the weights are one, the formula for the expectation of g is given by:

$$\mathbf{E}\{g(\mathbf{X})\} = \frac{\sum_i g(x_i)}{N}$$

where: N is the number of samples taken.

The expectation value is a linear operator thus:

For a constant, C, $\mathbf{E}\{C\} = C$ and

For two functions $g_1(\mathbf{X})$ and $g_2(\mathbf{X})$ and two constants C_1 and C_2,

$$\mathbf{E}\{C_1 g_1(\mathbf{X}) + C_2 g_2(\mathbf{X})\} = C_1\mathbf{E}\{g_1(\mathbf{X})\} + C_2\mathbf{E}\{g_2(\mathbf{X})\}$$

In general the expectation value of the product of two functions of a random variable does not equal the product of the expectation values, i.e.:

$$\mathbf{E}\{g_1(\mathbf{X})g_2(\mathbf{X})\} \neq \mathbf{E}\{g_1(\mathbf{X})\}\mathbf{E}\{g_2(\mathbf{X})\}$$

However, if $\mathbf{X}$ and $\mathbf{Y}$ are mutually independent random variables then:

$$\mathbf{E}\{g_1(\mathbf{X})g_2(\mathbf{Y})\} = \mathbf{E}\{g_1(\mathbf{X})\}\mathbf{E}\{g_2(\mathbf{Y})\}$$

The expectation value is the basis from which the usual measures (mean, variance, standard deviation, etc.) associated with the moments of a random variable are derived. For example $\mathbf{E}\{\mathbf{X}\}$ is the mean or average of $\mathbf{X}$.

Expected Frequencies *n* A predicted *frequency* obtained for an experiment based on theory, assumptions, models, etc. It is used in hypothesis tests and comparisons with the actual or observed frequency. In tests involving *contingency tables*, the expected frequencies are the frequencies that are predicted for each cell in the table, generally assuming that the variables and categories of the table are independent.

Expected Value See ▶ Expectation Value.

Experiment *n* A process through which data is acquired. The data could be measurements, observations, etc. An experiment is often a process carried out under controlled conditions.

Experimental Design *n* The structure or organization of an *experiment*. It usually defines the *experimental units*, the *treatments*, the structure of and methods for collecting the data, and the questions that are intended to be answered by the experiment. See also ▶ Completely Randomized Design, ▶ Factorial Design, ▶ Randomized Block Design, and ▶ Randomized Complete Block Design.

Exponential Smoothing *n* A process for the *smoothing* of *time series* data, by calculating a new time series whose values are calculated from the current value in the original series, x_i, and the previous value in the new series, y_{i-1} as follows:

$$y_0 = x_i \text{ and}$$

$$y_i = ax_i + (1 - \alpha)y_{i-1} \text{ for } i \text{ greater than } 0.$$

The factor α is referred to as the *smoothing factor.*

Extrapolation *n* The process of estimating the value of a dependent variable for a value of an independent variable which lies outside the range of the independent variable values contained in the set of known or observed independent-dependent value pairs. For a *time series* an *extrapolation* generally tries to estimate values in the series for times that have not yet occurred.

F

Factor *n* An independent variable or controllable situation or property of an experiment. The values associated or set for a factor are *levels.* If a factor takes on real number levels it is a *quantitative factor* otherwise it is a *qualitative factor.* See also ▶ Treatment.

Factorial Design *n* A type of *experimental design* in which the *treatments* consist of a multiple combinations of two or more *factors*, where the combination of *levels* for the factors is different for each treatment. In general there are finite sets of levels for each factor and the treatments consist of all possible combinations of levels for the factors. The most simple factorial design would consist of two factors each with two levels. This would create two times two or four treatments.

Factorial design is useful in that it is an efficient way to measure in one experiment the response due to multiple factors as well as the *interactions* between factors.

Five-Number Summary *n* Also denoted as *5-number summary*; an abbreviated representation or summary of a data set which states: the lower limit, the *first quartile*, the *median*, the *third quartile*, and the upper limit. The five-number summary is often displayed using a *box-and-whisker plot.*

Frequency *n* The frequency is the number of occurrences of a particular outcome, result, or observation in an experiment. It may be expressed as a ratio of the number of particular occurrences to the total number of observations, which is often referred to as the *relative frequency* when used refers to the actual count and not the ratio.

Frequency Table *n* A frequency table is a simple method of representing or summarizing a set of data. It is a list or table displaying a record of the *frequencies* of each particular result, value or observation (how often each result, value, or observation or a set of these occurs in an experiment). The observations are divided into categories that are usually independent with their occurrence frequencies listed in a tabular form. The frequencies may be listed as absolute, relative, or in percent, or sometimes more than one of these forms. The data may be discrete or continuous. When the display of the relationship of multiple variables is desired, a *contingency table* is used.

G

Gaussian Distribution *n* An alternate name for the *normal distribution.*

Geometric Distribution *n* A *probability distribution* of a *discrete random variable*, **X**, with non-negative integer values, *k*, that have a *probability*, **P**(*k*), of the form:

$$\mathbf{P}(k) = (1 - p)^k p$$

where: p is a real number greater than zero and less than or equal to 1
k is a non-negative integer (0,1,2, . . .)
The geometric distribution has:
$mean = (1-p)/p$
standard deviation $= (1-p)^{1/2}/p$
variance $= (1-p)/p^2$
$skewness = (2-p)/(1-p)^{1/2}$,
kurtosis $= 9 + p^2/(1-p)$, and
excess kurtosis $= 6 + p^2/(1-p)$.
The geometric distribution is often used to describe the probability that the first success in a sequence of *Bernoulli Trials* occurs on the kth trial where p is the probability of success for a trial.
Sometimes the geometric distribution is stated in the form that $\mathbf{P}(x)$ is the probability of k failures before the first success. In this case the geometric distribution takes the form of:

$$\mathbf{P}(k) = (1-p)^{k-1}p$$

where: k has the range of positive integers, (1,2, . . .)
This form of the geometric distribution has a *mean* equal to p^{-1} but the same values for the standard deviation, variance, skewness, and kurtosis as the first form.

H

Histogram *n* A type of quantitative graphical display used for displaying the distribution of *categorical data* by placing contiguous rectangles along the horizontal axis above the category range whose *frequency* they enumerate. The width of the rectangle represents *range* interval of the category and the rectangle area represents the total number of occurrences for a category. Histograms are a type of *bar chart*. Histograms are often used to visualize *frequency* tables.

Hypothesis *n* See ► Statistical Hypothesis.

Hypothesis Test *n* A process or reasoning used to test the validity of a statistical hypothesis. Formally it is a process that allows for the acceptance of the assumed or starting hypothesis, which is referred to as the null hypothesis, or its rejection in favor of the alternative hypothesis, which is the negation of the null hypothesis.

I

Independent Events *n* Two events, E_1 and E_2, are independent if either of their conditional probabilities given the other is the probability of the event by itself. That is, $\mathbf{P}(E_1|E_2) = \mathbf{P}(E_1)$ or $\mathbf{P}(E_2|E_1) = \mathbf{P}(E_2)$. Thus, if two events are independent, the probability of one event is the same whether the other event has occurred or has not occurred. If E_1 is independent of E_2 then E_2 is independent of E_1. Independent events are sometimes referred to as *statistically independent events*.

Independent Groups *n* An alternate term for *independent samples*.

Independent Random Variables *n* A set of *random variables*, $\mathbf{X}_1, \ldots, \mathbf{X}_N$, defined on the same sample space, is independent if all possible combinations of *events* involving the variables are *independent events*. This is true if and only if the *probability* of obtaining the values, $x_1, \ldots x_N$, $\mathbf{P}(x_1, \ldots, x_N)$ is equal the product of the individual probabilities, $\mathbf{P}(x_1) \bullet \mathbf{P}(x_2) \bullet \ldots \bullet \mathbf{P}(x_N)$, for all possible combinations of values of $x_1, \ldots, x_N$.

Independent Samples *n* Also referred to as *independent groups*, different samples that have no particular relation to each other. Theoretically there should be no correlation between the groups that is related to the *parameter* of interest with respect to the *population*.

Indicator Variable *n* An alternate name for a *dummy variable*.

Inference *n* See ► Statistical Inference.

Interaction *n* A change in the character of the relationship between a *parameter* and the *level* of one *factor* due to a second factor. More precisely the response to a *treatment* can be expressed as the relationship between the *mean*, μ_X of a *random variable*, $\mathbf{X}$, and the levels of factors A and B, which can be expressed as:

$$\mu_X = \mu_0 + f_A(A) + f_B(B) + f_{AB}(A, B)$$

where: μ_0 is a constant with respect to A and B,
f_A is a function of A alone,
f_B is a function of B alone, and
f_{AB} is a function of both A and B.
If f_{AB} is non-zero then there is an interaction and f_{AB} is the interaction, if it is zero then there is no interaction. f_A and f_B if non-zero are the *main effects* of A *and* B respectively.

Interquartile Range *n* The difference between the third and first *quartiles.* The interquartile range is often abbreviated as *IQR* and also referred to as *midspread* or *middle fifty*; since it is the range interval that contains the middle 50% of their observations. This corresponds to the rectangle in the *box-and-whisker plot.* The interquartile range is used as a measure of dispersion.

Interval *n* A bounded set of real numbers that contains the subset of all numbers of a set that lie between two numbers. Whether or not the bounding numbers are included in the interval depends on the type. A *closed interval* contains both of the bounding points. An *open interval* contains neither of the bounding points. A *semiopen interval* or *semiclosed interval* contains only one of the bounding points, the lower or upper but not both. For example the semiopen interval of points greater than or equal to 3 and less than 7, from the set, {0,1,3,4,7,8,10}, would be the set {3,4}, since it contains the lower bound 3 but not the upper bound 7.

Interval Scale *n* A quantitative scale that is divided by equal *intervals* but may have any arbitrary point as its zero or beginning. This creates a scale where sums and differences in values have useful meanings but products and ratios do not. The common temperature scales, Celsius and Fahrenheit, are examples.

IQR *n* The abbreviation for *interquartile range.*

Irregular Component *n* An alternate term for *residual component.*

Irregular Variations *n* An alternate term for *residual movement.*

K

K-S Test *n* An abbreviation for the *Kolmogorov-Smirnov test.*

Kolmogorov-Smirnov Test *n* Also known as the *K-S test* or the *Kolmogorov-Smirnov goodness-of-fit test,* a distribution free, *nonparametric test* used to test whether two measured distributions from two different *samples* which may or may not come from two different populations or a measured distribution and a theoretical distribution. The test is performed by calculating the maximum distance between two points on the *distribution functions,* $F_1(x)$ and $F_2(x)$, for the two distributions, to get one of three parameters:

$$D_{2S} = \sup_x |F_2(x) - F_1(x)|$$

the *two-sided test* for F_2 equal F_1,

$$D_{1S2} = \sup_x [F_2(x) - F_1(x)]$$

the *one-sided test* for F_2 less than or equal to F_1, or

$$D_{1S1} = \sup_x [F_1(x) - F_2(x)]$$

the *one-sided test* for $F1$ less than or equal to $F2$.
A *significance level,* α, is chosen and a value $t_{n,m,1-\alpha}$ where, n and m are the number of units in samples 1 and 2 respectively, or a value $t_{n,1-\alpha}$, where n is the number of units in the one measured sample, is selected from a *Smirnov table.* If the distance, D, under consideration is greater than the t value from the table, then the *null hypothesis* under consideration,
$F_2 = F_1$, $F_2 \le F_1$, or $F_1 \le F_2$, is rejected.

Kruskal-Wallis Test *n* A *nonparametric test* used to test whether the average ranks of two or more *samples* from different *populations* or the same population are the same. The *null hypothesis* is that the average or *mean* ranks for the samples are equal; therefore, the null hypothesis will be rejected if at least one average rank is different. The average rank is determined by ranking, based on their value, all observations, from one to N, where N is the total of all observations, and then calculating the mean rank for each sample. If the values of more than one sample are equal then all of these values would be given the same rank, which is the average of the ranks they would occupy if slightly different.
In mathematical terms: for M, samples, each sample having n_i units, N would be given by:

$$N = \sum_i n_i$$

and

$$R_i = \sum_j r_{ij}$$

where: R_i is the sum of the overall ranks of the ith sample and
r_{ij} is the overall rank of the jth unit in the ith sample.
The *statistic* used in the test, H, is given by:

$$H = \left[\frac{12}{N(N+1)} \sum_i \frac{R_i^2}{n_i}\right] - 3(N+1)$$

If H is greater than the value chosen from the Kruskal-Wallis table, $h_{n1,n2,...nM,1-\alpha}$, where M is the number of samples and α is the *significance level*, then the null hypothesis is rejected. The Kruskal-Wallis tables exist for small numbers of samples. If a large number of samples are included, H, the behavior of the statistic, H, is approximated by the *chi-square distribution* with $M-1$ degrees of freedom and the $h_{M-1,\ 1-\alpha}$ value from the chi-square table is used.

Kurtosis *n* A primary measure of shape or descriptive measure which describes the peakedness of a probability distribution, population or sample. It is ideally the 4th normalized central moment (or by many authors as the 4th normalized central moment minus 3, which is also called the *excess kurtosis* or simply the *excess*) and is often denoted as κ or γ_2. The precise mathematical definition of kurtosis for a random variable, **X**, defined on a probability space, **S**, is given by:

$$k = \frac{\int S\ (X-\mu)^4 f(X)dX}{\sigma^4}$$

for a continuous random variable with probability density function, $f(X)$, mean, μ, and standard deviation, σ; by:

$$\kappa = \frac{\sum_i (x_i-\mu)^4 \mathbf{P}_i}{\sigma^4}$$

for a discrete random variable, with discrete values, x_i, and associated probabilities, $\mathbf{P}_i$, and by:

$$\kappa = \frac{\sum_i (x_i-\mu)^4 \mathbf{W}_i}{\sigma^4 \sum_i \mathbf{W}_i}$$

for statistical samples, x_i, with weights, $\mathbf{W}_i$.

Many authors will define kurtosis as the normalized fourth central moment minus 3. This is done primarily to reference the kurtosis to the normal distribution which has a kurtosis of 3. This form is also known as and, in the authors' opinion, more properly known as the *excess kurtosis.* The excess kurtosis is primarily used to relate the flatness of a distribution to that of the normal distribution. A distribution that is sharper or more bunched in the middle than the normal distribution has a negative excess kurtosis and is referred to as a leptocurtical or leptokurtic distribution. A distribution that is flatter or more spread out from the normal distribution has a positive excess kurtosis and is referred to as a platicurtical or platykurtic distribution. A distribution that has the same peakedness as the normal distribution has a zero excess kurtosis and is referred to as a mesocurtical or mesokurtic distribution.

L

Law of Alternatives *n* An alternate name for Law of Total Probability.

Law of Total Probability *n* Also known as the *Theorem on Total Probability* or the *Law of Alternatives.* It states that the probability, $P(A)$ of an *event, A*, is equal to the sum of the *conditional probabilities* of A, given events, E_i, $P(A|E_i)$, *times* the probability of event, E_i, for $i = 1, 2, 3, \ldots, N$ where N is a positive integer or infinity, and where E_is are non-overlapping and form a partition of a *sample space* that covers the sample space of A. This can be expressed as:

$$\mathbf{P}(A) = \sum_i \mathbf{P}(A|E_i)\mathbf{P}(E_i)$$

This is often expressed in the form of set intersections as:

$$\mathbf{P}(A) = \sum_i \mathbf{P}(A \cap E_i)$$

which is more correctly known as the *law of alternatives.*

Least Squares Method *n* A method for finding the *best fit* of a function of variables to a set of observed variable relations. In the *least squares sense* a best fit is determined by minimizing the sum of the squares of the *residuals* of a model and corresponding set of observed variable relations. Its purpose is to determine the constant parameters in a function of one or more variables so that a good estimate of an additional variable can be obtained. More precisely, if a model that predicts the values of a *dependent random variable,* **Y**, as a function of one or more independent random variables, $\mathbf{X}_j$, for $j-1, 2, \ldots, N$ and of one or more constants, $a_1, a_2, \ldots a_M$, such that $y_i = f(x_{1i}, x_{2i}, \ldots, x_{Ni}, a_1, a_2, \ldots, a_M)$, then the *least squares method* chooses values for the a_i constants that will minimize the sum, S, given by:

$$S = \sum_i [y_i - f(x_{1i}, x_{2i}, \ldots, x_{Ni}, a_1, a_2, \ldots a_M)]^2$$

If the partial derivatives of f with respect to the a_k constants exist then the minimizing values for the constants can be found by setting the M partial derivatives, $\partial s/\partial a_k = 0$ This generates M equations with the a_k

constants as the M unknowns given for each $k = 1, 2, \ldots,$ M by:

$$0 = \sum_i [y_i - f(x_{1i}, x_{2i}, \ldots, x_{Ni}, a_1, a_2, \ldots a_M)]^2$$
$$\frac{\partial f(x_{1i}, x_{2i}, \ldots, x_{Ni}, a_2, \ldots a_M)}{\partial a_k}$$

The above M equations can be simultaneously solved for the a_k constants.

If the function f is linear with respect to the x_js, then f would be given by the form:

$$f(x_{1i}, x_{2i}, \ldots x_{Mi}, a_0, a_1, a_2, \ldots a_M) = a_0 + \sum_{j=1}^{M} a_j x_{ji}$$

and there would be $M+1$ equations which, if there were L data points, would have the form:

$$\sum_{i=1}^{L} y_i = La_0 + \sum_{j=1}^{M} a_j \sum_{i=1}^{L} x_{ji}$$

for $k = 0$ and

$$\sum_{i=1}^{L} x_{ki} y_i = a_0 \sum_{i=1}^{L} x_{ki} + \sum_{j=1}^{M} a_j \sum_{i=1}^{L} x_{ki} x_{ji}$$

for $k = 1, 2, \ldots, M$.

These equations generate the *least squares regression line.*

Leptocurtical Distribution *n* A distribution that is sharper or more bunched in the middle than the normal distribution. It has a negative excess kurtosis. See ► Kurtosis.

Leptokurtic Distribution *n* A distribution that is sharper or more bunched in the middle than the normal distribution. It has a negative excess kurtosis. See ► Kurtosis.

Level *n* The value taken on, associated with, or set for a *factor.* Factors can be numerical, e.g., a number between 0 and 25, or non-numerical, e.g., months of the year.

M

Main Effect *n* A change in the character of the relationship (effect) between a *parameter* and the *level* of one *factor* averaged across all levels of all other factors in the experiment. More precisely the response to a *treatment* can be expressed as the relationship between the *mean,* μ_X of a *random variable,* **X**, and the levels of factors A and B, which can be expressed as:

$$\mu_X = \mu_0 + f_A(A) + f_B(B) + f_{AB}(A, B)$$

where: μ_0 is a constant with respect to A and B,
f_A is a function of A alone,
f_B is a function of B alone, and
f_{AB} is a function of both A and B.

If f_{AB} is non-zero then there is an *interaction* and f_{AB} is the interaction, if it is zero then there is no interaction. Although technically not the case, f_A and f_B if non-zero are often referred to as the *main effects* of A and B respectively. In a true representation the *main effect* of the factor A, α_A, would be given by:

$$\alpha_A = f_A(A) + \langle f_B(B) + f_{AB}(A, B) \rangle_B$$

where: $< >_B$ is the average with respect to the factor, B.

Mann-Whitney U Test *n* Also known as the Wilcoxon-Mann-Whitney test, the *Mann-Whitney-Wilcoxon test,* the *MWW test,* or the *Wilcoxon rank-sum test* is a *nonparametric test* used to test whether two *samples* of different sizes come the same *population* or populations with the same distribution. The samples must be independent.

In mathematical terms: Given two samples **X** and **Y** with sizes N and M respectively, order all $N + M$ *units* from both samples together from 1 to $N + M$. If the value of more than one sample are equal then all of these values would be given the same rank, which is the average of the ranks they would occupy if slightly different. The test *statistic,* U is the sum over $j = 1$ to M for each y_j, of the number of x_i units that are ranked less than or equal to y_j with the equal units being assigned a count of only one half and all of the units strictly less than being assigned a value of one. If $N + M$ is large U can be approximated as follows:

$$U = NM + 0.5N(N+1) - \sum_i R(x_i)$$

where: $R(x_i)$ is the rank of x_i within the $N + M$ total units.

A *significance level,* α, is chosen and compared to the values, t, taken from the *Mann-Whitney table* for one of three *null hypothesis* cases: $\mathbf{P(X < Y)} = 1/2$ (the two sided case), $\mathbf{P(X < Y)} \leq 1/2$ (the $\mathbf{X > Y}$ one sided case) and $\mathbf{P(X < Y)} \geq 1/2$ (the $\mathbf{X < Y}$ one-sided case). If $t_{N,M,1-\alpha/2} < U < t_{N,M,\alpha/2}$ for the two-sided case, $U < t_{N,M,\alpha}$ for the $\mathbf{X > Y}$ one-sided case, and $U > t_{N,M,1-\alpha}$ for the $\mathbf{X < Y}$ one-sided case, the null hypothesis is rejected.

If N and M are larger than the values available in the tables a *normal distribution* approximation, with *mean,* $\mu = NM/2$ and a *variance,* $\sigma^2 = NM(N+M+1)/12$ provides good results.

Matched Group Design *n* A type of study or experiment design that employs *correlated samples*, where the *elements* in the two groups having different modifications of the variable under study are matched. The simplest and most often group division is the control group and the test group. Often the groups are not explicitly paired by subsets of equal numbered elements. For example, two groups each containing *M* females and *N* males. See ► Correlated Samples.

Matched Groups *n* An alternate term for *correlated samples.*

Matched Samples *n* An alternate term for *correlated samples.*

Mathematical Expectation See ► Expectation Value.

Mean *n* The primary descriptive measure of a probability distribution, population or sample. It is a measure of central tendency. The mean of a random variable is defined as the expectation value of that random variable. In statistics the mean of a sample or population is the *arithmetic mean* or simple average of the set of numbers, i.e., the sum of all numbers in the set divided by the number of numbers in the set. The arithmetic mean of the set **X**, ‹**X**›, is given by:

$$\langle \mathbf{X} \rangle \frac{\sum_i x_i}{N}$$

where: x_i are the elements of the set **X** and
N is the number of elements in **X**.
If the set is a sample of ‹**X**› is referred to as the *sample mean.* If the set is the population then ‹**X**› is often replaced by μ and referred to as the *population mean.* For a random variable, **X**, defined on a probability space, **S**, the mean is the expectation value of **X**, **E** {**X**}, if that expectation exists. For a continuous random variable, with a probability density function, $f(\mathbf{X})$, the mean (usually denoted by μ) is given by:

$$\mu = \int_S Xf(X)dX$$

and for a discrete random variable, with discrete probabilities, $\mathbf{P}_i$, the mean, μ is given by:

$$\mu = \sum_i x_i \mathbf{P}_i$$

This form of the mean is an example of a *weighted average* and is often referred to as a *weighted mean.*

Measure of Central Tendency *n* A single value that gives a measure of the "most typical" value of a variable within a set of observations. That is a single number that best describes or summarizes the variable. There are several *measures of central tendency* which define quantitative values related to the values taken on by a variable in a set of observations. These include: *mean, mode,* and *median.* A *measure of central tendency* along with a *measure of dispersion* give a simple quantitative description of shape of the distribution of values within the set of observations.

Measure of Dispersion *n* A quantitative number related to the *dispersion* of the set of values of a variable in a set of observations. See ► Dispersion.

Median *n* The value that is in the middle of an ordered set of values. For a finite set, if there is an odd number of values then the median is equal to the value of the element in the middle of the set and if there is an even number of values the median is the average of the two values in the center. The median is equal to the *second quartile* and the 50th *percentile.* The *median* is one of the *common measures of central tendency.*

Mesocurtical Distribution *n* A distribution that has the same peakedness as the normal distribution. It has a zero excess kurtosis. See ► Kurtosis.

Mesokurtic Distribution *n* A distribution that has the same peakedness as the normal distribution. It has a zero excess kurtosis. See ► Kurtosis.

Middle Fifty *n* An alternate term for *interquartile range.*

Midspread *n* An alternate term for *interquartile range.*

Mode *n* The value of the data or observation in a set that has the highest *frequency* of occurrence. For a continuous distribution, the mode is the point on the *density function* with the greatest magnitude. In some situations there may be more than one mode. This occurs when there are multiple values that have the same and highest frequency of occurrence. A set with a single mode is referred to as *unimodal* and a set with more than one mode is referred to as *multimodal.* The terms *bimodal* and *trimodal* are commonly used for sets with two and three modes respectively. The *mode* is one of the common *measures of central tendency.*

Moment *n* Formally the *moment of a probability distribution.* For a random variable, **X**, defined on a probability space, **S**, for an integer, *k* greater than or equal to 0, the kth moment about a constant, *C*, $\mathbf{M}_k\{\mathbf{X}-C\}$, are given by:

$$\mathbf{M}_k\{\mathbf{X} - C\} = \int_S (\mathbf{X} - C)k\, f(\mathbf{X})d\mathbf{X}$$

and for a discrete random variable, with discrete random variable, with discrete probabilities, $\mathbf{P}_i$, the moments about *C*, are given by:

$$\mathrm{M}_k\{\mathbf{X} - \mathbf{C}\} = \sum_i (\mathrm{x_i} - \mathbf{C})\mathbf{k}\mathbf{P}_i$$

For statistical samples, x_i, with weights, $\mathbf{W}_i$, the moments about C are given by:

$$M_k\{\mathbf{X}-C\} = \frac{\sum_i (x_i - C)\mathbf{k}\mathbf{W}_i}{\sum_i \mathbf{W}_i}$$

If the kth moment exists then all lower order moments below k also exist, i.e., if the kth moment exists then the moments from 0 through k exist. The moments about zero are most often simply referred to as *moments* or as *raw moments.*

Since the expectation value is linear, the 0th raw moment is $\mathbf{E}\{(\mathbf{X}-0)^0\}=\mathbf{E}\{1\}=1$.

The first moment about zero (the first raw moment) is $\mathbf{E}\{\mathbf{X}\}$ which is the mean, μ. The first moment about the mean is significant in that probability distributions are often characterized by their moments about their mean, μ. These moments are referred to as the *central moments.* The central moments are given by:

$$\mathbf{M}_k\{\mathbf{X}-\mu\} = \int_S (\mathbf{X}-\mu)kf(\mathbf{X})d\mathbf{X}$$

for a continuous random variable with probability density function, $f(\mathbf{X})$, by:

$$M_k\{\mathbf{X}-\mu\} = \sum_i (x_i-\mu)\mathbf{k}\mathbf{P}_i$$

for a discrete random variable, with discrete probabilities, $\mathbf{P}_i$, and by

$$\mathbf{M}_k\{\mathbf{X}-\mu\} = \frac{\sum_i (x_i-\mu)\mathbf{k}\mathbf{W}_i}{\sum_i \mathbf{W}_i}$$

for statistical samples, x_i, with weights, $\mathbf{W}_i$.

The second central moment is called the *variance* and is often denoted by σ^2 and sometimes by Var($\mathbf{X}$). The *standard deviation* is defined as the positive square root of the variance and is often denoted by σ.

There is one other often used class of moments. Since the standard deviation is related to the spread of the probability distribution, it is often used to normalize the central moments. The moments normalized in this manner are referred to as *normalized central moments, normalized moments,* or *standardized moments* and are given by:

$$\frac{\mathbf{M}_k\{\mathbf{X}-\mu\}}{\sigma^k} = \frac{\int_S (\mathbf{X}-\mu)\mathbf{k}f(\mathbf{X})d\mathbf{X}}{\sigma^k}$$

or

$$\frac{\mathbf{M}_k\{\mathbf{X}-\mu\}}{\sigma^k} = \frac{\sum_i (x_i-\mu)\mathbf{k}\mathbf{P}_i}{\sigma^k}$$

or

$$\frac{\mathbf{M}_k\{\mathbf{X}-\mu\}}{\sigma^k} = \frac{\sum_i (x_i-\mu)\mathbf{k}\mathbf{W}_i}{\sigma^k \sum_i \mathbf{W}_i}$$

The *skewness* which is defined as the normalized third central moment and is often denoted by γ or γ_1, is related to the asymmetry of the probability distribution about its mean. A probability distribution that is symmetric about its mean will have a skewness of zero.

The *kurtosis* which is related to the flatness of a distribution is ideally defined as the normalized fourth central moment and is often denoted by κ or γ_2; however, many authors will define kurtosis as the normalized fourth central moment minus 3. This is done primarily to reference the kurtosis to the normal distribution which has a kurtosis of 3. This form is also known as, and in the author's opinion, more properly known as the excess kurtosis.

Moving Average Smoothing *n* Also known as *running average smoothing,* it is a process for the *smoothing* of *time series* data or measured *density functions,* by calculating a new time series or density function whose values are equal to the *mean* of N consecutive points of the series or function located around the points in the original series. The *moving average* is said to be of order N since N points are used to calculate each mean. In the case where the means are weighted, i.e., any type of non-recursive finite linear filter, the process is referred to as *moving weighted average smoothing.*

Moving Medians Smoothing *n* An alternate term for *running medians smoothing.*

Multiple Linear Regression *n* A form of *regression* where the *dependent variable* is assumed to be a linear function of more than one *independent variable.* There are many methods for calculating the constants of the single variable linear function. *Least squares method* is the most widely used. The multiple linear regression model function has the form:

$$\mathbf{Y} = a_0 + \sum_{j=1}^{M} a_j\mathbf{X}_j$$

where: $\mathbf{Y}$ is the dependent variable,
$\mathbf{X}_j$ is the one of the M independent variables, and
a_0 and the a_js are the constants of the *best fit line* which is the *regression line.*

The regression equation given by:

$$\mathbf{Y} = a_0 + \sum_{j=1}^{M} a_j\mathbf{X}_j + err$$

where: *err* is the error term.

Multiple Regression Correlation Coefficient *n* An alternate term for *coefficient of determination.*

Multiplication Rule *n* Also called the *multiplicative law of probability* and is one of the primary rules of probability. The multiplication rule states that the probability that event 1, E_1, and event 2, E_2, both occur, $\mathbf{P}(E_1 \cap E_2)$, is given by:

$$\mathbf{P}(E_1 \cap E_2) = \mathbf{P}(E_1)\mathbf{P}(E_2|E_1)$$

where: $\mathbf{P}(E_1)$ is the probability that E_1 occurs and $\mathbf{P}(E_2|E_1)$ is the conditional probability of E_2 given E_1.
or:

$$\mathbf{P}(E_1 \cap E_2) = \mathbf{P}(E_2)\mathbf{P}(E_1|E_2)$$

where: $\mathbf{P}(E2)$ is the probability that $E2$ occurs and $\mathbf{P}(E_1|E_2)$ is the conditional probability of E_1 given E_2. If the two events 1 and 2 are mutually exclusive then the probability that both event 1 and 2 occur, $\mathbf{P}(E_1 \cap E_2)$, is zero. If the two events are independent then the conditional probability of E_1 given E_2 is merely $\mathbf{P}(E_1)$. Similarly for independent events 1 and 2, $\mathbf{P}(E_2|E_1) = \mathrm{P}(E_2)$, and the multiplication rule reduces to:

$$\mathbf{P}(E_1 \cap E_2) = \mathbf{P}(E_1)\mathbf{P}(E_2)$$

Multiplicative Law of Probability See ▶ Multiplication Rule

Mutually Exclusive Events *n* Two events, E_1 and E_2, are mutually exclusive if their intersection in the sample space, **S**, of a random experiment contains no points (is the empty set). This means that the occurrence of one precludes the occurrence of the other; therefore, the probability of their both occurring is zero, i.e., $\mathbf{P}(E_1 \cap E_2) = 0$. The conditional probabilities for the mutually exclusive events are also zero, i.e., $\mathbf{P}(E_1|E_2) = 0$ and $\mathbf{P}(E_2|E_1) = 0$. Mutually exclusive events are not necessarily complementary but complementary events are mutually exclusive.

MWW Test *n* short for the *Mann-Whitney-Wilcoxon Test,* and alternate name for the *Mann-Whitney U Test.*

N

Nominal Data *n* A type of *categorical data* in which the property of the *population* has only a finite number of more than two states. The states can be named but cannot be ordered by value; for example, species of birds present in a given area. Also see ▶ Ordinal Data.

Non-Linear Regression *n* A form of *regression* where the *dependent variable* is assumed to be a non-linear function of one or more *independent variables.* There are many methods for calculating the constants of the non-linear function. *Least squares method* is the most widely used even though it is not as straight forward as with the linear regression models.

Nonparametric Test *n* A type of *hypothesis test* which does not assume a particular form of the *probability distribution* of the subject *population.* Nonparametric tests are more useful than tests that require a certain probability to the population under study, when the characteristics of the subject population are uncertain or unknown.

Normal Distribution *n* A *probability distribution* of a *continuous random variable,* **X**, with values, *x*, that have a *density function, f(x)*, of the form:

$$f(x) = \frac{1}{\sigma\sqrt{2\pi}} e^{-(x-\mu)^2/2\sigma^2}$$

where: σ is greater than zero
The normal distribution has:
mean = μ
standard deviation = σ
variance = σ^2
skewness = 0
kurtosis = 3, and
excess kurtosis = 0.
The normal distribution is often standardized, especially when given in tabular form. The *standardized normal distribution* is the normal distribution with a zero mean and a standard deviation of one.

Null Hypothesis *n* The starting hypothesis in hypothesis testing. It is normally symbolized by H_0. Traditionally the hypothesis in hypothesis testing was stated in the form that "..." has no effect, hence the "null". The alternative or negation of the null hypothesis is referred to as the alternative hypothesis.

O

Observed Frequencies *n* In contingency table problems, the observed frequencies are the frequencies actually obtained in each cell of the table, from our random sample. When conducting a chi-squared test, the term "observed frequencies" is used to describe the actual data in the contingency table.

One-Sample T-Test *n* See ▶ T-Test.

One-Sided Test *n* Also known as a *one-tailed test*, a *hypothesis test* in which the values of the *population parameter* for which the *null hypothesis* is rejected are contained in one tail of the *probability distribution* of the parameter. See also ▶ Two-Sided Test.

One-Tailed Test *n* An alternate term for a *one-sided test.*

One-Way Analysis of Variance *n* A *hypothesis test* used to test the *null hypothesis* that the *means* of two or more *samples* are equal. The samples must be independent, the *populations* from which the samples were obtained are assumed to be *normally distributed* with equal *variances*, and there is a single factor with M levels of classification or *treatment* used for classification or *treatment* used for classification of each sample. There are thus M samples, $\mathbf{Y}_1, \mathbf{Y}_2, \ldots, \mathbf{Y}_M$, each with N_j elements, $y_{j1}, y_{j2}, \ldots, y_{jN_j}$. In a simple linear model, y_{ji} would be given by:

$$y_{ji} = \mu + \tau_j + \varepsilon_{ji}$$

where: μ is the *general* or *grand* mean over all of the elements in all samples,

τ_j is the deviation of the jth sample's mean from the grand mean, and

ε_{ji} is the error associated with y_{ji}.

The null hypothesis that all of the sample means are equal becomes that all of the τ_js are equal. The null hypothesis is tested using the between group variation and the within group variation and the *Fisher test.*

The between group variation, S_B^2, is the weighted sum of the squares of the differences between the individual sample means, μ_j, and the grand mean, μ, divided by the degrees of freedom associated with the number of samples, $M-1$. This is given by:

$$S_B^2 = \frac{\sum_j N_j(\mu_j - \mu)^2}{M-1}$$

The within group variation, S_E^2, is the sum of the squares over all elements, y_{ji}, minus their respective means, μ_j, divided by the degrees of freedom associated with the total number of elements, $N-M$, where N is equal to the sum of the N_js for j=1,2, . . ., M. This within group variation is given by:

$$S_E^2 = \frac{\sum\limits_j \sum\limits_{i=1}^{N_j} (y_{ji} - \mu_j)^2}{N-M}$$

The *Fisher statistic, F*, is given by the ratio of S_B^2 to S_E^2 which gives:

$$F = \frac{(N-M)\sum\limits_j N_j(\mu_j\mu)^2}{(M-1)\sum\limits_j \sum\limits_{i=1}^{N_j} (y_{ji} - \mu_j)^2}$$

A *significance level*, α, is chosen and F is compared to the values, $f_{M-1,N-M,\alpha}$, taken from the *Fisher table* for use in testing the *null hypothesis.* If $F \geq f_{M-1,N-M,\alpha}$, the null hypothesis is rejected and the differences are most likely due to the treatments or differences in classes of the samples.

Ordinal Data *n* A type of *categorical data* in which the property of the *population* has more than two states and the states can be ordered by value; for example, ages in years of the people in Spain. See ▶ Nominal Data.

Outcome *n* In statistics, the result of a *random experiment.* A set of outcomes is referred to as an *event.*

Outlier *n* An observation or data set which far separated from the rest of the observations. It may be much smaller or much larger than the rest of the observations. Often times it is due to measurement errors. If a data set's values are due to errors it can skew or *bias* the results, in this case a determination may be made to discard that particular measurement. In general, outliers should not be discarded simply because they are outliers.

P

P-P Plot *n* An alternate term for probability to probability plot.

P-Value *n* The *P-value* in *hypothesis testing* is the probability, assuming that the *null hypothesis* is true, of obtaining an outcome as extreme or more extreme through chance alone as the observed outcome.

The *P-value* is used in conjunction with the *significance level* in statistical tests, e.g., *chi-square tests*, to determine the validity of the hypothesis.

Paired Sample T-Test *n* See ▶ T-Test.

Parameter *n* A parameter is a characteristic or property of a population. Often described in a more specific sense as a numerical property of a population. Usually a characteristic of a sample of a population is referred to as a statistic of the sample.

Pearson Correlation Coefficient *n* A short name for *Pearson's product-moment correlation coefficient.*

Pearson's Product-Moment Correlation Coefficient *n* Also known as the *Pearson Correlation Coefficient, Pearson's r*, or *PMCC* is the most widely used measure of the linear correlation between two variables. For a *sample* of pairs, (x_i, y_i), of observations or measurements of two variables, **X** and **Y**, Pearson's product-moment correlation coefficient is equal to the un-weighted *covariance* of the variable pairs divided by the product of the individual variable sample *standard deviations*. For N pairs, (x_i, y_i), with individual *sample means*, μ_x and μ_y, Pearson's product-moment correlation coefficient, r, is given by:

$$r = \frac{\sum_{i=1}^{N}(x_i - \mu_x)(y_i - \mu_y)}{\sqrt{\sum_{i=1}^{N}(x_i - \mu_x)^2}\sqrt{\sum_{i=1}^{N}(y_i - \mu_y)^2}}$$

The value of r is greater than or equal to -1 and less than or equal to 1. If **Y** is an exact linear function of **X** then r will be exactly equal to one if the linear term is positive and exactly equal to minus one if the linear term is negative. A correlation coefficient value near zero indicates that there is little if any linear correlation.

Percentile *n* Also known as *centile*, is an element in the set of location values that divide the set of observations (data sample) into one hundred groups containing equal numbers of observations. The nth percentile is the value that would divide the sample so that the n percent of the observations lie below the value and $(100-n)$ percent of the values lie above the value.

Pie Chart *n* A type of quantitative graphical display used for displaying and summarizing *categorical data* in the form of a circle divided by radii into sectors whose included angle represents the proportion of the data in that category.

Placebo *n* An inert object, process, or procedure which has no effect on the outcome of an experiment. It is often used in studies in an attempt to remove the subject's or observer's *bias* as to the expected outcome of the experiment. Studies involving *blinding* often involve the use of placebos, which are given to the control group which is not receiving the procedure under test, so that individuals involved in the study, both subjects and observers, are not aware of which subjects are in which group.

Platicurtical Distribution *n* A distribution that is flatter or more spread out toward the edges than the normal distribution. It has a positive excess kurtosis. See ▶ Kurtosis.

Platykurtic Distribution *n* A distribution that is flatter or more spread out toward the edges than the normal distribution. It has a positive excess kurtosis. See ▶ Kurtosis.

Poisson Distribution *n* A *probability distribution* of a *discrete random variable*, **X**, with non-negative integer values, x, that have a *probability*, $\mathbf{P}(x)$, of the form:

$$\mathbf{P}(x) = \frac{\lambda^x e^{-\lambda}}{x!}$$

where: λ is a real number greater than zero and x is a non-negative integer $(0, 1, 2, \ldots)$

The Poisson distribution has:

mean $= \lambda$,

standard deviation $= \lambda^{1/2}$,

variance $= \lambda$

skewness $= \lambda^{-1/2}$

kurtosis $= 3 + \lambda^{-1}$, and

excess kurtosis $= \lambda^{-1}$.

The Poisson distribution is often used to describe the number of occurrences of an event in a fixed period of time, if the events meet certain criteria: the events occur with a constant average rate, R, that is constant over time and the occurrence of any event is independent of the occurrence of any other event. In this case, λ would be the expected number of occurrences in the time interval, Δt, that is $\lambda = R\,\Delta t$. x is the number of occurrences during a time period, Δt.

In the limit as λ gets larger and goes toward infinity, the Poisson distribution goes to the *normal distribution*, with a *mean* and *variance* equal to λ.

Population *n* A collection of units or statistical units being studied or the set of all measurements of interest to a study. Each individual unit is referred to as an element of the population. Any subset of the units in a population is referred to as a sample.

Power *n* Also known as the *power of the test*, it is the probability of rejecting the *null hypothesis* when it is false. In *hypothesis testing* this is equal to one minus the probability of a *type II error*, which is often represented by β. The *power* is, therefore, equal to $1-\beta$. See also ▶ Type I Error.

Power of the Test *n* See ▶ Power.

Precision *n* The degree to which future estimates will show the same value for a *statistic* or the degree to which the statistic for the current *sample* agrees with the actual *parameter* of the *population*. *Precision* is often given in terms of the *standard error*.

Probability *n* A *probability distribution* is a function, **P**, that assigns a value to each event in a sample space, **S**, of a random experiment, which obeys the following three laws:

1. **P**(E_i) is greater than or equal to zero, for all events E_i in **S**,
2. **P**(**S**) = 1, and
3. **P** (union of any two events, E_j and E_k) equals **P**(E_j) + **P**(E_k), whenever the intersection of the two events is the empty set, i.e., $E_j \cap E_k = \emptyset$.

If **P** is a probability distribution over a sample space and E is an event in that sample space, then **P**(E) is the *probability* of E in **S**. The three laws shown above are known as the axioms of probability.

The probability provides a quantitative description of the likely occurrence of a particular event and of the relative frequency of outcomes.

Probability Density Function *n* An alternate term for the *density function.*

Probability Distribution *n* See ▶ Probability.

Probability to Probability Plot *n* Also known as a *P-P Plot,* a plot of the measured *statistic* samples to the theoretical or model *distribution function* for the experiment, that is used to determine whether the measured data fits the model or theoretical distribution. The plot is constructed by ordering the N statistic sample values, x_k, from the least value to the maximum value as, x_1, x_2, ... x_N. Then $(k-1\backslash 2)/N$ is plotted against the theoretical or model distribution function, $F(x_k)$ for all of the values in the set. The closer the plot is to linear the better the model fits the data. See also ▶ Quantile to Quantile Plot.

Q

Q-Q Plot *n* An alternate term for *quantile to quantile plot.*

Quantile *n* An element in the set of location values that divide the set of observations, data sample, population, or probability distribution into groups containing equal numbers of observations. In the case of discrete sets, equal means as equal as possible. There are several common types of *quantiles* based on the number of groups: *quartile* = four groups, *quintile* = *five groups, decile* = ten groups, *percentile* or *centile* = one hundred groups.

Quantile to Quantile Plot *n* Also known as *Q-Q plot,* it is a plot of the measured *statistic* samples to the theoretical or model *distribution function* for the experiment, that is used to determine whether the measured data fits the model or theoretical distribution. The plot is constructed by ordering the N statistic sample values, x_k, from the least value to the maximum value as, x_1, x_2, ... x_N. Then x_k is plotted against the theoretical or model calculated values of x for $(k-1/2)/N$ which are calculated from the model distribution function inverse, F^{-1}, as $F^{-1}(k-1/2/N)$, for all of the values of k in the set. The closer the plot is to linear the better the model fits the data. See also ▶ Probability to Probability Plot.

Quartile *n* A location value of a sample data, set of observations, or probability distribution, that locates the division of the sample, etc., into four parts each containing equal numbers of observations. In the case of discrete data the equal numbers means as nearly equal as possible. There are three division values which are termed:

first quartile or *lower quartile,* where 25% of the observations lie before or below the value and 75% lie after or above the value.

second quartile or *median,* where 50% of the observations lie before or below the value and 50% lie after or above the value.

third quartile or *upper quartile,* where 75% of the observations lie before or below the value and 25% lie after or above the value.

Quintile *n* An element in the set of location values that divides the set of observations (data sample) into five groups containing equal numbers of observations. A *quintile* is a common type of *quantile.*

Quota Sampling *n* A sampling technique in which a fixed number of *elements* are chosen from groups of the *population.* The groups are chosen based on some characteristic of the population usually in an attempt to represent the population as closely as possible. The *samples* from each group are not necessarily chosen randomly and are often chosen for convenience, which differentiates *quota sampling* from *stratified sampling.*

R

Random Experiment *n* An observed phenomenon or experiment in which the *outcome* is not predictable in advance. More formally a random experiment is an experiment which follows three rules:

1. The sample space is a set (or list) in which all the possible outcomes can be specified.

2. The experiment can be repeated or repeatedly sampled.
3. The various outcomes of the experiment may not always have the same result.

As a general rule, a probability can be assigned to each event in the sample space of the random experiment.

Random Sampling *n* A process or technique for choosing a *sample*, where each *element* is chosen by chance. This does not mean that all elements are equally likely to be chosen; however, this is generally the desire. The case in which each is equally likely to be chosen throughout the sampling is referred to as *simple random sampling.*

Random Variable *n* Formally a function defined on a sample space or variable determined by the outcome of a random experiment. See also ► Continuous Random Variable and ► Discrete Random Variable.

Randomization *n* In statistics, the process or procedure for randomly assigning *treatments* to *experimental units* in an *experiment.* The main purpose of using *randomization* is to remove or reduce *bias* in the experiment.

Randomized Block Design *n* A type of *experimental design* in which the *experimental units* are divided into *blocks* and the *treatments* are randomly assigned to each block's units. The blocks are generally chosen so that the units within each block are homogenous with respect to some set of *parameters.*

If the treatments are required to appear at least once in each group, then the design is referred to as a *completely randomized block design.* This should not be confused with the more formal term *randomized complete block design* which each block has the same number of units as treatments and the treatments are randomly assigned within each block so that each treatment appears once in each block.

Randomized Complete Block Design *n* A type of *randomized block design* in which each *block* has the same number of *units* as the number of *treatments*, each treatment occurs exactly once within each block, and the treatments are assigned randomly within each block. As usual with randomized block design, the groups are generally chosen to be as homogenous as possible with respect to some set of *parameters.*

There is an alternate form in which each block contains only one unit which receives all of the treatments but in random orders.

There is a third alternative form in which each block is an exact replication of the experiment.

Range *n* The difference between the highest and lowest value of a variable or quantitative characteristic in a data sample or set of observations. It is one of the simplest *measures of dispersion* to obtain; however, it is strongly affected by *outliers.*

Rectangular Distribution *n* An alternate name for the *uniform distribution.*

Regression *n* Also known as *regression analysis*, it is the process of estimating the relationship between two or more variables by assuming that one variable, the *dependent variable*, is a function of the other variables, the *independent variables.* This analysis uses a measure of total *residual* error to calculate the function's constants to give the *best estimate* of the dependent variable over a set of observations or measurements. The *least squares method* is one of the most common methods for determining the function's constants.

If there is only one independent variable, the analysis is referred to as *simple regression.* If the model function for dependent variable includes more than one independent variable, the analysis is referred to as multiple regression.

If the regression assumes a linear relationship then it is referred to as *simple linear regression* if there is one independent variable or, *multiple linear regression* if there is more than one independent variable. If the regression model has a non-linear function then the regression is referred to as *non-linear regression.*

Regression Correlation Coefficient *n* An alternate term for *coefficient of determination.*

Regression Equation *n* The equation developed by the *regression analysis* for a set of measurements or observations from the regression model equation and the *residual* errors. For a regression model for a dependent variable $\mathbf{Y}$ as a function, f, of N independent variables, $\mathbf{X}_j$, for $j = 1, 2, \ldots, N$, and M functional constants, $a_1, a_2, \ldots, a_M$, it has the form:

$$\mathbf{Y} = f(\mathbf{X}_1, \mathbf{X}_2, \ldots, \mathbf{X}_N, a_1, a_2, \ldots, a_M) + err$$

where: err is the residual error.

If the regression function is linear then the regression equation is the *regression line* plus the residual error.

Regression Line *n* The line given by the equations developed by the *regression analysis* for either *simple linear regression* or *multiple linear regression.* This line represents the *best fit* line which minimizes the *residual* for a data set.

Rejection Region *n* An alternate term for the *critical region.*

Relative Frequency of Outcomes *n* The relative frequency of outcomes, F_E, is the number of outcomes,

n_E, falling in an event, E, divided by the total number of outcomes in the random experiment, n_s:

$$F_E = \frac{n_E}{n_s}$$

The limit of F_E as n_s goes to infinity is equal to the probability, $\mathbf{P}(E)$, of event, E.

Residual *n* The difference between the estimated value and the actual observed value for a variable. It is generally used to mean the difference or errors in a *regression* model, which is the difference between the observed value and the value calculated from the *regression equation.*

Residual Component *n* The component of the values of a *time series* that are due to *residual movement.* This is the component that is remaining when the other components of the values have been removed. The other components are due to *trends, seasonal variation,* and *cyclical fluctuation.* Techniques such as *smoothing* are often applied in order to remove the residual component that may obscure other features, from a time series.

Residual Movement *n* Also known as *irregular variations,* it is an erratic or irregular movement or fluctuations in a *time series,* that is not identifiable long term movement or cyclic movements associated with *trend, seasonal variation,* or *cyclical fluctuation.* The component of the values of the time series associated with the residual movement is referred to as the *residual component.* This component often appears as random noise.

Rules of Probability *n* The rules of probability traditionally include the *addition rule* which states that the probability that event 1, E_1, occurs, event 2, E_2, occurs, or both events 1 and 2 occur is given by:

$$\mathbf{P}(E_1 \cup E_2) = \mathbf{P}(E_1) + \mathbf{P}(E_2) - \mathbf{P}(E_1 \cap E_2)$$

where: $\mathbf{P}(E_1)$ is the probability that E_1 occurs,
$\mathbf{P}(E_2)$ is the probability that E_2 occurs, and
$\mathbf{P}(E_1 \cap E_2)$ is the probability that both E_1 and E_2 occur.
and the *multiplication rule* which states that the probability that event 1, E_1, and event 2, E_2, both occur, $\mathbf{P}(E_1 \cap E_2)$, is given by:

$$\mathbf{P}(E_1 \cap E_2) = \mathbf{P}(E_1)\mathbf{P}(E_2|E_1)$$

where: $\mathbf{P}(E_1)$ is the probability that E_1 occurs, and
$\mathbf{P}(E_2|E_1)$ is the conditional probability of E_2 given E_1.
or:

$$\mathbf{P}(E_1 \cap E_2) = \mathbf{P}(E_2)\mathbf{P}(E_1|E_2)$$

where: $\mathbf{P}(E_2)$ is the probability that E_2 occurs, and
$\mathbf{P}(E_1|E_2)$ is the conditional probability of E_1 given E_2.

The traditional rules of probability are sometimes expanded to include the two additional rules; the probability of an impossible event is zero and the probability of the complement of an event is one minus the probability of that event or $\mathbf{P}(\sim E) = 1 - \mathbf{P}(E)$.

Running Average Smoothing *n* An alternate term for *moving average smoothing.*

Running Medians Smoothing *n* Also known as *moving medians smoothing,* it is a process for the *smoothing* of *time series* data or measured *density functions,* by calculating a new time series or density function whose values are equal to the *median* of *N* consecutive points of the series or function located around the points in the original series. The *running median* is said to be of order *N* since *N* points are used to calculate each median.

Runs Test *n* Also known as the *Wald-Wolfowitz* test is a *nonparametric test* used to test for randomness in a *sequence.* The test sequence must have only two classifications of values, such as hot or cold. The test uses the *probability* of the number of runs of consecutive same elements within the sequence. If the probability of getting the exact number of runs that occurred in the sequence is either too high or too low the *null hypothesis,* that the events are random, is rejected.
The probability of getting exactly k runs, $p\ (k, N, M)$, where N is the number of elements of the first classification in the sequence and M is the number of elements of the second classification, is given by:

$$p(k, N, M) = \frac{N!(N-1)!M!(M-1)!}{(N+M)(i-1)!(j-1)!}\left[\frac{1}{(N-i)!(M-j)!} + \frac{1}{(N-j)!(M-i)!}\right]$$

where: $i = j = k/2$ for k even and
$i = (k+1)/2$ and $j = (k-1)/2$ for k odd.
For a test with r number of runs, a *significance level,* α, is chosen and compared to the values of the probabilities of k less than or equal to r, $\mathbf{P}(k \le r, N, M)$, and the probability of k greater than or equal to r, $\mathbf{P}(k \ge r, N, M)$. The two-tailed test is used, thus if $\mathbf{P}(k \le r, N, M) \le \alpha/2$ or $\mathbf{P}(k \ge r, N, M) \le \alpha/2$ then the *null hypothesis* is rejected. One-tailed tests would compare either of the probabilities to α to test if the number of runs was too small (which is the most common usage) or too large.
If N and M are larger (usually greater than 10), a *normal distribution* approximation, with *mean,* $\mu = 1+2NM/(N+M)$ and *variance,* $\sigma^2 = 2NM(2NM-N-M)/[(N+M)^2(N+M-1)]$ provides good results.

S

Sample *n* Any subset of the units in a population. A sample is usually a subset chosen for statistical analysis in order to make predictions or to draw conclusions about the whole population.

Sample Mean *n* The *arithmetic mean* of the *elements* in the *sample*. For a sample, **X**, consisting of $x_1, x_2, \ldots x_N$, the *sample mean* would be given by:

$$\langle X \rangle \frac{\sum_i x_i}{N}$$

Sample Space *n* The sample space is a set (or list) of all the possible outcomes of a random experiment. The set of all possible samples of an experiment is the sample space for the experiment. The set is referred to as *exhaustive* since it represents all possible outcomes of the experiment. The sample space is usually denoted by **S** or Ω.

Sample Standard Deviation *n* The positive square root of the *sample variance*. This is equivalent to the *standard deviation* of a *sample* of the *population*.

Sample Variance *n* The uniformly weighted *variance* of the *elements* in the *sample* with respect to the *sample mean*. For a sample, **X**, consisting of $x_1, x_2, \ldots x_N$, the *sample mean* would be given by:

$$\langle X \rangle \frac{\sum_i x_i}{N}$$

The *sample standard deviation* is the positive square root of the sample variance.

Sampling Distribution *n* A probability distribution or density function of a sample statistic.

Sampling Variability *n* The set of differences for a *sample statistic* that are obtained from different samples chosen from the same *population*.

Scatter Plot *n* A graphical representation of the relationship between two different variables associated with a set of *categorical data*, where pairs of observations of the two variables are placed as points on a two axis plot with the position of the point in two dimensions is determined by the values of the variables in each pair of observations, with the value of the first variable determining the distance along the horizontal axis and the value of the second variable determining the distance along the second axis. The primary function of the scatter plot is to illustrate the relation or lack thereof between the two variables. The tighter the grouping of the points along a curve, the greater the indication of a relationship between the two variables. A straight line would indicate a linear relationship.

Seasonal Component *n* The component of the values of a *time series* that are due to *seasonal variation.*

Seasonal Variation *n* An identifiable (generally periodic) movement in a *time series* that is associated with seasons of the year, etc. Climate and cultural customs are often associated with these movements. The component of the values of the time series associated with the seasonal variations is referred to as the *seasonal component.* Other periodic movement in a time series is referred to as *cyclical fluctuation.*

Secular Trend *n* An alternate term for *trend.*

Sequence *n* An ordered set of elements where the first element can be identified and the element following each element except for the last element can be identified. This definition would allow one to write down the elements of the set in the order of their occurrence and implies that this can be done in only one way.

Sign Test *n* A *nonparametric test* used to test whether there is no difference between the distributions of two *random variables*. It mathematically tests the *hypothesis* that the *probability* that the *element* from the first distribution is greater than the element from the second distribution in a pair of random observations or measurements is one half. The test is performed by measuring *N* sets of pairs, (x_i, y_i), from the random variables, **X**, and **Y**. The pairs where the values are equal are eliminated leaving *M* pairs. Of the *M* pairs there are *T* pairs where, $y_i - x_i$ is positive (or greater than zero). If the hypothesis in this situation is true, the distribution of *T* is *binomial*, $\mathbf{P}(T,M,0.5)$ with probability per pair of 0.5. A *significance level*, α, is chosen and compared to the values, *t*, taken from the binomial distribution for one of three *null hypothesis* cases: $\mathbf{P}(\mathbf{X} < \mathbf{Y}) = \mathbf{P}(\mathbf{X} > \mathbf{Y})$ (the two sided case), $\mathbf{P}(X < Y) \leq \mathbf{P}(X > Y)$ (the $\mathbf{X} > \mathbf{Y}$ one-sided case), and $\mathbf{P}(\mathbf{X} < \mathbf{Y}) \geq \mathbf{P}(\mathbf{X} > \mathbf{Y})$(the $\mathbf{X} < \mathbf{Y}$ one-sided case). If $t_{\alpha/2} < T < M - t_{\alpha/2}$ for the two-sided case, $T < M - t_\alpha$for the $\mathbf{X} > \mathbf{Y}$ one-sided case, and $T > t_\alpha$for the $X < \mathbf{Y}$ one-sided case, the null hypothesis is accepted.

If *M* is larger (usually greater than 20), a *normal distribution* approximation, with *mean*, $\mu = M/2$ and *variance*, $\sigma^2 = M/4$ provides good results.

Significance Level *n* Also known as the *level of significance, alpha level*, or *the probability of type I error* and is often represented by α. The significance level is the probability of rejecting the null hypothesis when it is true. This is equivalent to the probability of a type I error.

Simple Hypothesis *n* A hypothesis that completely describes the distribution of the *outcomes* of a *random experiment.* If the outcomes involve more than one variable, then the distributions of all outcomes must be exactly described. If the hypothesis only partially describes and does not completely describe the distribution, it is referred to as a *composite hypothesis.*

Simple Linear Regression *n* A form of *regression* where the *dependent variable* is assumed to be a linear function of only one *independent variable.* There are many methods for calculating the two constants of the single variable linear function. *Least squares method* is the most widely used. The simple linear regression model function has the form:

$$\mathbf{Y} = a_0 + a_1\mathbf{X}$$

where: **Y** is the dependent variable,
X is the single independent variable, and
a_0 and a_1 are the constants of the *best fit line* which is the *regression line.*
The regression equation given by:

$$\mathbf{Y} = a_0 + a_1\mathbf{X} + err$$

where: *err* is the error term.

Simple Random Sampling *n* A *random sampling* technique in which any *sample* of a given number of *elements* has the same probability of being chosen from a fixed *population.* This implies that all elements are equally likely to be chosen at any time during the sampling.

Single-Blind Study *n* A *single-blind study* is a type of study or experiment which attempts to eliminate *bias* induced by human observers or subjects involved with the experiment, by allowing only the observers or the subjects, but not both, to be aware of which group is associated with which situation. A study where neither the observers or subjects are allowed to know the group associations is referred to as a *double-blind study.* The double-blind study is usually preferred over the single-blind study; however, there are many situations, due to the logistics of the study, in which the observers or researchers or the subjects must know which individuals are in each group, making only a single-blind study possible. See ▶ Blinding.

Skewness *n* The most often used descriptive measure of the deviation from symmetry of a probability distribution, population or sample. It is the 3rd normalized central moment and is often denoted as γ or γ_1. The precise mathematical definition of skewness for a random variable, **X**, defined on a probability space, **S**, is given by:

$$\gamma = \frac{\int_{\mathbf{S}} (\mathbf{X} - \mu)^3 f(\mathbf{X})\, d\mathbf{X}}{\sigma^3}$$

for a continuous random variable with probability density function, $f(\mathbf{X})$, mean, μ, and standard deviation, σ; by:

$$\gamma = \frac{\sum_i (x_i - \mu)^3 \mathbf{P}_i}{\sigma^3}$$

for a discrete random variable, with discrete values, x_i, and associated probabilities, $\mathbf{P}_i$, and by:

$$\gamma = \frac{\sum_i (x_i - \mu)^3 \mathbf{W}_i}{\sigma^3 \sum_i \mathbf{W}_i}$$

for statistical samples, x_i, with weights, $\mathbf{W}_i$.
For a symmetric distribution the skewness is 0. If the tail of the distribution extends more to the left the skewness is negative and if the tail extends more to the right the skewness is positive.

Smoothing *n* The process of removing random fluctuations or irregularities from *time series* data or *density functions.* The purpose is generally to remove these features that may mask the parameters of interest in the data. Smoothing techniques make no strong assumptions about the parameters of the data, they meekly assume that the major features should be smooth, especially in the case of large numbers. To this extent smoothing techniques attempt to remove the random noise and irregular variations or in the case of *time series* data the *residual movement* and leave unaffected or weakly affected the features due to actual parameters of the data. The primary technique for the smoothing of time series data is to use some type of lowpass filtering technique. The most common type of lowpass filter smoothing is *moving average smoothing* or a *weighted average* version of it.

Smoothing Factor *n* The factor greater than zero and less than one that is used in *exponential smoothing.*

Spatial Sampling *n* The sampling of *spatial data,* that is data that is distributed over two or more dimensions. For example, the number of people per square mile in different states within the United States.

Spearman Rank Correlation Coefficient *n* Also known as *Spearman's rho* or *SRCC* and usually designated by r_s or ρ, is a widely used nonparametric measure of the correlation between two variables. It is generally used when the values of the variables are ambiguous or hard to obtain. For a *sample* of pairs, (x_i, y_i), of observations or measurements of two variables, **X** and **Y**, the Spearman rank correlation coefficient is equivalent to *Pearson's product-moment correlation coefficient* with the values of the pairs (x_i, y_i) replaced by each value's ranking within its own set. That is the values of x_i are ordered from 1 to N, where N is the number of pairs. Each x_i is replaced by its number (rank) in the order. If multiple x_is have the same value they all are given the average of the ranks that they would have had if their values varied slightly. The Spearman rank correlation coefficient is thus equal to the un-weighted *covariance* of the variable rank pairs divided by the product of the individual variable sample rank *standard deviations.* For N rank pairs, (x_i, y_i), with individual rank *sample means*, μ_x and μ_y, the Spearman rank correlation coefficient, r_s, is given by:

$$r_s = \frac{\sum_{i=1}^{N}(x_i - \mu_x)(y_i - \mu_y)}{\sqrt{\sum_{i=1}^{N}(x_i - \mu_x)^2}\sqrt{\sum_{i=1}^{N}(y_i - \mu_y)^2}}$$

where: x_i and y_i are the ranks of the elements of the ith pair.

If there are no rank ties then the expression for r_s can be reduced to:

$$r_s = 1 - \frac{6\sum_{i=1}^{N} d_i^2}{N(N^2 - 1)}$$

where: $d_i = x_i - y_i$, the difference in the ith pair's elements' ranks.

The value of r_s is greater than or equal to -1 and less than or equal to 1. If **Y** ranks are an exact linear function of **X** ranks then r_s will be exactly equal to 1 if the linear term is positive and exactly equal to -1 if the linear term is negative. A correlation coefficient value near zero indicates that there is little if any linear correlation between the ranks of **X** and **Y.**

Standard Deviation *n* A measure of the spread or dispersion of a random variable, probability distribution, population, or sample from the mean. It is defined as the positive square root of the variance and is one of the most used descriptive measures of a random variable, probability distribution, population or sample. The variance of a random variable is defined as the second central moment of the random variable and is often denoted by σ^2; where σ is the standard deviation. The precise mathematical definition of the standard deviation for a random variable, **X**, defined on a probability space, **S**, is given by:

$$\sigma = \sqrt{\int_S (X - \mu)^2 f(X)\,dX}$$

for a continuous random variable with probability density function, $f(\mathbf{X})$ and mean, μ, by:

$$\sigma = \sqrt{\sum_i (x_i - \mu)^2 \mathbf{P}_i}$$

for a discrete random variable, with discrete values, x_i, and associated probabilities, $\mathbf{P}_i$, and by:

$$\sigma = \sqrt{\frac{\sum_i (x_i - \mu)^2 \mathbf{W}_i}{\sum_i \mathbf{W}_i}}$$

for statistical samples, x_i, with weights, $\mathbf{W}_i$.

Standard Error *n* Also known as the *standard error of the estimate* or *standard error of the mean* and is the *standard deviation* of the *statistic* of the *sample* with respect to the *population.* This means that if one were to calculate the statistic for all possible samples of a given size over the entire population, then the standard deviation of this set of values would be the *standard error.* The *standard error,* σ_S, for a sample, S, of a finite population is given by:

$$\sigma_S = \frac{\sigma_P}{\sqrt{N_S}}\sqrt{\frac{N_P - N_S}{N_P}}$$

where: σ_P is the standard deviation of the *parameter* for the entire population,

N_P is the number of *elements* in the population, and

N_S is the number of elements in the sample, S.

When the population has an infinite number of elements, the *standard error* becomes:

$$\sigma_S = \frac{\sigma_P}{\sqrt{N_S}}$$

Standardized Normal Distribution *n* A *normal distribution* which has a zero *mean* and a *standard deviation* equal to one. See also ▶ Normal Distribution.

Statistic *n* A numerical parameter or characteristic of a sample of a population is referred to as a statistic of the sample.

Statistical Hypothesis *n* A statistical hypothesis is a statement about a population or a parameter of a population. The hypothesis may involve the distribution of one or more random variables and their distributions. There is no implication as to whether the statement is true or false. Hypothesis testing is employed to examine a hypothesis in order to verify it.

Statistical Inference *n* The process of making estimates of or drawing conclusions about a *population* based on knowledge about a *sample* associated with the population.

Statistically Independent Events See ► Independent Events.

Stem-and-Leaf Diagram *n* An alternate name for *stemplot.*

Stemplot *n* Also known as a *stem-and-leaf diagram, stem-and leaf plot,* and *stem-and-leaf graph,* it is a semi-graphical representation of the *frequency* distribution of *categorical data* in a vertical tabular representation of a *histogram,* where there are two columns of data separated by a line or bar; the left column (stems) represents the category range and the right column (leaves) represents the *elements* within the range, with each element listed in an abbreviated form, usually by a single number. In this way the number of leaves shows the frequency in the range with each leaf showing the value of a single element in the range. Because each element is represented, stemplots are used to represent only small sets of data. An example stemplot is shown below:

1|04
2|114
3|02446
4|00147899
5|012389
6|4789
7|33
8|8
9|12

Key: R|d represents R.d; i.e., 2|7 is 2.7

This stemplot represents the set {1.0, 1.4, 2.1, 2.1, 2.4, 3.0, 3.2, 3.4, 3.4, 3.6, 4.0, 4.0, 4.1, 4.4, 4.7, 4.8, 4.9, 4.9, 5.0, 5.1, 5.2, 5.3, 5.8, 5.9, 6.4, 6.7, 6.8, 6.9, 7.3, 7.3, 8.8, 9.1, 9.2).

As can be seen, the number of elements in each range is indicated by the number of digits on the right of the bar and the value of each element can be derived from each digit's value.

Stepwise Regression *n* A type of *regression* where the regression is calculated multiple times and in which a set of *predictor variables* are added or removed from the regression model as individuals or groups at each step or stage where the regression analysis is run and evaluated by some measure such as the *coefficient of determination* or the *t-test* in order to determine which variables are important to the model and what effect the variables have on the model.

Stratified Sampling *n* A *random sampling* technique in which the entire *population* is divided into non-intersecting subsets of *elements* based on some feature of the population (such as age, etc.,) and each group or *strata* is *randomly sampled.* The sample sizes are often but not always proportional to the strata size. This sampling method is often used when the population is extremely heterogenous as an attempt to insure that the total sampling is representative of the population. This sampling method is also used when it is desired to see the effect of each strata on the *statistic* of interest or when it is expected that the value of the statistic of interest varies among the strata.

Student's T-Distribution *n* An alternate name for the *t-distribution.*

Student's T-Test *n* An alternate name for the *t-test.*

Subjective Probability *n* A *probability* based on the judgment of an individual or a group of individuals. Subjective probability can be based on experience, judgment, based on a set of values without using quantitative statistical inference, etc.

Symmetry *n* The property of a set, a *probability distribution,* or a *density function* of being reflectively equal about its *median* or center value. This means that if the lower half of the set, probability distribution, or density function is known, the upper half can be constructed by reflecting the lower half about the median and similarly if the upper half is known the lower half can be constructed by reflection about the median.
A symmetric distribution has a *skewness* equal to zero.

T

T-Distribution *n* Also known as the *Student's T-Distribution,* it is a probability distribution that describes the variation of the *sample mean,* μ_X, of a *random variable,* **Y**, of a number of randomly chosen samples of size, *n*, about the *mean,* μ, of the *population,* when the probability distribution of the random variable is the *normal distribution* and the *sample standard deviation, s,* is independent from the sample mean. More formally the t-distribution is the distribution of the random variable, **T**, where its values, *t*, are given by:

$$t = \frac{\mu_X - \mu}{sn^{-1/2}}$$

The *density function*, $f(t)$, of the t-distribution is given by:

$$f(t) = \frac{\Gamma(n/2)[1 + t^2/(n-1)]^{-n/2}}{(n-1)^{1/2}(\pi)^{1/2}\Gamma[(n-1)/2]}$$

where: Γ () is the gamma function.
$n-1$ is the number of *degrees of freedom* of the t-distribution.
The t-distribution has:
mean = 0 for $n > 2$,
standard deviation = $(n-1)^{1/2}/(n-3)^{1/2}$ for $n > 3$,
variance = $(n-1)/(n-3)$ for $n > 3$,
skewness = 0 for $n > 4$,
kurtosis = $(3n-9)/(n-5)$ for $n > 5$, and
excess kurtosis = $6/(n-5)$ for $n > 5$.

T-Test *n* Also known as the *Student's T-Test* is a *hypothesis test* as to the validity of the *sample* that was taken from a *population* with *mean*, μ_0 of a *normally distributed random variable*, **X**, (the *one-sample t-test*), the equivalence of the means of two samples (the *two-sample t-test*), or the equivalence of the means of the same measurement made under different conditions on the same sample, where each unit is measured under both conditions (the *paired sample t-test*).
For the *one-sample t-test* the *null hypothesis* is that the mean, μ, of **X** for the population from which the sample was selected is equal to μ_0. The test is performed by choosing a *significance level*, α, and calculating using the *t-distribution* the *critical values* on both tails needed to give a *critical region* whose volume is equal to α. This is the *two-sided test* version of the one-sample t-test. The use of slightly different null hypotheses of $\mu > \mu_0$ or $\mu < \mu_0$ leads to *one-sided test* versions of the one-sample t-test.
For the *two-sample t-test* the null hypothesis is that the mean of the population from which first sample was selected, μ_A, and the mean of the population from which the second sample was selected, μ_B, are equal. The test is performed by selecting a significance level, α, and calculating using the t-distribution random variable, **T**, with values, *t*, given by (assuming the two samples are independent and the two population distributions have the same variance):

$$t = \frac{\mu_{XA} - \mu_{XB}}{\left[\frac{(n_A-1)s_A^2 + (n_B-1)s_B^2}{n_A+n_B-2}\right]^{1/2}\left[\frac{1}{n_A} + \frac{1}{n_B}\right]^{1/2}}$$

where: μ_{XA} and μ_{XB} are the *sample means of the two samples A* and *B*,
n_A and n_B are the sample sizes, and
s_A and s_B are the sample standard deviations.
Note: n in the t-distribution is replaced with $(n_A + n_B - 1)$ in the t-distribution *density function* as the number of *degrees of freedom* is (n_A+n_B-2); since, the two means are not known. This is the two-sided test version. The one-sided test versions would have null hypotheses of $\mu_A > \mu_B$ and $\mu_A < \mu_B$.

Target Population *n* The *population* that is the desired focus of the experiment or study. It is generally considered to mean the entire population of interest.

Test Statistic *n* Also known as a *decision maker*, it is a *statistic* calculated from the test information in the sample, which is used to determine if the *null hypothesis* should be rejected. See also ▶ Hypothesis Test.

Theorem on Total Probability *n* An alternate name for *law of total probability*.

Time Series *n* A *sequence* of measurements or observations made in chronological order or at successive times. The time spacings are most often equal but do not have to be. The time series must describe the changes in a variable over the period of time of the observation or measurement.

Transformation *n* A functional replacement or change in variables of one or more variables. *Transformations* are often used to simplify the behavior or shape of a *random variable's probability distribution*, to create a constant variable, or simplify the relationship between variables. *Transformations* are generally named by either their purpose or by their functional transformation. Two common transformations named for their purpose are the *transformation to normality*, whose purpose is to change the shape of a probability distribution so that it is closer to that of the *normal distribution*, and the *transformation to linearity* whose purpose is to change the relation between two variables so that it is linear. Common types of transformations named for their functional transformation are the *logarithmic transformation*, which replaces a variable by its logarithm, the *1/x transformation*, which replaces a variable by its inverse, and the *x^2 transformation*, which replaces a variable by its square.

Transformation to Linearity *n* A *transformation* whose purpose is to change the relation between two variables so that it is linear. The transformation may be a single type of mathematical transformation applied to one of the variables or more than one mathematical process applied to both variables to generate two new variables that have a linear relationship.

Transformation to Normality *n* A *transformation* whose purpose is to change the shape of a probability distribution so that it is closer to that of the *normal distribution.* In many cases this is done to remove the *skewness* of the distribution to prepare the data for certain types of analysis or to make the data analysis easier to visualize and understand.

Treatment *n* A defined combination of *levels* for one or more *factors.* Treatments are usually given to some part of the *experimental units* in an *experiment.* Treatments are generally defined in the *experimental design.*

Trend *n* Also known as *secular trend,* it is an identifiable long term movement in a *time series.* The component of the values of the time series associated with the trend is referred to as the *trend component.*

Trend Component *n* The component of the values of a *time series* that are due to *trends.* Techniques such as *differencing* are often applied in order to remove the trend component, that may obscure other features, from a time series.

Two-Sample T-Test *n* See ▶ T-Test.

Two-Sided Test *n* Also known as a *two-tailed test,* it is a *hypothesis test* in which the values of the *population parameter* for which the *null hypothesis* is rejected are contained in both tails of the *probability distribution* of the parameter. See also ▶ One-Sided Test.

Two-Tailed Test *n* An alternate team for a *two-sided test.*

Two-Way Analysis of Variance *n* A *hypothesis test* used to test the *null hypothesis* that the *means* of two or more *samples* are equal. The samples must be independent, the *populations* from which the samples were obtained are assumed to be *normally distributed* with equal *variances,* and there are two factors with M_1 and M_2 respective levels of classification of each sample. There are thus $M = M_1M_2$ samples, $\mathbf{Y}_{jk}$, for $j = 1, 2, \ldots, M_1$ and $k = 1,2, \ldots M_2$, each with N_{jk} elements, $y_{jk1}, y_{jk2}, \ldots, y_{jkNjk}$. In a simple linear model, y_{jki} would be given by:

$$y_{jki} = \mu + \tau_{1j} + \tau_{2k} + \tau_{12jk} + \varepsilon_{jki}$$

where: μ is the *general* or *grand* mean over all of the elements in all samples,
τ_{1j} is the deviation due to factor 1 level j,
τ_{2k} is the deviation due to factor 2 level k,
τ_{12jk} is the deviation due to the interaction of factor 1 level j and factor 2
level k, and
ε_{jki} is the error associated with y_{jki}.
The null hypothesis that all of the sample means are equal is specified as three separate null hypotheses; all of the τ_{1j}s are equal, all of the τ_{2k}s are equal, and all of the τ_{12jk}s are equal. The null hypotheses are tested separately using the between group variation and the within group variation and the *Fisher test.*

For factor 1, the between group variation, $s_{B1}{}^2$, is the weighted sum of the squares of the differences between the individual sample means, μ_{1j} (where μ_{1j} is the mean over all samples with factor 1 level j), and the grand mean, μ, divided by the degrees of freedom associated with the number of samples, M_1-1. This is given by:

$$s_{B1}{}^2 = \frac{\sum_j \left[(\mu_{1j} - \mu)^2 \sum_k N_{jk} \right]}{M_1 - 1}$$

Similarly for factor 2, the between group variation, $s_{B2}{}^2$, is given by:

$$s_{B2}{}^2 = \frac{\sum_j \left[(\mu_{2k} - \mu)^2 \sum_j N_{jk} \right]}{M_2 - 1}$$

For the interaction, the between group variation, $s_{B12}{}^2$, is given by:

$$s_{B12}{}^2 = \frac{\sum_j \sum_k N_{jk}(\mu_{jk} - \mu_{1j} - \mu_{2k} + \mu)^2}{(M_1 - 1)(M_2 - 1)}$$

where: μ_{jk} is the mean of the $\mathbf{Y}_{jk}$ sample.
The within group variation, $s_E{}^2$, is the sum of the squares over all elements, y_{jki}, minus their respective means, μ_{jk}, divided by the degrees of freedom associated with the total number of elements, $N-M_1M_2$, where, N is equal to the sum of the N_{jk}s. This within group variation is given by:

$$S_E^2 = \frac{\sum_j \sum_k \sum_{i-1}^{N_{jk}} (y_{jki} - \mu_{jk})^2}{N - M_1 M_2}$$

The *Fisher* statistics, F_1, F_2, an F_{12}, for factor 1, factor 2, and the interactions, are given by the ratios of $s_B{}^2$s to $s_E{}^2$ which gives:

$$F_1 = \frac{s_{B1}{}^2}{s_E{}^2}$$

$$F_2 = \frac{s_{B2}{}^2}{s_E{}^2}$$

$$F_1 = \frac{s_{B12}{}^2}{s_E{}^2}$$

A *significance level,* α, is chosen the Fs are compared to the values, $f\ (M_1-1,\ N-M_1M_2,\ \alpha)$, $f\ (M_2-1,\ N-M_1M_2,\ \alpha)$, and $f((M_1-1)(M_2-1),\ N-M_1M_2,\ \alpha)$

taken from the *Fisher table* for use in testing the *null hypotheses.* If $F_1 \geq f(M_1-1,\ N-M_1M_2,\ \alpha)$ the factor 1 null hypothesis is rejected and the τ_{1j}s are not all equal. If $F_2 \geq f(M_2-1,\ N-M_1M_2,\ \alpha)$ the factor 2 null hypothesis is rejected and the τ_{2k}s are not all equal. If $F_{12} \geq f((M_1-1)(M_2-1),\ N-M_1M_2,\ \alpha)$ the interactions null hypothesis is rejected and the τ_{12jk}s are not all equal. See also ▶ One-Way Analysis of Variance.

Type I Error *n* The error of rejecting the *null hypothesis* when it is true. In *hypothesis testing* the probability of a type I error is referred to as the *significance level,* which is often represented as α. See also ▶ Type II Error.

Type II Error *n* The error of accepting the *null hypothesis* when it is false. In *hypothesis testing* the probability of a type II error is often represented by β. The probability of rejecting the null hypothesis when it is false is referred to as the *power* or *power of the test,* which is equal to 1- β. See also ▶ Type I Error.

U

Uniform Distribution *n* A *probability distribution* where the *probability* in the case of a *discrete random variable* or the *density function* in the case of a *continuous random variable,* **X**, with values, *x*, are constant (equal) over an interval, (a,b), where x is greater than or equal to a and x less than or equal to b and x is zero outside the interval. The *uniform distribution* is sometimes referred to as the *rectangular distribution*; since, a plot of its probability or density function resembles a rectangle. When the random variable, **X**, is discrete, the uniform distribution is referred to as the *discrete uniform distribution* and has the *probability,* $\mathbf{P}(x_k)$, with the form of:

$$\mathbf{P}(x_k) = \frac{1}{b+1-a}$$

for: k an integer, $a \leq k \geq b$ and $\mathbf{P}(x_k) = 0$ otherwise. The discrete uniform distribution has:

mean $= (a+b)/2$,

standard deviation $=[(b=1-a)^2 - 1]^{1/2}/12$ ½,

variance $= [(b+1-a)^2 - 1]/12$,

skewness $= 0$,

kurtosis $= [9(b+1-a)^2-21]/[5(b+1-a)^2 - 5]$, and

excess kurtosis $= [6(b+1-a)^2 + 6]/[5-5(b+1-a)^2]$.

When the random variable, **X**, is continuous, the uniform distribution is referred to as the *continuous uniform distribution* and has the *density function,* $f(x)$, with the form:

$$f(x) = \frac{1}{b-a}$$

for: x a real number, $a \leq x \geq b$ and $f(x) = 0$ otherwise. The continuous uniform distribution has:

mean $= (a+b)/2$,

standard deviation $= (b-a)/12^{1/2}$,

variance $= (b-a)^2/12$,

skewness $= 0$,

kurtosis $= 9/5$, and

excess kurtosis $= -6/5$

Unit *n* A fundamental, individual item on which an experiment or study is based. If the study concerns sampling the unit is referred to as a *sampling unit.* If the study is an experiment then the units are referred to as *experimental units.* See also ▶ Element.

Variance *n* One of the most used descriptive measures of a probability distribution, population or sample. It is equal to the square of the *standard deviation.* It is a measure of deviation from the expectation value or mean. It is a measure of statistical dispersion which quantifies the deviations from the expectation value as the mean of the squares of the distance from the mean. The variance of a random variable is defined as the second central moment of the random variable and is often denoted by σ^2 and sometimes by Var(**X**), defined on a probability space, **S**, is given by:

$$\sigma^2 = \int_S = (\mathbf{X}-\mu)^2 f(\mathbf{X})d\mathbf{X}$$

for a continuous random variable with probability density function, $f(\mathrm{X})$ and mean, μ, by:

$$\sigma^2 = \sum_i (x_i - \mu)^2 - \mathbf{P}_i$$

for a discrete random variable, with discrete values, x_i, and associated probabilities, $\mathbf{P}i$, and by:

$$\sigma^2 = \frac{\sum_i (x_i-\mu)^2 \mathbf{W}_i}{\sum_i \mathbf{W}_i}$$

for statistical samples, x_i, with weights, $\mathbf{W}_i$.

The *standard deviation* is defined as the positive square root of the variance and is often denoted by σ.

Wald-Wolfowitz Test *n* An alternate name for a *runs test.*

Weighted Average *n* Also referred to as *weighted arithmetic mean* or *weighted arithmetic average* and is defined as for a set of values and non-negative associated weights as the sum of all values times their associated weights divided by the sum of the weights. The weights average of the set **X**, ‹**X**›, with elements x_i and associated weights, $\mathbf{W}_i$, is given by:

$$\langle \mathbf{X} \rangle = \frac{\sum_i x_i \mathbf{W}_i}{\sum_i \mathbf{W}_i}$$

where: x_i are the elements of the set **X** and
$\mathbf{W}_i$ is the associated weight for x_i.
If the weights are the probabilities, $\mathbf{P}_i$, associated with x_i, then ‹**X**› is the mean, μ of **X** and the sum of the $\mathbf{P}_i$ is zero. See also ▶ Arithmetic Mean.

Weighted Mean *n* See ▶ Weighted Average and ▶ Mean.

Wilcoxon Rank-Sum Test *n* An alternate name for the *Mann-Whitney U Test.*

Wilcoxon Signed Test *n* Also known as the *Wilcoxon Signed Rank Test*, the *Wilcoxon Matched-Pairs Signed-Rank Test*, or the *Wilcoxon Rank Sum Test*, and is a *nonparametric test* used to test the *null hypothesis* that the *median* of the difference between the values of pairs of *elements* from related *samples* or repeated measurements on the same sample, is zero. It is assumed that the observations are independent. The differences, z_i, between the pairs of *random variables*, (x_i, y_i) are calculated as $z_i = y_i - x_i$ and the zero differences are disguarded. The remaining, M, differences are ranked from 1 to M according to their absolute values. When two or more differences have the same absolute value they are all given the average range of the ranks they would have occupied if slightly different. The *statistic* is the sum of the ranks and calculated for the negative differences, T. It is convention to take the smaller of the two T statistics to use in the test.
A *significance level*, α, is chosen and compared to the values, t, taken from the *Wilcoxon signed table* for use in testing one of three *null hypothesis* cases: the median of the z_is is equal to zero (the two sided case), the median of the z_is is equal to or less than zero (the negative median one-sided case), and the median of the z_is is greater than or equal to zero (the positive median one-sided case). If the $T < t_{\alpha/2}$ or $T < t_{1-\alpha/2}$ for the two-sided case, $T > {}_{1-\alpha}$ for the negative median one-sided case, or $T < t_\alpha$for the positive median one-sided case, the null hypothesis is rejected.
The tables are usually good for values of M less than 50. If M is large (in fact for M greater than 20), a *normal distribution* approximation, with *mean*, $\mu = M(M+1)/4$ and *variance*, $\sigma^2 = M(M+1)(2M+1)/24$ provides good results.

Wilcoxon-Mann-Whitney Test *n* An alternate name for the *Mann-Whitney U Test.*

References

Clelland RC, deCani JS, Brown FE (1973) Basic statistics with business applications, 2nd edn. Wiley, New York
Dodge Y (2008) The concise encyclopedia of statistics. Springer, New York
Feller W (1968) An introduction to probability theory and its applications, vol 1, 3rd edn. Wiley, New York
Hoel PG, Port SC, Stone CJ (1972) Introduction to stochastic processes. Houghton Mifflin, Boston
Mendenhall W (1975) Introduction to probability and statistics, 4th edn. Wadsworth, Belmont
Mills TC (1998) Time series techniques for economists. Cambridge University Press, Cambridge
Papoulis A (1965) Probability, random variables, and stochastic processes. McGraw-Hill, New York
Runyon RP, Haber A (1971) Fundamentals of behavioral statistics, 2nd edn. Addison-Wesley, Reading
Simonoff JS (1998) Smoothing methods in statistics. Springer, Heidelberg

Appendix F

Nomenclature for Organic Polymers

The IUPAC system of polymer nomenclature has aided the generation of unambiguous names that reflect the historical development of chemistry. However, the explosion in the circulation of information and the globalization of human activities mean that it is now necessary to have a common language for use in legal situations, patents, export-import regulations, and environmental health and safety information. Rather than recommending a unique name for each structure, rules have been developed for assigning IUPAC names, while continuing to allow alternative names in order to preserve the diversity and adaptability of nomenclature (Jones et al., 2009).

A section on nomenclature of polymers is included in Physical Properties of Polymers (Mark, 1996), and the basic nomenclature principles are mentioned in the following pages.

The CRC Handbook (Lide, 2004) contains a section by R. B. Fox and E. S. Wilks on the nomenclature and gives examples of the naming process of polymers and copolymers, and some of these comments and examples are discussed below.

Organic polymers have traditionally been named on the basis of the monomer used, a hypothetical monomer, or a semi-systematic structure. Alternatively, they may be named in the same way as organic compounds, i.e., on the basis of a structure as drawn. The former method, often called "source-based nomenclature" or "monomer-based nomenclature," sometimes results in ambiguity and multiple names for a single material. The latter method, termed "structure-based nomenclature," generates a sometimes cumbersome unique name for a given polymer, independent of its source. Within their limitations, both types of names are acceptable and well-documented (IUPAC, 1991). The use of stereochemical descriptors with both types of polymer nomenclature has been published.

Traditional Polymer Names: Monomer-Based Names

"Polystyrene" is the name of a homopolymer made from the single monomer styrene. When the name of a monomer comprises two or more words, the name should be enclosed in parentheses, as in "poly(methyl methacrylate)" or "poly(4-bromostyrene)" to identify the monomer more clearly. This method can result in several names for a given polymer: thus, "poly(ethylene glycol)," "poly(ethylene oxide)," and "poly(oxirane)" describe the same polymer. Sometimes, the name of a hypothetical monomer is used, as in "poly(vinyl alcohol)." Even though a name like "polyethylene" covers a multitude of materials, the system does provide understandable names when a single monomer is involved in the synthesis of a single polymer. When one monomer can yield more than one polymer, e.g., 1,3-butadiene or acrolein, some sort of structural notation must be used to identify the product, and one is not far from a formal structure-based name.

Copolymers, Block Polymers, and Graft Polymers

When more than one monomer is involved, monomer-based names are more complex. Some common polymers have been given names based on an apparent structure, as with "poly(ethylene terephthalate)." A better system has been approved by the IUPAC (1991). With this method, the arrangement of the monomeric units is introduced

IUPAC source-based copolymer classification

No.	Copolymer type	Connective	Example
1	Unspecified or unknown	*-co-*	Poly(A-*co*-B)
2	Random (obeys Bernoullian distribution)	*-ran-*	Poly(A-*ran*-B)
3	Statistical (obeys known statistical laws)	*-stat-*	Poly(A-*stat*-B)
4	Alternating (for two monomeric units)	*-alt-*	Poly(A-*alt*-B)
5	Periodic (ordered sequence for two or more monomeric units)	*-per-*	Poly(A-*per*-B-*per*-C)
6	Block (linear block arrangement)	*-block-*	PolyA-*block*-polyB
7	Graft (side chains connected to main chains)	*-graft-*	PolyA-*graft*-polyB

through use of an italicized connective placed between the names of the monomers. For monomer names represented by A, B, and C, the various types of arrangements are shown in table IUPAC source-Based Copolymer Classification.

The information in the following table, Source-Based Copolymer Nomenclature, contains examples of common or semi-systematic names of copolymers. The systematic names of comonomers may also be used; thus, the polyacrylonitrile-*block*-polybutadiene-*block*-polystyrene polymer in Source-Based Copolymer Nomenclature may also be named poly(prop-2-enenitrile)-*block*-polybuta-1,3-diene-*block*-poly(ethenylbenzene). IUPAC does not require alphabetized names of comonomers within a polymer name; many names are thus possible for some copolymers.

These connectives may be used in combination and with small, non-repeating (i.e., non-polymeric) junction units; see, for example, Source-Based Copolymer Nomenclature, line 8. Along dash may be used in place of the connective -*block*-; thus, in Source-Based Copolymer Nomenclature, the polymers of lines 7 and 8 may also be written as shown on lines 9 and 10.

Source-based copolymer nomenclature

No.	Copolymer name
1	Poly(propene-*co*-methacrylonitrile)
2	Poly[(acrylic acid)-*ran*-(ethyl acrylate)]
3	Poly(butene-*stat*-ethylene-*stat*-styrene)
4	Poly[(sebacic acid)-*alt*-butanediol]
5	Poly[(ethylene oxide)-*per*-(ethylene oxide)-*per*-tetrahydrofuran]
6	Polyisoprene-*graft*-poly(methacrylic acid)
7	Polyacrylonitrile-*block*-polybutadiene-*block*-polystyrene
8	Polystyrene-*block*-dimethylsilylene-*block*-polybutadiene
9	Polyacrylonitrile—polybutadiene—polystyrene
10	Polystyrene—dimethylsilylene—polybutadiene

IUPAC also recommends an alternative scheme for naming copolymers that comprises use of "copoly" as a prefix followed by the names of the comonomers, a solidus (an oblique stroke) to separate comonomer names, and addition before "copoly" of any applicable connectives listed in Source-Based Copolymer Nomenclature except -*co*-.

The information in Source-Based Copolymer Nomenclature (Alternative Format) gives the same examples shown in Source-Based Copolymer Nomenclature but with the alternative format. Comonomer names need not be parenthesized.

Source-based copolymer nomenclature (Alternative format)

No.	Polymer name
1	Copoly(propene/methacrylonitrile)
2	-copoly(acrylic acid/ethyl acrylate)
3	-copoly(butene/ethylene/styrene)
4	-copoly(sebacic acid/butanediol)
5	-copoly(acrylonitrile/butadiene/styrene)
6	-copoly(ethylene oxide/ethylene oxide/tetrahydrofuran)
7	-copoly(isoprene/methacrylic acid)

Connectives for non-linear macromolecules and macromolecular assemblies

No.	Type	Connective
1	Branched (type unspecified)	Branch
2	Branched with branch point of functionality f	f-Branch
3	Comb	Comb
4	Cross-link	ι (Greek iota)
5	Cyclic	Cyclo
6	Interpenetrating polymer network	ipn
7	Long-chain branched	l-Branch
8	Network	Net
9	Polymer blend	Blend
10	Polymer-polymer complex	Compl
11	Semi-interpenetrating polymer network	Sipn
12	Short-chain branched	sh-Branch
13	Star	Star
14	Star with f arms	f-Star

Non-linear polymers are named by using the italicized connective as a *prefix* to the source-based name of the polymer component or components to which the prefix applies; some examples are listed in the Non-Linear Macromolecules table.

Non-linear macromolecules

No.	Polymer name	Polymer structural features
1	Poly(methacrylic acid)-*comb*-polyacrylonitrile	Comb polymer with a poly (methacrylic acid) backbone and polyacrylonitrile side chains
2	*Comb*-poly [ethylene-*stat*-(vinyl chloride)]	Comb polymer with unspecified backbone composition and statistical ethylene/vinyl chloride copolymer side chains
3	Polybutadiene-*comb*-(polyethylene; polypropene)	Comb polymer with butadiene backbone and side chains of polyethylene and polypropene
4	*Star*-(polyA; polyB; polyC; polyD; polyE)	Star polymer with arms derived from monomers A, B, C, D, and E, respectively
5	*Star*-(polyA-*block*-polyB-*block*-polyC)	Star polymer with every arm comprising a tri-block segment derived from comonomers A, B, and C
6	*Star*-poly (propylene oxide)	A star polymer prepared from propylene oxide
7	5-*Star*-poly (propylene oxide)	A five-arm star polymer prepared from propylene oxide
8	*Star*-(polyacrylonitrile; polypropylene) (Mr 10000:25000)	A star polymer containing polyacrylonitrile arms of MW 10000 and polypropylene arms of MW 25000

Macromolecular assemblies held together by forces other than covalent bonds are named by inserting the appropriate italicized connective between names of individual components as described in the following table, Examples of Polymer Blends and Nets.

Examples of polymer blends and nets

No.	Polymer name
1	Polyethylene-*blend*-polypropene
2	Poly(methacrylic acid)-*blend*-poly(ethyl acrylate)
3	*Net*-poly(4-methylstyrene-ι-divinylbenzene)
4	*Net*-poly[styrene-*alt*-(maleic anhydride)]-ι-(polyethylene glycol; polypropylene glycol)
5	*Net*-poly(ethyl methacrylate)-*sipn*-polyethylene
6	[*Net*-poly(butadiene-*stat*-styrene)]-*ipn*-[*net*-poly(4-methylstyrene-ι-divinylbenzene)]

Structure-Based Polymer Nomenclature

Regular Single-Strand Polymers

Structure-based nomenclature has been approved by the IUPAC4 and is currently being updated; it is used by *Chemical Abstracts* (1999). Monomer names are not used. To the extent that a polymer chain can be described by a repeating unit in the chain, it can be named "poly (repeating unit)." For regular single-strand polymers, "repeating unit" is a bivalent group; for regular double-strand (ladder and spiro) polymers, "repeating unit" is usually a tetravalent group (IUPAC, 1993).

Since there are usually many possible repeating units in a given chain, it is necessary to select one, called the "constitutional repeating unit" (CRU) to provide a unique and unambiguous name, "poly(CRU)," where "CRU" is a recitation of the names of successive units as one proceeds through the CRU from left to right. For this purpose, a portion of the main chain structure that includes at least two repeating sequences is written out. These sequences will typically be composed of bivalent subunits such as –CH2–, –O–, and groups from ring systems, each of which can be named by the usual nomenclature rules (IUPAC, 1979, 1993).

Where a chain is simply one long sequence comprising repetition of a single subunit, that subunit is itself the CRU, as in "poly(methylene)" or "poly(1,4-phenylene)." In chains having more than one kind of subunit, a seniority system is used to determine the beginning of the CRU and the direction in which to move along the main chain atoms (following the shortest path in rings) to complete the CRU. Determination of the first, most senior, subunit, is based on a descending order of seniority: (1) heterocyclic rings, (2) hetero atoms, (3) carbocyclic rings, and, lowest, (4) acyclic carbon chains.

Within each of these classes, there is a further order of seniority that follows the usual rules of nomenclature.
Heterocycles: A nitrogen-containing ring system is senior to a ring system not containing nitrogen (IUPAC, 1976, 1993). Further descending order of seniority is determined by:

1. The highest number of rings in the ring system
2. The largest individual ring in the ring system
3. The largest number of hetero atoms
4. The greatest variety of hetero atoms

Hetero atoms: The senior bivalent subunit is the one nearest the top, right-hand corner of the Periodic Table; the order of seniority is: O, S, Se, Te, N, P, As, Sb, Bi, Si, Ge, Sn, Pb, B, Hg.

Carbocycles: Seniority (IUPAC, 1976) is determined by:

1. The highest number of rings in the ring system
2. The largest individual ring in the ring system
3. Degree of ring saturation; an unsaturated ring is senior to a saturated ring of the same size

Carbon chains: Descending order of seniority is determined by:

1. Chain length (longer is senior to shorter)
2. Highest degree of unsaturation
3. Number of substituents (higher number is senior to lower number)
4. Ascending order of locants
5. Alphabetical order of names of substituent groups

Among equivalent ring systems, preference is given to the one having lowest locants for the free valences in the subunit, and among otherwise identical ring systems, the one having least hydrogenation is senior. Lowest locants in unsaturated chains are also given preference. Lowest locants for substituents are the final determinant of seniority. Poly (1,3-butadiene) obtained by polymerization of 1,3-butadiene in the so-called 1,4- mode is frequently drawn incorrectly in publications as – (CH2–CH=CH–CH2)*n*-; the double bond should be assigned the lowest locant possible, i.e., the structure should be drawn as – (CH=CH–CH2–CH2)*n*-.

Direction within the repeating unit depends upon the shortest path, which is determined by counting main chain atoms, both cyclic and acyclic, from the most senior subunit to another subunit of the same kind or to a subunit next lower in seniority. When identification and orientation of the CRU have been accomplished, the CRU is named by writing, in sequence, the names of the largest possible subunits within the CRU from left to right. For example, the main chain of the polymer traditionally named "poly(ethylene terephthalate)" has the structure shown in the following figure, Structure-based name: poly(oxyethyleneoxyterephthaloyl); traditional name: poly(ethylene terephthalate).

$$\left[-O-CH_2CH_2-O-\overset{O}{\overset{\|}{C}}-C_6H_4-\overset{O}{\overset{\|}{C}}-\right]_n$$

Polyethylene terephalate

The CRU in the above figure, Structure-based name: poly(oxyethyleneoxyterephthaloyl); traditional name: poly(ethylene terephthalate), is enclosed in brackets and read from left to right. It is selected because (1) either backbone oxygen atom qualifies as the "most senior subunit," (2) the shortest path length from either –O– to the other –O– is via the ethylene subunit. Orientation of the CRU is thus defined by (1) beginning at the –O– marked with an asterisk, and (2) reading in the direction of the arrow. The structure-based name of this polymer is therefore "poly(oxyethyleneoxyterephthaloyl)," not much longer than the traditional name and much more adaptable to the complexities of substitution. As organic nomenclature evolves, more systematic names may be used for subunits, e.g., "ethane-1,2-diyl" instead of "ethylene." IUPAC still prefers "ethylene" for the –CH2–CH2– unit, however, but also accepts "ethane-1,2-diyl."

Structure-based nomenclature can also be used when the CRU backbone has no carbon atoms. An example is the polymer traditionally named "poly(dimethylsiloxane)," which on the basis of structure would be named "poly (oxydimethylsilylene)" or "poly(oxydimethylsilanediyl)." This nomenclature method has also been applied to inorganic and coordination polymers and to double-strand (ladder and spiro) organic polymers (IUPAC, 1993).

Irregular Single-Strand Polymers

Polymers that cannot be described by the repetition of a single CRU or comprise units not all connected identically in a directional sense can also be named on a structure basis. These include copolymers, block and graft polymers, and star polymers. They are given names of the type "poly(A/B/C...)," where A, B, C, etc. are the names of the component constitutional units, the number of which are minimized. The constitutional units may include regular or irregular blocks as well as atoms or atomic groupings, and each is named by the method described above or by the rules of organic nomenclature.

The solidus denotes an unspecified arrangement of the units within the main chain. For example, a statistical copolymer derived from styrene and vinyl chloride with the monomeric units joined head-to-tail is named "poly (l-chloroethylene/l-phenylethylene)." A polymer obtained by 1,4- polymerization and both head-to-head and head-to-tail 1,2- polymerization of 1,3-butadiene would

$$\left(-\underset{Cl}{\underset{|}{CH}}-CH_2-/-\underset{C_6H_5}{\underset{|}{CH}}-CH_2-\right)_n-\left(-CH=CH-CH_2-CH_2-/-\underset{CH=CH_2}{\underset{|}{CH}}-CH_2-/-CH_2-\underset{H_2C=CH}{\underset{|}{CH}}-\right)_{ni}$$

be named "poly(but-1-ene-l,4-diyl/l-vinylethylene/2-vinylethylene)" (Note: Poly(1,3 butadiene) is usually incorrectly expressed as $-CH_2-CH=CH-CH_2)n-$ should be expression with the = in the *lowest* locant possible or $-CH=CH-CH_2-CH_2)n-$). In graphic representations of these polymers, shown in the following figure, Graphic Representations of Copolymers, the hyphens or dashes at each end of each CRU depiction are shown *completely within* the enclosing parentheses; this indicates that they are not necessarily the terminal bonds of the macromolecule.

A long hyphen is used to separate components in names of block polymers, as in "poly(A)—poly(B)—poly(C)," or "poly(A)—X—poly(B)" in which X is a non-polymeric junction unit, e.g., dimethylsilylene.

In graphic representations of these polymers, the blocks are shown connected when the bonding is known (see the figure Polystyrene—polyethylene—polystyrene),

$$\left(\underset{C_6H_5}{CH}-CH_2\right)_p\left(CH_2-CH_2\right)_q\left(\underset{C_6H_5}{CH}-CH_2\right)_r$$

when the bonding between the blocks is unknown, the blocks are separated by solid and are shown *completely within* the outer set of enclosing parentheses (see the figure Poly[poly(methyl methacrylate)—polystyrene—poly (methyl acrylate)]), see IUPAC (1994).

$$\left[\left(\overset{CH_3}{\underset{C(=O)OCH_3}{C}}-CH_2\right)_p/\left(\underset{C_6H_5}{CH}-CH_2\right)_q/\left(\underset{C(=O)OCH_3}{CH}-CH_2\right)_r\right]_{ni}$$

Graft polymers are named in the same way as a substituted polymer but without the ending "yl" for the grafted chain; the name of a regular polymer, comprising Z units in which some have grafts of "poly (A)," is "poly[Z/poly(A)Z]." Star polymers are treated as a central unit with substituent blocks, as in "tetrakis (polymethylene)silane" (IUPAC, 1994).

Other Nomenclature Articles and Publications

In addition to the *Chemical Abstracts* and IUPAC documents cited above and listed below, other articles on polymer nomenclature are available. A 1999 article lists significant documents on polymer nomenclature published during the last 50 years in books, encyclopedias, and journals by *Chemical Abstracts*, IUPAC, and individual authors (Wilks, 1999). A comprehensive review of source-based and structure-based nomenclature for all of the major classes of polymers (Wilks, 2000a) and a short tutorial on the correct identification, orientation, and naming of most commonly encountered constitutional repeating units were both published in 2000 (Wilks, 2000b).

References

Chemical Abstracts Service. Naming and indexing of chemical substances for chemical abstracts, Appendix IV. Chemical Abstracts 1999 Index Guide

International Union of Pure and Applied Chemistry (1976) Nomenclature of regular single-strand organic polymers (Recommendations 1975). Pure Appl Chem 48:373–385

International Union of Pure and Applied Chemistry (1979) Nomenclature of organic chemistry, Sections A, B, C, D, E, F, and H. Pergamon Press, Oxford, UK

International Union of Pure and Applied Chemistry (1981) Stereochemical definitions and notations relating to polymers (Recommendations 1980). Pure Appl Chem 53:733–752

International Union of Pure and Applied Chemistry (1985a) Source-based nomenclature for copolymers (Recommendations 1985). Pure Appl Chem 57:1427–1440

International Union of Pure and Applied Chemistry (1985b) Nomenclature of regular double-strand and quasi-single-strand inorganic and coordination polymers (Recommendations 1984). Pure Appl Chem 57:149–168

International Union of Pure and Applied Chemistry (1991) Compendium of macromolecular nomenclature. Blackwell Scientific Publications, Oxford, UK

International Union of Pure and Applied Chemistry (1993a) A guide to IUPAC nomenclature of organic compounds. Blackwell Scientific Publications, Oxford, UK

International Union of Pure and Applied Chemistry (1993b) Nomenclature of regular double-strand (ladder and spiro) organic polymers (Recommendations 1993). Pure Appl Chem 65:1561–1580

International Union of Pure and Applied Chemistry (1994a) Structure-based nomenclature for irregular single-strand organic polymers (Recommendations 1994). Pure Appl Chem 66:873–889

International Union of Pure and Applied Chemistry (1994b) Graphic representations (chemical formulae) of macromolecules (Recommendations 1994). Pure Appl Chem 66:2469–2482

International Union of Pure and Applied Chemistry (1997) Source-based nomenclature for non-linear macromolecules and macromolecular assemblies (Recommendations 1997). Pure Appl Chem 69:2511–2521

Jones RG, Rahovec J, Stepto RFTR (2009) Compendium of polymer terminology and nomenclature: IUPAC (Recommendations 2008). Royal Society of Chemistry, London

Lide DR (ed) (2004) CRC handbook of chemistry and physics. Taylor & Francis, Boca Raton, FL

Mark JE (ed) (1996) Physical properties of polymers. Springer, New York

Wilks ES (1999) Macromolecular nomenclature note No. 17: "Whither Nomenclature?" Polym Prep 40(2):6–11

Wilks ES (2000a) Polymer nomenclature: the controversy between source-based and structure-based representations (a personal perspective). Prog Polym Sci 25:9–100

Wilks ES (2000b) Macromolecular nomenclature note No. 18: "SRUs: Using the Rules." Polym Prep 41(1):6a–11a

Appendix G

Interactive Polymer Technology Programs

Dear Reader,

We are happy to provide a collection of unique and useful tools and interactive programs along with this Springer Reference. You will find short descriptions of the different functions below. Please download the software at the following website:

http://extras.springer.com/2011/978-1-4419-6247-8

Please note that the file is more than 200 MB. Download the ZIP file and unzip it. It is strongly recommended to read the ReadMe.txt before installing. The software is started by opening the file InPolyTech.pdf and following the instructions. Detailed instructions can be found under "Help Instructions."

The software consists of the following 15 programs and tools we are sure you will find most helpful in your daily work:

Abbreviations and Symbols

This section lists some of the common abbreviations and symbols used in polymer science. The symbols are in alphabetical order and in tabular form in a standard PDF document.

Pronunciation of Terms – Audio

This section allows the user to hear the English pronunciation of selected polymer terms. To hear a term type the first few letters of the term in the first box and the window list will scroll automatically so that the term is in view. A term can then be played by double-clicking the left mouse button on the term.

These instructions can be viewed in the section by pressing the "HELP" button located in the upper right of the list window.

Greek – Russian – English Alphabets

This section lists the letters and numbers of the Greek, Russian, and English alphabets and numbers. They are in tabular form in a standard PDF document.

Mathematical Symbols

This section lists some of the common mathematical symbols for operations, constants, and trigonometric functions. The symbols are in tabular form in a standard PDF document.

Polymer Science Equations – Interactive

This section contains interactive polymer science equations and their definitions. The user can enter values for most of the variables and the program will calculate the rest. In addition many of the equations contain tabulated lists of material parameters from which the user can select a value.

Periodic Table – Interactive

This section contains an interactive periodic table of the elements.

Molecular Imaging – 3D Models

This section allows the user to display and manipulate a 3-dimensional model of the listed molecules. IT MAY TAKE A LONG TIME TO LOAD AND APPEAR!

Molecular Structure Diagrams – 2D

This section allows the user to display and manipulate a 2-dimensional structure diagram, 3-dimensional simple ball-and-stick model, or both of the listed molecules.

Solubility of Polymers – Interactive

This section allows the user to calculate the RED fraction from the Hansen solubility parameters for a selected polymer and a mixture of up to three solvents. If three solvents are selected the program calculates and displays the RED for each solvent, the best RED and mixture formulation for every combination of two of the three solvents, and the best RED and mixture formulation for a mixture of all three solvents. The results are displayed in a results window.

Polymer Solution Rheology – Interactive

This section allows the user to calculate the viscosity of a single polymer single solvent solution for a polymer of specified molecular weight. The results are displayed in the lower results window.

Chlorine Aqueous Solutions – Interactive

This section allows the user to calculate the concentrations for the various ion species in a chlorine water solution from the knowledge of the temperature and the concentration of only one species. The results are displayed in the lower results window.

International Standards – Units

This section lists the most common ISO units, abbreviations, and symbols. They are grouped in tables by type or category in a standard PDF document.

Units Converter – Interactive

This section allows the user to calculate the value of a quantity in one type of unit when it is given in another.

Algebraic Calculator with Complex Numbers

This section provides J-Complex Calculator, which is an algebraic calculator that handles complex numbers.

Probability and Statistics Equations – Interactive

This section contains interactive probability and statistics equations and terms along with their definitions. The user can enter values for most of the variables and the program will calculate the rest. In addition many of the equations can contain lists of values for calculating statistics. These lists can be saved and read in and modified at a later date.

Disclaimer

▸ "This software is provided by the copyright holder(s) and contributors "As Is" and any express or implied warranties, including, but not limited to, the implied warranties of merchantability and fitness for a particular purpose are disclaimed. In no event shall the copyright owner or contributors be liable for any direct, indirect, incidental, special, exemplary, or consequential damages (including, but not limited to, procurement of substitute goods or services; loss of use, data, or profits; or business interruption) however caused and on any theory of liability, whether in contract, strict liability, or tort (including negligence or otherwise) arising in any way out of the use of this software, even if advised of the possibility of such damage."

References

(1976) Encyclopedia of PVC. Marcel Dekker, New York

(1978) Paint/coatings dictionary. Federation of Societies for Coatings Technology, Blue Bell, PA

(1981) Magnetic materials – a glossary. Pfizer Corporation, Easton

(1985) Systems and specifications. Steel Structures Paint Council

(1990) Encyclopedia of polymer science and engineering, 2nd edn. Wiley, New York

(1995) Paint and coating testing manual (Gardner-Sward Handbook) MNL 17, 14th edn. ASTM, Conshohocken

(1996) Kirk-Othmer encyclopedia of chemical technology: pigments-powders. Wiley, New York

(1997) Tests for comparative flammability of liquids, UI 340, Laboratories Incorporated Underwriters, New York

(1998) Kirk-Othmer encyclopedia of chemical technology, vols 1–19. Wiley, New York

(2000) McCutcheon's emulsifiers and detergent: North American edition, vol 1. McCutcheon

(2000) Stedman's medical dictionary, 27th edn. Lippincott Williams and Wilkins, New York

(2001) Merck Index, 13th edn. Merc, Whitehouse Station

(2001) Paint: pigment, drying oils, polymers, resins, naval stores, cellulosics esters, and ink vehicles, vol 3. American Society for Testing and Material

(2002) Intelligent macromolecules for smart devices. Springer, New York

(2002) McGraw Hill dictionary of geology and mineralogy. McGraw Hill, New York

(2002) Microsoft computer dictionary. Microsoft Press

(2003) Industrial dye: chemistry, properties and applications, Wiley, New York

(2004) Merriam-Webster's Collegiate Dictionary, 11th edn. Merriam-Webster, Springfield

Adamaon AW, Gast AP (1997) Physical chemistry of surfaces. Wiley, New York

Akin RB (1962) Acetal resins. Reinhold, New York

Aldrich Chemical Co (1990) Catalog handbook of fine chemicals

Alfrey T, Price CC (1947) J Polymer Sci 2:101

Allcock HR, Mark J, Lampe F (2003) Contemporary polymer chemistry. Prentice Hall, New York

American Society for Testing and Materials, 100 Barr Harbor Drive, West Conshohocken, PA 19428–2959, www.astm.org

Appleman BR, Hower HE (1985) Surface preparation: the state of the art. Steel Structures Paint Council, Pittsburgh, PA

Ash M, Ash I (1982–1983) Encyclopedia of plastics, polymers, and resins, vols I–III. Chemical Publishing, New York

Ash M, Ash I (1996) Handbook of paint and coating raw materials: trade name products – chemical products dictionary with trade name cross-references. Ashgate, New York

Ash M, Ash I (1998) Handbook of fillers, extenders and dilutents. Synapse Information Resources, New York

Ash M, Ash I (2000) Handbook of industrial surfactants. Synapse Information Resources, New York

Ash M, Ash I (2000) Industrial surfacants. Synapse Information Resourses, New York

Ashford NA, Miller CS (1997) Chemical exposures: low levels and high stakes. Wiley, New York

Atkins PW, De Paula J (2001) Physical chemistry, 7th edn. Freeman, New York

Avrami M (1939–1941) Kinetics of phase change I: general theory. J Chem Phys 7:1103; Kinetics of phase change II: transformation-time relations for random distribution of nuclei. 8:212; Kinetics of phase change III: granulation, phase change and microstructures. 9:177

Baboian R (2002) Corrosion engineer's handbook, 3rd edn. NACE International – The Corrosion Society, Houston

Babrauskas V (2003) Ignition handbook. Fie Science, New York

Bar-Cohen Y (ed) (2001) Electroactive polymer (EAP) actuators as artificial muscles. SPIE, Bellingham

Barner L (2003) Surface grafting via the reversible addition-fragmentation chain-transfer (RAFT) process: from polypropylene beads to core shell microspheres. Aust J Chem 56:1091

Bart J (2005) Additives in polymers: industrial analysis and applications. Wiley, New York

Barton AFM (1983) Handbook of solubility parameters and other cohesion parameters. CRC Press, Boca Raton

Becher P (1989) Dictionary of colloid and surface science. Marcel Dekker, New York

Becher P (2001) Emulsions: theory and practice. American Chemical Society, Washington DC

Belfield KD, Crivello JV (eds) (2003) Photoinitiated polymerization. American Chemical Society Publications, Washington, DC

Bergey's manual of systematic bacteriology, 2nd edn. Springer, New York, April 2001

Berne BJ (2000) Dynamic light scattering: applications to chemistry, biology and physics. Dover, New York

Bhushan B (ed) (2004) Springer handbook of nanotechnology. Springer, New York

Bhushan B, Fuchs H, Hosaka S (2004) Applied scanning probe methods. Springer, New York

Biederman H (2004) Plasma polymer films. Imperial College Press, London

Billmeyer FW Jr (1984) Textbook of polymer science, 3rd edn. Wiley-Interscience, New York

Billmeyer FW, Saltzman M (1966) Principles of color technology. Wiley, New York

Black JG (2002) Microbiology, 5th edn. Wiley, New York

Bovey FAA, Mirau PA (1996) NMR of polymers. Elsevier, New York

Box GE, Hunter WG, Hunter JS (2005) Statistics for experimenters: innovation, and discovery, 2nd edn. Wiley, New York

Braga PC, Rici D (2004) Atomic force microscopy: biomedical methods and applications. Humana, Italy

Brandrup J, Immergut EH (eds) (1989) Polymer handbook, 3rd ed. Wiley, New York

Breitmaier E, Voelter W (1986) Carbon-13 NMR spectroscopy. Wiley, New York

Brown R (1999) Handbook of physical polymer testing, vol 50. Marcel Dekker, New York

Brown W (1996) Light scattering: principles and development. Oxford University Press, New York
Bruno TJ, Paris S (2005) CRC handbook of fundamental spectroscopic correlation charts. CRC, Boca Raton
Burell R, Faires JD (2004) Numerical analysis. Brookes/Cole, New York
Cacioli P, Hawthorne DG, Laslett RL, Rizzardo E, Solomen DH (1986) Copolymerization of o-unsaturated oligo (methyl methacrylate): new macromonomers. J Macromolecular Sci (1998), Part A: Pure Appl Chem 23(7):839–852
Callister WD (2002) Materials science and engineering. Wiley, New York
Carley JF (ed) (1993) Whittington's dictionary of plastics. Technomic Publishing, Lancaster, PA
Chiefari J, Chong YK, Ercole F, Krstina J, Jeffery J, Le TPT, Mayadunne RTA, Meijs GF, Moad CL, Moad G, Rizzardo E, Thang SH (1998) Living free-radical polymerization by reversible addition-fragmentation chain transfer: the RAFT process. Macromolecules 31(16):5559–5562
Chung DD (1994) Carbon fiber compositers. Elsevier, New York
Coleman MM, Strauss S (1998) Fundamentals of polymer science: an introductory text. CRC, Boca Raton
Collins EA, Bares J, Billmeyer FW Jr (1973) Experiments in polymer science. Wiley-Interscience, New York
Collins PJ (1997) Introduction to liquid crystals: chemistry and physics, vol 1. Taylor & Francis, New York
Color photographic standards for surface preparation (SSPC-VIS 1) (1982) Steel Structures Paint Council: The Society for Protective Coatings, 40 24th Street, 6th Floor, Pittsburg, PA 15222–4656 USA, www.sspc.org
Commission Internationale De L'Eclairage (CIE), CIE Central Bureau, Kegelgasse 27, A-1030, Wien, Austria. [English: International Commission on Illumination, Vienna, Austria], www.cie.co.at/headerbasic.html
Complete textile glossary (2000) Celanese Acetate LLC, Three Park Avenue, New York
Connors KA (1990) Chemical kinetics. Wiley, New York
Coussot P (2005) Rheometry of pastes, suspensions and grannular materials: applications in industry and environment. Wiley, New York
Coyard H, Tuck N, Deligny P, Oldring PKT (2001) Resins for surface coatings: acrylics and epoxies, vol 12. Wiley, New York
Cowie JMG, Valeria A (2008) Polymers: chemistry and physics of modern materials, 3rd edn. CRC Press, Boca Raton, FL
Czeslaw K, Harvey S, Wilusz E (1990) Encyclopedia of tribology. Elsevier, New York
D'Agostino R, Favia P, Fracassi F (eds) (1997) Plasma processing of polymers. Kluwer, New York
Dainth J (2004) Dictionary of chemistry. Oxford University Press, Oxford
DeLevie R (1996) Principles of quantitative analysis. McGraw Hill, New York
Deligny P, Oldring PKT, Tuck N (eds) (2001) Resins for surface coatings, alkyds and polyester, vol 22. Wiley, New York
Dissado LA, Fothergill CJ (eds) (1992) Electrical degradation and breakdown of polymers. Institution of Electrical Engineering (IEE), London
Donnet J-B, Wang M-J (1993) Carbon black. Marcel Dekker, New York
Driggers RC, Cox P, Edwards T (1998) Introduction to infrared and electro-optical systems. Artech House, Aylesford
Elias H-G (1977) Macromolecules, vol 1–2. Plenum, New York
Elias H-G (2003) An introduction to plastics. Wiley, New York
Elmas M (1973) Powder technology publication, series No 5: Fluidized – bed powder coatings, Powder Advisory Center, London
Emerson JA, Torkelson JM (eds) (1991) Optical and electrical properties of polymers: materials research society symposium proceedings, vol 24. Materials Research Society, Warrendale
Finzel WA (1996) Silicones in coatings. Federation of Societies for Coatings Technology, Blue Bell
Flick EW (1975) Water-based paint formulations. Noyes, Park Ridge
Flick EW (1991) Industrial synthetic resins handbook. Williams Andrews/Noyes, New York
Flick WE (1977) Solvent-based paint formulation. Noyes, Park Ridge
Flory PJ (1953) Principles of polymer science. The Cornell University Press, Ithaca
Flory PJ (1960) The statistical thermodynamics of solutions. Wiley, New York
Flory PJ (1969) Statistical mechanics of chain molecules. Interscience, New York
Fouassier J-P (1995) Photoinitiation, photopolymerization and photocuring. Hanser-Gardner, New York
Fox AM (2001) Optical properties of solids. Oxford University Press, Oxford
Franklin AJ (1976) Cement and mortar additives. Noyes, Park Ridge
Freir GD (1965) University physics. Appleton-Century-Crofts, New York
Froment GF (1990) Chemical reactor analysis and design. Wiley, New York
Gair A (1996) Artist's manual. Chronicle, San Francisco
Galina H, Spiegel S, Meisel I, Kniep CS, Grieve K (2001) Polymer networks. Wiley, New York
Giambattista A, Richardson R, Richardson RC, Richardson B (2003) College physics. McGraw Hill, New York
Glenz W (ed) (2001) A glossary of plastics terminology in 5 languages, 5th edn. Hanser Gardner, Cinicinnati
Godlly EW (1995) Naming organic compounds: a systematic instruction manual. Pearson, New York
Goldberg DE (2003) Fundamentals of chemistry. McGraw Hill Science/Engineering/Math, New York
Gooch JW (1988) Development of high solids magnetic dispersions and coatings. J Coating Tech 60(757):37–44
Gooch JW (1993) Lead based paint handbook. Plenum, New York
Gooch JW (1997) Analysis and deformulation of polymeric materials. Plenum, New York
Gooch JW (2002) Emulsification and polymerization of alkyd resins. Kluwer Academic/Plenum, New York
Goodwin JW, Goodwin J, Hughes RW (2000) Rheology for chemists. Royal Society of Chemistry, London
Grellman W, Seidler S (eds) (2001) Deformation and fracture behavior of polymers. Springer, New York
Groenewoud WM (2001) Characterization of polymers by thermal analysis. Elsevier, New York
Hansen CM, Beerbower A (1971) Solubility parameters. In: Standen A (ed) Kirk-Othmer encyclopedia of chemical technology (suppl vol), 2nd edn. Interscience, New York, p 889
Hare CH (2001) Paint film degradation – mechanisms and control. Steel Structures Paint Council, Pittsburgh
Harper CA (2000) Modern plastics encyclopedia. McGraw Hill, New York
Harper CA (ed) (2002) Handbook of plastics, elastomers and composites, 4th edn. McGraw Hill, New York
Harrington BJ (2001) Industrial cleaning technology. Kluwer Academic, New York
Harris CM (2005) Dictionary of architecture and construction. McGraw Hill, New York
Harris DC (2002) Quantitative chemical analysis. W.H. Freeman, New York

Hartland S (ed) (2004) Surface and interfacial tension. CRC Press, Boca Raton

Heeger AJ, Ulrich DR, Orenstein J (eds) (1988) Nonlinear optical properties of polymers. Materials Research Society, Pittsburgh

Herbst W, Hunger K (2004) Industrial organic pigments. Wiley, Weinheim

Hess M, Morgans WM (1979) Paint film defects. Wiley, New York

Hibbard MJ (2001) Mineralogy. McGraw Hill, New York

Higham T, Ramsey B, Owen C (2004) Radiocarbon and archaeology. Oxford University School of Archaeology, Oxford

Hildebrand JH, Scott RL (1962) Regular solutions. Prentice-Hall, Englewood Cliffs, NJ

Hoadley RB (2000) Understanding wood. The Taunton, Newtown

Hollander AP, Hatton PV (2004) Biopolymer methods in tissue engineering. Humana, Totowa

Holst GC (2003) Electro-optical imaging system performance. JDC Publishing, Winter Park, FLorida

Houston PL (2001) Chemical kinetics and reaction dynamics. McGraw Hill, New York

Houwink R, Salomon G (1967) Adhesion and adhesives, vol 2, 2nd edn. Elsevier, Amsterdam

Huggins ML (1958) Physical chemistry of high polymers. Wiley, New York

Hull D (1999) Fractography: observing, measuring and interpreting fracture surface topography. Cambridge University Press, Cambridge

Humphries M (2000) Fabric glossary. Prentice-Hall, Upper-Saddle River, NJ

Iaac MD, Ishal O (2005) Engineering mechanics of composite materials. Oxford University Press, Oxford

Isobe Y, Okuyama K, Hosaka A, Kubota Y (1983) Effect of the pigment vehicle interaction on the properties of magnetic coatings. J Coating Tech 55(698):23

IUPAC Handbook (2000) 2000–2001, International Union of Pure and Applied Chemistry, Wiley, New York

IUPAC World Polymer Congress (2002) Progress in polymer science and technology: 2002 IUPAC World Polymer Congress, Beijing, China, July 7–12, 2002. Wiley, New York

Johnson SF (2001) History of light and colour measurement: a science in the shadows. Taylor & Francis, London

Kadolph SJJ, Langford AL (2001) Textiles. Pearson, New York

Kakudo M (2005) X-ray diffraction of macromolecules. Springer, New York

Kamide K, Dobashi T (2000) Physical chemistry of polymer solutions. Elsevier, New York

Keane JD (ed) (1993) Steel structures painting manual. Steel Structures Paint Council, Pittsburgh

Keane JD (ed) (1993) Steel structures painting manual: good painting practice, vol 1. Steel Structures Painting Council, Pittsburgh

Kemp RB (1999) Handbook of thermal analysis and calorimetry. Elsevier, New York

Kidder RC (1994) Handbook of fire retardant coatings and fire testing services. CRC, Boca Raton

Kirk PM, Cannon PF (2001) Fungi, 9th edn. CABI, Wallingford, UK

Klempner D, Frisch KC (eds) (2001) Advances in urethane science and technology. Rapra Technology, Shropshire, UK

Klocek P (ed) (1991) Handbook of infrared materials. Marcel Dekker, New York

Kokhanovsky AA (2004) Light scattering media optics. Springer, New York

Krause A, Lange A, Ezrin M (1988) Plastics analysis guide: chemical and instrumental methods. Oxford University Press, Oxford

Kricheldorf HR, Swift G, Nuyken O, Huang SJ (2004) Handbook of polymer synthesis. CRC, Boca Raton

Kroschwitz JI (ed) (1990) Concise dictionary of polymer science and engineering. Wiley, New York

Kroschwitz JI (ed) (1990) Polymers: polymer characterization and analysis. Wiley, New York

Ku CC, Liepins R (1987) Electrical properties of polymers. Hanser, New York

Langenheim JH (2003) Plant resins: chemistry, evolution ecology and ethnobotany. Timber, Portland

Leach RH, Pierce RJ, Hickman EP, Mackenzie MJ, Smith HG (eds) (1993) Printing ink manual, 5th edn. Blueprint, New York

Lee SM (1989) Dictionary of composite materials technology. Technomic, Lancaster, PA

Lenz RW (1967) Organic chemistry of synthetic high polymers. Interscience, New York

Leonard EC (ed) (1970–1971) Vinyl and diene monomers, 3 vols. Wiley, New York

Levenspiel O (1998) Chemical reaction engineering. Wiley, New York

Levin HL (2005) The earth through time. Wiley, New York

Lewis PA (ed) (1985–1990) Pigment handbook, vols 1–4, 2nd edn. Wiley, New York

Licari JJ (2003) Coating materials for electronics applications: polymers, processes, reliability and testing. Noyes Data Corporation/Noyes Publications, New York

Lide DR (ed) (2004) CRC handbook of chemistry and physics. CRC Press, Boca Raton

Loveland RP (1981) Photomicrography. Krieger, New York

Lovell PA, El-Aasser M (eds) (1997) Emulsion polymerization and emulsion polymers. Wiley, New York

Lowe JJ (1997) Radiocarbon dating. Wiley, New York

Madox DM (1998) Handbook of physical vapor deposition (PVD) processing. Noyes Data Corporation, New York

Mark JE (ed) (1996) Physical properties of polymers handbook. Springer, New York

Martens CR (1961) Alkyd resins. Reinhold, New York

Martens CR (1964) Emulsion and water-soluble paints and coatings. Reinhold, New York

Martens CR (1968) Technology of paints, varnishes and lacquers. Reinhold, New York

Mathews CK, van Holde KE, Ahern KG (2000) Biochemistry, 3rd edn. Addison Wesley, San Francisco

Matyjaszewski K, Scott Gaynor, Jin-Shan Wang (1995) Controlled radical polymerizations: the use of alkyl iodides in degenerative transfer. Macromolecules 28(6):2093–2095

Mayer R, Sheehan S (1991) Artist's handbook of materials and techniques. Viking Adult, New York

McCleverty JA, Connelly NG (2001) Nomenclature of inorganic chemistry II: recommendations 2000. The Royal Society of Chemistry, London

McDonald R (ed) (1997) Colour physics for industry, 2nd edn. Society of Dyers and Colourists. West Yorkshire, England

Meeten GH (1986) Optical properties of polymers. Springer, New York

Miller ML (1966) The structure of polymers. Reinhold, New York

Mittal KL (2003) Contact angle, wettability and adhesion, vols 1–3. VPS International Science Publishers, Zeist, The Netherlands

Mittal KL (ed) (2004) Silanes and other coupling agents, vol 3. VSP International Science, New York

Moad G, Chong YK, Almar P, Ezio R, San HT (2004) Advances in RAFT polymerization: the synthesis of polymers with defined end-groups. Polymers (Elsevier) 46:8458–8468

Moad G, Chong YK, Postma A, Rizzardo E, Thang SH (2004) Advances in RAFT polymerization: the synthesis of polymers with defined end-groups CSIRO Molecular Science, Bayview Ave, Clayton, Vic. 3168, Australia
Moad G, Rizzardo E, Thang SH (2008) Radical addition fragmentation chemistry in polymer synthesis. Polymer 49(5):1079–1131
Modern plastics encyclopedia. McGraw Hill/Modern Plastics, New York
Moller KD (2003) Optics. Springer, New York
Morrison GR (2004) X-ray microscopy: techniques and applications. World Scientific, Singapore
Morrison RT, Boyd RN (1992) Organic chemistry, 6th edn. Prentice Hall, Englewood Cliffs, NJ
Muizebelt WJ, Donkerbroek JJ, Nielsen MWF, Hussem JB, Biedmond MEF, Klaasen RP, Zabel KH (1998) Oxidative crosslinking of alkyd resins studied with mass spectroscopy and NMR using model compounds. J Coating Techn 70(876):83–92
Munson BR, Young DF, Okiishi TH (2005) Fundamentals of fluid mechanics. Wiley, New York
Murphy J (1998) Reinforced plastics handbook. Elsevier, New York
National Association of Printing Ink Manufacturers, Inc. (1999) Printing ink handbook, 5th edn. Kluwer Academic, London
Nasongkla N, Bey E, Ren J, Ai H, Khemtong C, Guthi JS, Chin SF, Sherry AD, Boothman DA, Gao J (2006) Multifuntional polymeric micelles as cancer targeted, MRI-ultrasensitive drug delivery systems. Nano Letters 6(11):2427–2430
Nelson G (1990) Fire and polymers: hazards identification and prevention. Oxford University Press, Oxford
Nesse WD (2003) Introduction to optical mineralogy. Oxford University Press, New York
NRC (1987) Radiogenic age and isotopic studies: report 1. National Research Council of Canada NRC Research, Ottawa
O'Conner DJJ, Smart RS, Sexton BA (2003) Surface analysis methods in materials science. Springer, New York
Odian G (2004) Principles of polymerization. Wiley, New York
Oss CJ (1994) Interfacial forces in aqueous media. Marcel Dekker, New York
Parfitt CD, Sing KSW (1976) Characterization of powder surfaces. Academic, London
Parfitt GD (1969) Dispersion of powders in liquids. Elsevier, New York
Patton TC (1979) Paint flow and pigment dispersion: a rheological approach to coating and ink technology. Wiley, New York
Paul N. Gardner, Company, Inc., 316 N. E. Fifth Street, Pompano Beach, Fl, www.gardco.com
Pecora R, Pecora R (1985) Dynamic scattering: applications of photon correlation spectroscopy. Kluwer Academic, New York
Perkins D (2001) Mineralogy. Prentice Hall, New York
Perrier S, Takolpuckdee P (2005) Macromolecular design via reversible addition–fragmentation chain transfer (RAFT)/xanthates (MADIX) polymerization. J Polym Sci Part A 43:5347–5393
Perry RH, Green DW (1997) Perry's chemical engineer's handbook, 7th edn. McGraw Hill, New York
Pethrick RA, Pethrick RA (eds) (1999) Modern techniques for polymer characterization. Wiley, New York
Pierson HO (1994) Handbook of carbon, graphite, diamond and fullerenes. Noyes Data Corporation/Noyes Corporation, New York
Pittance JC (ed) (1990) Engineering plastics and composites. SAM International, Materials Park
Provder T (ed) Particle size distribution II: assessment and characterization. Oxford University Press, New York
Provder T, Texter J (eds) (2004) Particle sizing and chacterization. American Chemical Society, Washington DC
Prutton M, El Gomati M (eds) (2006) Scanning Auger electron microscopy. Wiley, New York
Ready RG (1996) Thermodynamics. Plenum, New York
Rhodes G (1999) Crystallography made crystal clear: a guide for users of macromolecular models. Elsevier, New York
Rizzardo E, Chiefari J, Mayadunne R, Moad G, Thang S (2008) Tailored polymer architectures by reversible additionfragmentation chain transfer. Macromol Symp 174:209–212
Rosato DV (ed) (1992) Rosato's plastics encyclopedia and dictionary. Hanser-Gardner, New York
Rosin J (1967) Reagent chemicals and standards. Van Nostrand Reinhold, New York
Russell JB (1980) General chemistry. McGraw Hill, New York
Salamone JC (ed) (1996) Polymeric materials encyclopedia. CRC, Boca Raton
Saleh BEA, Teich MC (1991) Fundamentals of photonics. Wiley, New York
Schoeser M (2003) World textiles: a concise history. Thames & Hudson
Seanor DA (1982) Electrical conduction in polymers. Academic, New York
Sepe MP (1998) Dynamic mechanical analysis, plastics design library. Norwich, New York
Shah V (1998) Handbook of plastics testing technology. Wiley, New York
Shahidi F, Bailey AE (eds) (2005) Bailey's industrial oil and fat products. Wiley
Shenoy AV (1996) Thermoplastics melt rheology and processing. Marcel Dekker
Sigma-Aldrich Catalog, www.sigma-aldrich.com, USA, 2006–2007
Skeist I (ed) (1990) Handbook of adhesives. Van Nostrand Reinhold, New York
Skoog DA, Holler FJ, Nieman TA (1997) Principles of instrumental analysis. Brooks/Coles, New York
Slade PE (2001) Polymer molecular weights, vol 4. Marcel Dekker, New York
Smith E, Dent G (2004) Modern Raman spectroscopy. Wiley, New York
Smith JM (1981) Chemical engineering kinetics. McGraw Hill, New York
Smith JM, Abbott MM, Abbot M, Van Ness HC, Van Ness HC (2004) Introduction to chemical engineering thermodynamics. McGraw Hill, New York
Smith MB, March J (2001) Advanced organic chemistry, 5th edn. Wiley, New York
Soderberg GA, Karcher PW (1958) Restoring and maintaining finishes – spot finishing. Bruce, Milwaukee
Solomon DH (1969) Kinetics and mechanisms of polymerization series, vol 2 – ring opening; vol 3 – step growth. Marcel Dekker, New York
Solomon DH, Hawthorne DG (1991) Chemistry of pigments and fillers. Krieger, New York
Staniforth M, Goldstein J, Echlin P, Lifshin E, Newbury DA (2003) Scanning electron microscopy and X-ray microanalysis. Springer, New York
Starr TF (1993) Data book of thermoset resins for composites. Elsevier, New York
Staudinger H, Heuer W (1930) A relationship between the viscosity and the molecular weight of polystyrene (German). Ber 63B:222–234
Strong AB (2000) Plastics materials and processing. Prentice Hall, Columbus
Stuart BH (2002) Polymer analysis. Wiley, New York
Stuart BH (2004) Infrared spectroscopy. Wiley, New York
Sudduth RD (2003) J Appl Sci 48(1):25–36
Suryanarayana C, Norton MG (2003) X-ray diffraction: a practical approach. Plenum, New York

Tanford C (1961) Physical chemistry of macromolecules. Wiley, New York
Tess RW, Poehlein GW (eds) (1985) Applied polymer science, 2nd edn. American Chemical Society, Washington, DC
Thompson, Michael, Baker MD, Tyson JF, Christie A (1985) Auger electron spectroscopy. Wiley, New York
Tooke WR Jr (1963) Film thickness testing: a paint inspection gage. Official Digest 35(462):691–698
Tortora PG (ed) (1997) Fairchild's dictionary of textiles. Fairchild, New York
Tortora PG, Merkel RS (2000) Fairchild's dictionary of, 7th edn. Fairchild, New York
Tracton AA (ed) (2005) Coatings technology handbook. Taylor & Francis, New York
Troitzsch J (2004) Plastics flammability handbook: principle, regulations, testing and approval. Hanser-Gardner, New York
Tuck N (2001) Waterborne and solvent based, alkyds and there end user application, vol 6. Wiley, New York
Uhlig HH (2000) Corrosion and corrosion control. Wiley, New York
Usmani AM (ed) (1997) Asphalt science and technology. Marcel Dekker
Van Wazer JR, Lyons JW, Kim KY, Colwell RE (1963) Viscosity and flow measurement. Interscience, New York
Vanderhoff JW, Gurnee EF (1956) Motion picture investigation of polymer latex phenomena. TAPPI 39(2):71–77
Vanderhoff JW, Tarkowski HL, Jenkins MC, Bradford EG (1966) Theoretical considerations of the interfacial forces involved in the coalescence of latex particles. J Macromol Chem 1(2):361–397
Vigo TL (1994) Textile processing: dyeing, finishing and performance. Elsevier, New York
Vincenti R (ed) (1994) Elsevier's textile dictionary. Elsevier, New York
Wallenberger FT, Weston NE (eds) (2003) Natural fibers, plastics and composites. Springer, New York
Wang X, Zhou Q (2004) Liquid crystalline polymers. World Scientific, Singapore
Watson P (1997) Physical chemistry. Wiley, New York
Watt IM (1997) Principles and practice of electron microscopy. Cambridge University Press, Cambridge
Weast RC (ed) Handbook of chemistry and physics, 52nd edn. The Chemical Rubber, Boca Raton
Weismantal GF (1981) Paint handbook. McGraw Hill, New York
Weldon DG (2001) Failure analysis of paints and coatings. Wiley, New York
Wells K, Beal S, Woodburn C, Durant J, Brandimane J (1997) Fabric dyeing and printing. Interweave, London
Whistler JN, BeMiller JN (eds) (1992) Industrial gums: polysaccharides and their derivatives. Elsevier
Whitten KW, Davis RE, Davis E, Peck LM, Stanley GG (2003) General chemistry. Brookes/Cole, New York
Wicks ZN, Jones FN, Pappas SP (1999) Organic coatings science and technology, 2nd edn. Wiley, New York
Wickson EJ (ed) (1993) Handbook of polyvinyl chloride formulating. Wiley, New York
Wijnekus FJM (1967) Elsevier's dictionary of the printing and allied industries in four languages. Research Institute for the Graphic and Allied Industries, TNO, Amsterdam, Elsevier, New York
Willard HH, Merritt LL, Dean JA (1974) Instrumental methods of analysis. Van Nostrand, New York
Witte RSS, Witte JS, Smith GS (2003) Statistics. Wiley, New York
Wolf WL, Zeissis GJ (1985) Infrared handbook. SPIE International Society for Optical Engineering
Wray HA (ed) (1991) Manual for flash point standards and their use. American Society for Testing and Materials, Philadelphia
Wypych A (ed) (2003) Plasticizer's database. Noyes, New York
Wypych G (ed) (2001) Handbook of solvents. Chemtec, New York
Yates M (2002) Fabrics. Norton, New York
Yau WW, Kirkland JJ, Bly DD (2001) Modern size-exclusion chromatography. Wiley, New York
Zaccaria VK, Utracki L (2003) Polymer blends. Springer, New York
Zaiko GE (ed) (1995) Degradation and stabilization of polymers. Nova Science, New York
Zhelyazova B, Kovacheva S (2002) Elsevier's dictionary of plastics and polymer in English-German-French-Spanish-Russian. Elsevier, New York

Characterization:

Summerscales J (1987) Shear modulus testing of composites. In: Proceedings 4th international conference on composite structures, vol 2, Paisley, 27–29 July 1987. Elsevier, pp 305–316
Summerscales J (ed) (1987/1990) Non-destructive testing of fibre reinforced plastics composites, 2 vols. Kluwer Academic, Dordrecht, ISBN 1-85166-093-3; ISBN 1-85166-468-8
Summerscales J (1990) NDT of advanced composites – an overview of the possibilities. Br J Non-Destr Test 32(11):568–577
Summerscales J (1994) Non-destructive measurement of the moisture content in fibre-reinforced plastics. Br J Non-Destr Test 36(2):64–72
Summerscales J (1994) Manufacturing defects in fibre-reinforced plastics composites. Insight 36(12):936–942